Proteins, Transmitters and Synapses

Proteins, Transmitters and Synapses

DAVID G. NICHOLLS
Department of Biochemistry, University of Dundee
Dundee, Scotland, UK

b
Blackwell
Science

To Solvig

Editorial Offices:
Osney Mead, Oxford OX2 0EL
25 John Street, London WC1N 2BL
23 Ainslie Place, Edinburgh EH3 6AJ
238 Main Street, Cambridge
Massachusetts 02142, USA
54 University Street, Carlton
Victoria 3053, Australia

Other Editorial Offices:
Arnette Blackwell SA
1, rue de Lille
75007 Paris
France

Blackwell Wissenschafts-Verlag GmbH
Kurfürstendamm 57
10707 Berlin
Germany

Feldgasse 13
A-1238 Wien
Austria

First published 1994
Reprinted 1995

Set by Setrite Typesetters Ltd, Hong Kong
Printed and bound in Great Britain
at the University Press, Cambridge

DISTRIBUTORS

Marston Book Services Ltd
PO Box 87
Oxford OX2 0DT
(*Orders*: Tel: 01865 791155
Fax: 01865 791927
Telex: 837515)

USA
Blackwell Science, Inc.
238 Main Street
Cambridge, MA 02142
(*Orders*: Tel: 800 215-1000
617 876-7000
Fax: 617 492-5263)

Canada
Oxford University Press
70 Wynford Drive
Don Mills
Ontario M3C 1J9
(*Orders*: Tel: 416 441-2941)

Australia
Blackwell Science Pty Ltd
54 University Street
Carlton, Victoria 3053
(*Orders*: Tel: 03 347-5552)

A catalogue record for this title
is available from the British Library

ISBN 0-632-03661-3

Library of Congress
Cataloging-in-Publication Data

Nicholls, David G.
Proteins, transmitters and synapses
David G. Nicholls.
p. cm.
Includes bibliographical references
and index.
1. Neurotransmitters
2. Neurotransmission. 3. Synapses.
I. Title.
[DNLM: 1. Neuroregulators – physiology.
2. Neurones – physiology.
3. Neurophysiology.
WL 102 N613m 1994]
QP364.7.N53 1994
599′.0188 – dc20
DNLM/DLC
for Library of Congress

Contents

Preface

The intellectual challenge posed by the attempt to obtain a molecular understanding of brain function has excited scientists from many disciplines of the life sciences, and has rendered the traditional divisions between biochemist, physiologist, anatomist and pharmacologist obsolete. Nevertheless, the majority of brain research is still performed by scientists, such as the author, whose training was in one of these major contributory disciplines rather than the global field of 'neuroscience'. This is apparent for example in the mutual difficulty which biochemists and electrophysiologists frequently experience when attempting to appreciate the subtleties of the other's approach in the primary literature.

In the last year or two a number of excellent introductory neuroscience texts have appeared, written predominantly from the standpoint of electrophysiology or molecular biology. The purpose of the present book, written by a neurochemist, is to provide an overview of current research into the molecular events which occur either side of the synaptic cleft across which information transfer occurs. This 'biochemistry of the synapse' has undergone dramatic advances in the past few years and has transformed neurochemistry, which was classically primarily concerned with the metabolic transformations of transmitters and related compounds in neuronal tissue. This book has been written in an attempt to achieve a synthesis of current understanding of synaptic biochemistry and to make the approach more accessible to neuroscientists with other backgrounds.

List of abbreviations

$\Delta\psi$	membrane potential across vesicular and mitochondrial membranes
AA	arachidonic acid
AcCoA	acetyl-CoA
ACh	acetylcholine
AChE	acetylcholinesterase
1S,3R-ACPD	1S,3R-aminocyclopentane-1,3-dicarboxylate
ACTH	adrenocorticotrophic hormone
ADP	adenosine diphosphate
Adr	adrenaline
AMPA	α-amino-5-methyl-4-isoxazole-propionic acid
AOAA	aminooxyacetate
4AP	4-aminopyridine
AP5/APV	2-amino-5-phosphonopentanoic acid
ARF	ADP ribosylation factor
βARK	β-adrenoceptor kinase
ATPase	adenosine triphosphatase
AVP	arginine-vasopressin
BAPTA	1,2-bis(2-aminophenoxy)-ethane-*N,N,N′,N′*-tetraacetic acid
BCECF	bis-carboxyethyl-carboxyfluorescein
BDZ	benzodiazepine
BoNT	botulin neurotoxin
αBTx	α-bungarotoxin
βBTx	β-bungarotoxin
$[Ca^{2+}]_c$	concentration of free Ca^{2+} in the cytoplasm
CAD	cadherin
cADPR	cyclic ADP-ribose
CaM	calmodulin
CAM	cell adhesion molecule
CaMKII	Ca^{2+}/calmodulin-dependent protein kinase II
cAMP	3′,5′-cyclic adenosine phosphate
CAT	choline acetyltransferase
CCh	carbachol
CCK	cholecystokinin
cDNA	complementary deoxyribonucleic acid
cGMP	3′,5′-cyclic guanosine triphosphate
CGRP	calcitonin gene-related peptide
Ch	choline
CHA	cyclohexyladenosine
CNQX	6-cyano-7-nitroquinoxaline-2,3-dione
CNS	central nervous system
COMT	catechol-*O*-methyltransferase
CPP	3-((RS)-carboxypiperazin-4-yl)-propyl-1-phosphonic acid
CSF	cerebrospinal fluid
DA	dopamine
DABA	diaminobutyrate
DAG	diacylglycerol
DARPP-32	dopamine and cyclic AMP-regulated phosphoprotein
DBH	dopamine β-hydroxylase
DCCD	dicyclohexylcarbodiimide
DFP	diisopropylflurophosphate
DHP	dihydropyridine
DTx	dendrotoxin
EGF	epidermal growth factor
EGTA	ethyleneglycol-bis-(β-aminoethyl ether)-*N,N,N′,N′*-tetraacetic acid
Enk-LI	enkephalin-like immunoreactivity
epsp	excitory postsynaptic potential

ER	endoplasmic reticulum
F	Farad, unit of capacitance
FCCP	carbonylcyanide-*p*-trifluoromethoxyphenylhydrazone
FGF	fibroblast growth factor
ΔG	Gibbs free energy
GΩ	giga ohm
G_s,G_i, etc.	trimeric G-proteins
$G\alpha_s$, etc.	α-subunit of a trimeric G-protein
GABA	γ-aminobutyric acid
GABA-A	(ionotropic) GABA-A receptor
GABA-B	(metabotropic) GABA-B receptor
GABA-T	GABA-2-oxyglutarate transaminase
GAD	glutamate decarboxylase
GAP	GTPase-activating protein
GAP-43	growth-associated protein 43 (other synonyms are neuromodulin, B-50)
GDH	glutamate dehydrogenase
GDP	guanosine diphosphate
G_t	transducin
GTP-γ-S	guanosine diphosphate sulphophosphate
Hb	haemoglobin
HCA	homocysteic acid
HPLC	high pressure liquid chromatography
5-HT	5-hydroxytryptamine, serotonin
$(1,4)IP_2$	(1,4)-inositol bisphosphate
$(1,4,5)IP_3$	(1,4,5)-inositol trisphosphate
$(1,3,4,5)IP_4$	(1,3,4,5)-inositol tetrakisphosphate
ipsp	inhibitory postsynaptic potential
K	equilibrium constant
K_d	dissociation constant
KA	kainate
kb	kilobase
kDa	kilodalton, molecular mass
L-AP4	L(+)-2-amino-4-phosphonobutyrate
LDCV	large dense-core (synaptic) vesicle
L-DOPA	L-3,4 dihydroxyphenylalanine
L-NOARG	L-*N*-nitroarginine
LTD	long-term depression
LTP	long-term potentiation
M_r	apparent molecular weight, e.g. from polyacrylamide gel
MAO	monoamine oxidase
MARCKS	myristoylated alanine-rich C-kinase substrate
MCDP	mast cell degranulating peptide
mepps	miniature end-plate potentials
mGluR	metabotropic glutamate receptor
MnAChR	muscle nicotinic acetylcholine receptor
MPTP	1, methyl-4-phenyl-1,2,3,6-tetrahydropyridine
α-MSH	α_1-melanocyte-stimulating hormone
NA	noradrenaline
N-CAD	neural cadherin
N-CAM	neural cell adhesion molecule
NEM	*N*-ethylmaleimide
NGF	nerve growth factor
nm	nanometre
NMDA	*N*-methyl-D-aspartate
NMJ	neuromuscular junction
NnAChR	neuronal nicotinic ACh receptor
NO	nitric oxide
NOS	nitric oxide synthase
NPY	neuropeptide Y
NSF	*N*-ethylmaleimide-sensitive factor
OXY	oxytocin
PC	phosphatidylcholine
PC12 cells	phaeochromocytoma cells, derived from chromaffin cell tumours
PCR	polymerase chain reaction
PDE	phosphodiesterase
PDGF	platelet-derived growth factor
PG	prostaglandin
ΔpH	pH gradient across a membrane
pH_i	internal pH
PI	phosphatidylinositol
PIP	phosphatidylinositol 4 phosphate
PIP_2	phosphatidylinositol bisphosphate

PIP_3	phosphatidylinositol trisphosphate
PI-PLC	phosphatidylinositol-specific phospholipase C
PKA	protein kinase A
PKC	protein kinase C
PLA_2	phosphalipase A_2
PLC	phospholipase C
PLD	phospholipase D
pmf	proton-motive force
PNMT	phenylethanolamine *N*-methyltransferase
PNS	peripheral nervous system
POMC	pro-opiomelanocortin
ProEnk	proenkephalin
pS	picoSiemen, unit of channel conductance
PSD	postsynaptic density
PTx	pertussis toxin
R	gas constant
S	Svedburg, unit of sedimentation
SAM	*S*-adenosylmethionine
SLMV	synaptic-like microvesicle
SNAP	soluble NSF attachment protein
SNAP-25	synaptosomal-associated protein 25
SNARE	SNAP receptor
SOM	somatostatin
SP	substance P
SSV	small synaptic vesicle
T	temperature (degrees Kelvin)
TEA	tetraethylammonium
TeTx	tetanus toxin
TRH	thyrotropin-releasing hormone
TTX	tetrodoxin
VACC	voltage-activated Ca^{2+}-channel
V-ATPase	V-type H^+ translocating ATPase
VIP	vaso-intestinal peptide
V_m	membrane potential across the plasma membrane

Part 1
Overview

1

Introduction: principles of neurotransmission

1.1 A neuroscience primer

1.1.1 Neurones and synapses

The vertebrate nervous system comprises the *central nervous system* or *CNS* (i.e. the brain together with the spinal cord), and the *peripheral nervous system* or *PNS* which relays impulses to and from the remainder of the body. Nervous tissue contains *neurones* (nerve cells) which are responsible for information transfer and processing, and *glia*, which are ten times more numerous than neurones and whose diverse functions include transmitter and metabolite supply, support, insulation, axonal guidance during development and regulation of the ionic environment of the neurones. The human brain has about 10^{11} neurones. One heroic quantification of the cells in a single milligram of monkey cerebral cortex revealed 120 000 neurones, 60 000 glial cells and no less than 280 000 000 synapses! There are far more neurones in an animal's nervous system than there are genes, or even base-pairs, in its nuclei.

There are several hundred types of neurones, some of which are summarized in Fig. 1.1. Neurones do, however, have certain generic features summarized in Fig. 1.2a. For clarity most diagrams of neurones greatly understate the enormous extent of the dendritic or axonal trees and the actual shape of a typical neurone is closer to that shown in Fig. 1.2b. The neuronal cytoskeleton, comprised of *microtubules*, *microfilaments* and *neurofilaments* (or *intermediate filaments*) with associated proteins, controls the complex and highly polarized shape of the neurone. The *cell body*, also called the *soma* or *perikaryon*, contains the nucleus and is the site of protein synthesis. Cell bodies can vary in diameter from 5 to 60 μm.

Growing out from the cell body are the *dendrites*, which can be regarded as an extension of the cell body. Some neurones have many dendrites; frequently one of these is particularly dominant — an *apical dendrite* — which extends in a line away from the cell body, giving off branches. In other cells, including those with the most highly branched dendrites, the *Purkinje cells* (Fig. 1.1), a single dendrite leaves the cell body and then branches.

Dendrites and cell bodies receive incoming (or *afferent*) signals from other neurones. Most signals are chemical and in the form of *neurotransmitters* which bind to *receptors* and generate or modulate changes in electrical potential across the dendritic membrane. Certain neurones have dendrites with projecting *dendritic spines* on which the receptors are localized.

A neurone integrates the signals received from many other neurones. In one extreme example, the cerebellar Purkinje cell (Fig. 1.1d), the number of inputs can reach 100 000. If the end result of this integration is to depolarize the cell body beyond a critical level, a nerve impulse in the form of an *action potential*, a transient (1 ms) change of 100 mV in the electrical potential

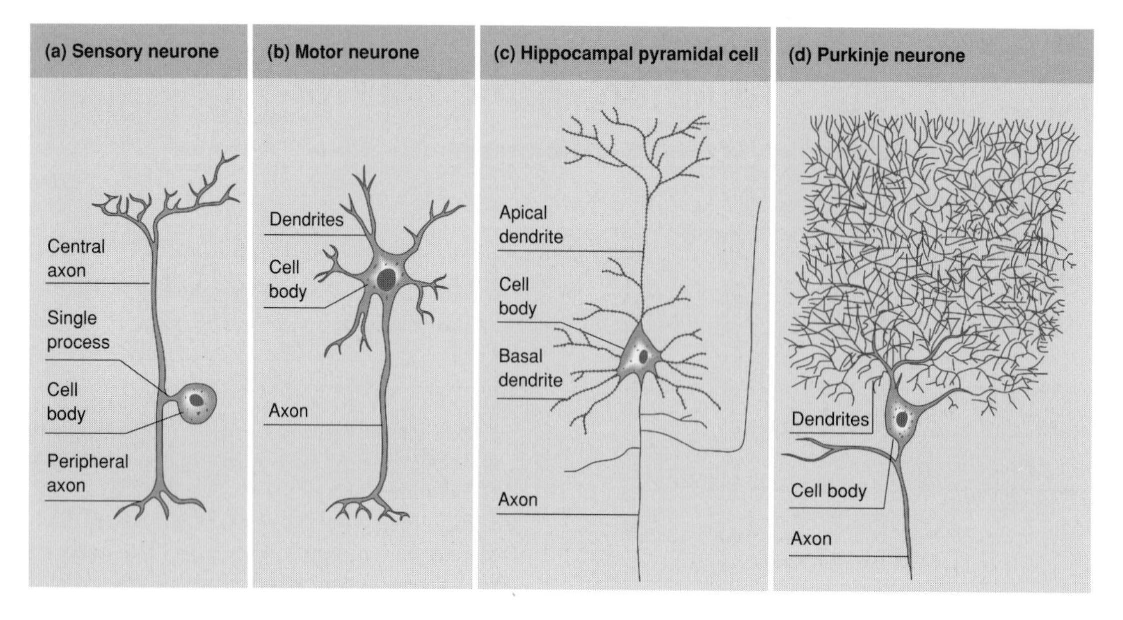

Fig. 1.1 Schematic representation of some common neuronal configurations. (a) Sensory neurones from dorsal root ganglia possess a single process emerging from their cell bodies. This bifurcates to give two axons, one of which receives sensory information from the periphery while the other transmits this information to the CNS. (b) Motor neurones innervate skeletal muscle. Dendrites emerge from the cell body while a branching axon transmits information to a group of muscle fibres. (c) Pyramidal cells of the hippocampus are so-called because of the shape of their cell body. Dendrites are covered with *dendritic spines* which form the postsynaptic elements of the synaptic inputs. (d) Cerebellar Purkinje neurones have an enormous, two-dimensional dendritic tree allowing an input from some 100 000 synapses.

across the plasma membrane, is propagated down a single *axon*, which is usually attached to the cell body at a conical projection of the cell body called the *axon hillock*. Axons are usually thinner and much longer than dendrites. The most common axons in the CNS are a few millimetres in length and branch profusely (Fig. 1.2b). Other axons may extend as far as 1 m.

At the end of each axon branch is a *nerve terminal* from which transmitter is released in response to Ca^{2+} entry triggered by the arrival of the action potential; some classes of neurones also release transmitter from a series of swellings (variously termed *varicosities*, *en passant terminals* or *boutons*) spaced along the branched axon. A typical neurone might possess 1000–2000 terminals or varicosities and receive synaptic inputs from a comparable number of other neurones, although these numbers can vary enormously. Plate 1 (facing page 132) shows a neurone from the hippocampus where the terminals synapsing onto the dendrites and cell body are stained by fluorescent antibody to a presynaptic protein, synaptotagmin.

Most release of neurotransmitter occurs into the narrow *synaptic cleft* which separates the *presynaptic* nerve terminal from the receptors present on the *postsynaptic* membrane of the subsequent neurone. The presynaptic membrane, synaptic cleft and postsynaptic membrane together make up the *synapse* (Fig. 1.3). While most receptors are located postsynaptically, there are also presynaptic receptors present on the terminals themselves. These may be *autoreceptors* responding to the released neurotransmitter and usually, but not invariably, providing a negative feedback, or *heteroreceptors* responding to a different receptor usually released at an *axo-axonal synapse* onto the terminal itself. Finally, bidirectional *dendro-dendritic* synapses occur when synaptic vesicles are present on both sides of a symmetrical synapse allowing for bidirectional signalling.

In addition to this synaptic transmission, some classes of neurotransmitter, notably the neuro-

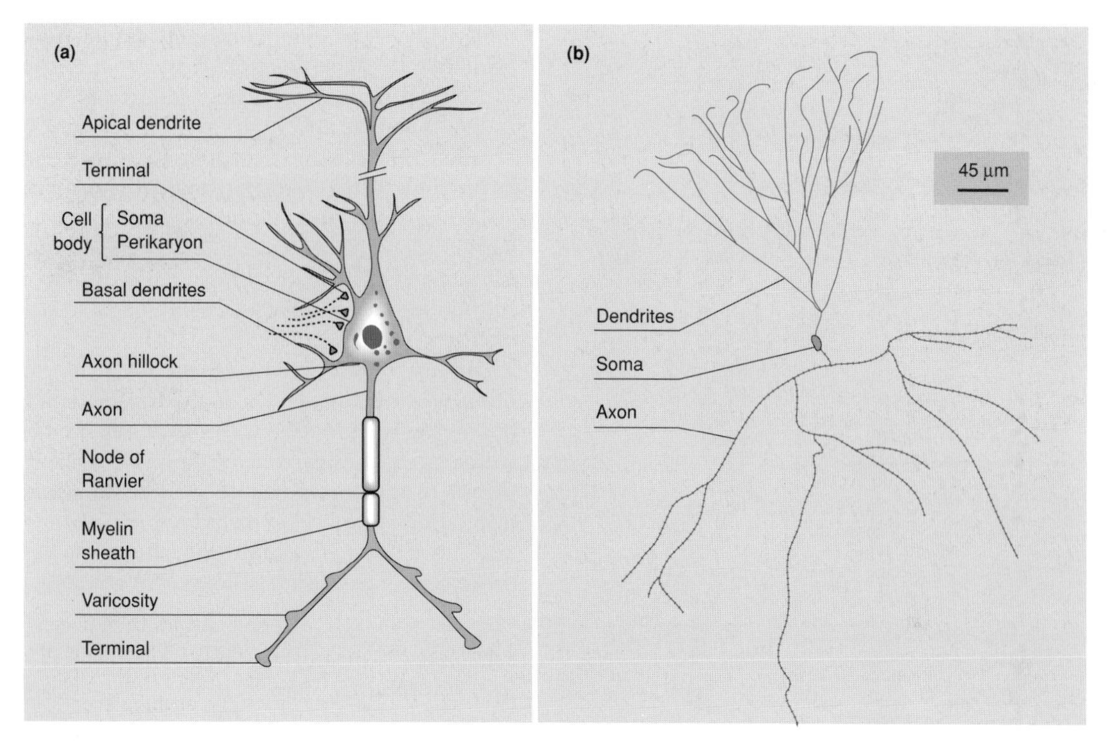

Fig. 1.2 Characteristic components of a vertebrate neurone. (a) The *cell body* (*soma* or *perikaryon*) contains the nucleus. The surface of the soma is greatly extended by *dendrites*. The surface of the soma and dendrites are densely covered with *presynaptic terminals* from other neurones which may be either excitatory or inhibitory. The *axon* extends from the *axon hillock* and may be up to 1 m in length but only a few micron in diameter. Long axons are usually covered by *myelin sheaths*, derived (in the CNS) from oligodendrocyte glial cells. At gaps between myelin sheaths, Na^+-channels are concentrated in *nodes of Ranvier*. The short axons of interneurones may be unmyelinated. Axons divide and may typically make 1000 synaptic contacts with other neurones, either via *presynaptic terminals* or *en passant varicosities*. The terminals contain small synaptic vesicles closely aligned to active zones which are the sites of transmitter release. After crossing the narrow (20 nm) synaptic cleft, transmitter binds to receptors on the postsynaptic membrane. (b) Camera lucida drawing of a dentate granule cell in the rat hippocampus showing the actual dimensions of the soma, dendritic tree and axon arbor of a typical neurone. The varicosities on the latter can just be distinguished. From Isokawa *et al.* (1993).

peptides, may be released at non-synaptic sites and diffuse a significant distance before binding to a receptor; in this way they can act as local *neuromodulators*. Synapses may occur not only between neurones but also between neurones and effector cells such as muscle cells, as at the *neuromuscular junction*.

While this book deals with the chemical synapse, it must also be borne in mind that *electrical synapses* exist where neurones are coupled via *gap junctions*, protein pores linking two closely apposed plasma membranes. Gap junctions allow cells to undergo synchronous bidirectional ionic and electrical changes. This means of communication is used by astrocytes (*see* Section 1.1.2) to form a continuous *syncytium*, and by some neurones where extremely fast communication is required, for example the horizontal cells in the retina discussed in Chapter 6.

1.1.2 Glia

The CNS contains many more glia than neurones. The main classes of glia in the CNS are *oligodendrocytes* and *astrocytes*. Oligodendrocytes have small-diameter (3–5 μm) cell bodies; those in white matter wrap their plasma membranes around a number of adjacent axons in order to provide a *myelin sheath*. In the peripheral nervous system, individual *Schwann cells* provide the myelin sheath for single axons

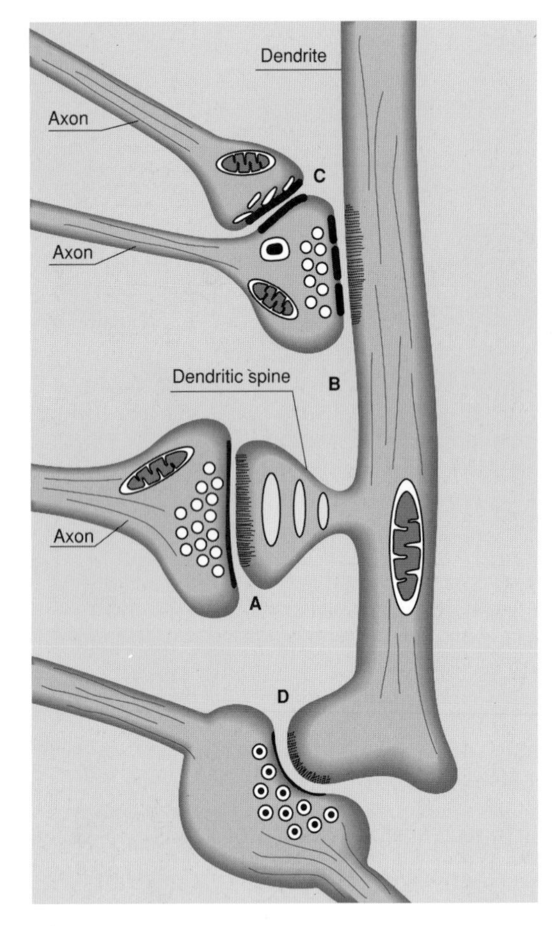

Fig. 1.3 The chemical synapse. (A) Excitatory synapse between a terminal, containing small electron-lucid synaptic vesicles, and a dendritic spine containing large Ca^{2+}-sequestering compartments (spine apparatus). (B) Excitatory synapse containing small synaptic vesicles and one large dense-cored vesicle, synapsing directly onto a dendrite, and regulated by an inhibitory axo-axonal synapse (C) which under the electron micrograph frequently contain flattened synaptic vesicles; (D) is a catecholamine-secreting *en passant* varicosity, or bouton, containing small vesicles which show a dense core in the electron microscope.

(Fig. 1.4). It should be noted that many short axons of interneurones are unmyelinated, although still surrounded by oligodendrocytes.

Astrocytes have a large (20 μm) cell body from which a number of processes radiate, sometimes terminating in thickenings, or end-feet, which surround capillaries in the CNS. It should be noted however that the *blood–brain barrier*, the uniquely effective barrier to the diffusion of large molecular weight compounds into brain, is due primarily to tight junctions between capillary endothelial cells. Astrocytes additionally surround synapses and assist in the removal of transmitters from the synaptic cleft. Thus they have highly active transport systems for amino acids and other transmitters. They also have a high K^+ conductance and may be able to buffer K^+, preventing an increase in the cation's concentration in the small extracellular space after intense neuronal activity.

During the embryonic development of the brain, macrophages invade and differentiate into *microglia*, with 2–3 μm diameter cell bodies, which can be considered to be the resident macrophages of brain, responsible for defence against infection. In the developing brain *radial glia* such as the *Bergmann glia* of the cerebellum form long processes which can assist in axonal guidance.

1.2 Neurotransmitter storage and release

Eukaryotic cells possess two distinct pathways for the export of secretory products by *exocytosis*, the process by which cytoplasmic vesicles fuse with the plasma membrane with the resulting release of their contents and the temporary or permanent incorporation of the vesicle membrane into the plasma membrane. All cells have a *constitutive* pathway, through which a continuous stream of small vesicles from the trans-Golgi network move to the plasma membrane and become incorporated into the plasma membrane. These vesicles do not contain secretory material but are instead the means whereby lipid and membrane protein components can be delivered to the plasma membrane. In addition to the constitutive pathway, many cells possess two additional *regulated* pathways for secretion, in which the secretory material is concentrated and packaged within vesicles, which are then stored near the site of ultimate release and only released in response to a stimulus, frequently an increase in cytoplasmic free Ca^{2+}.

Hormones and neurotransmitters can both be released by regulated pathways in which vesicles originate from the trans-Golgi. The vesicles are characteristically >70 nm in diameter, appear to contain a dense core under the electron microscope, and are consequently referred to as *large dense-core vesicles (LDCVs), secretory vesicles*

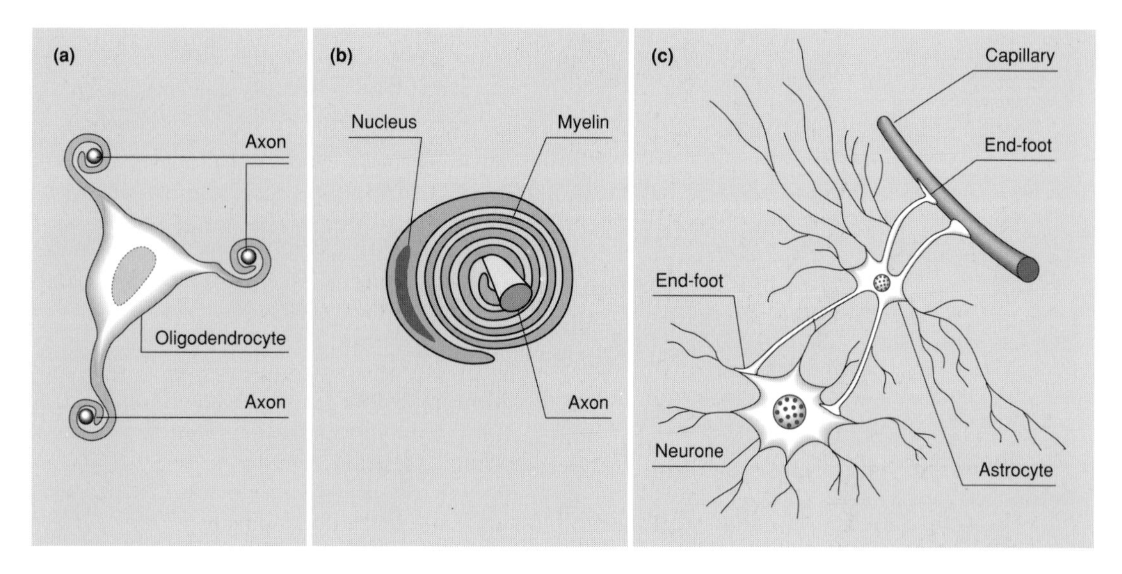

Fig. 1.4 Glia. (a) In white matter, oligodendrocyte processes wrap around a number of adjacent axons to provide a myelin sheath. (b) In the peripheral nervous system Schwann cells myelinate a single axon. (c) Astrocytes have extended 'star-like' processes which can form end-feet onto both neurones and capillaries where they assist in the formation of the blood–brain barrier.

or *secretory granules* (Fig. 1.5). A second type of regulated pathway is adopted by vesicles containing non-peptide neurotransmitters, which undergo local recycling and refilling after exocytosis without the need to return to the Golgi. In view of the relatively enormous distance a vesicle would have to travel to return to the Golgi in the cell body (in the extreme case up to 1 m), such local recycling has distinct advantages. The synaptic vesicles involved in this type of secretion are typically 50 nm in diameter and are referred to as *small synaptic vesicles (SSVs)*. LDCVs and SSVs can coexist in the same neurone.

Small synaptic vesicles are clustered in the region of the terminal adjacent to the synaptic cleft. This *active zone* contains a high concentration of voltage-activated Ca^{2+}-channels. Ca^{2+} entering through these channels binds to a so far unidentified *Ca^{2+} trigger*, which will be discussed in Chapter 7, to initiate exocytosis. At the neuromuscular junction active zones are arranged linearly, whereas in CNS terminals there is a dense two-dimensional array of release sites.

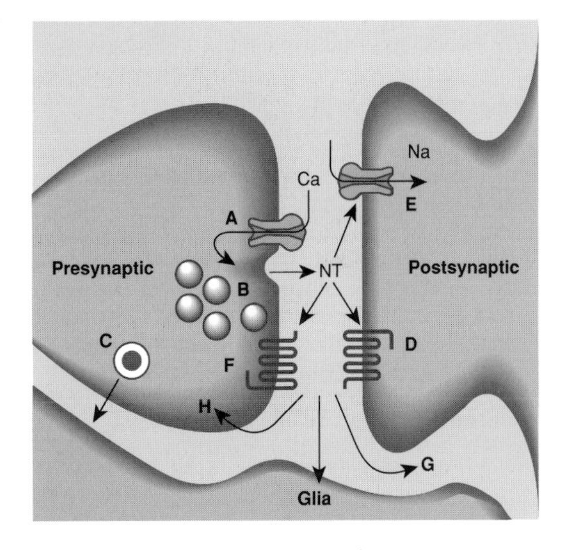

Fig. 1.5 Neurotransmission. Depolarization triggers Ca^{2+} entry at presynaptic Ca^{2+}-channels (A) which causes the release of small synaptic vesicles (B) containing non-peptide transmitters (NT). Neuropeptides are stored in large dense-core vesicles (C) and are released in response to repetitive stimulation of the terminal. Released transmitter interacts with either metabotropic (G-protein-coupled) receptors (D) or ionotropic receptors (E) on the postsynaptic membrane. Receptors may also be located on the presynaptic terminal (F) where they regulate release. Transmission can be terminated by transmitter degradation (G) or re-uptake (H).

1.2.1 Neurotransmitters

Neurotransmitters are very diverse (Fig. 1.6) but

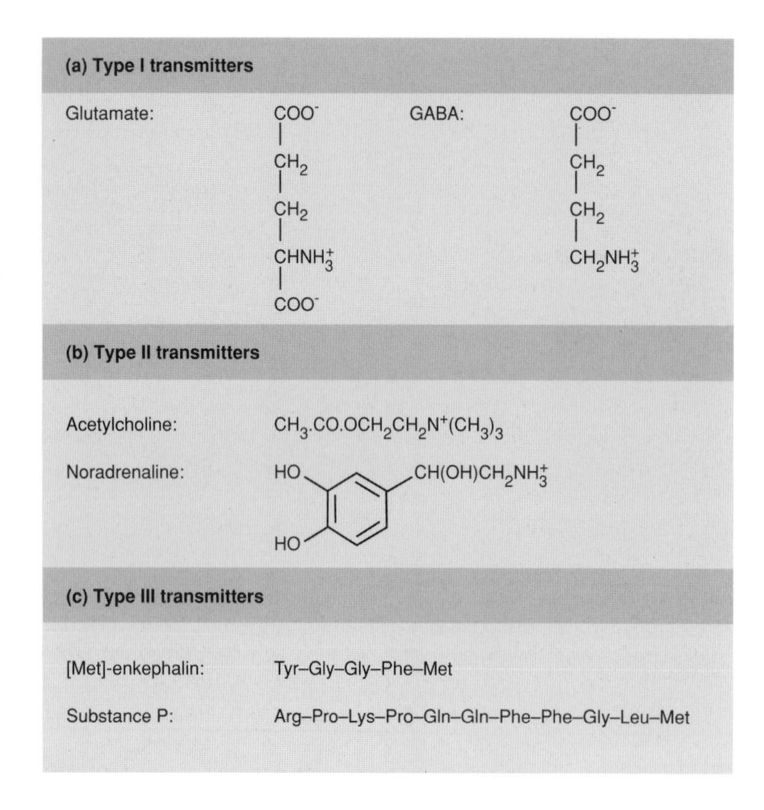

Fig. 1.6 Representative neurotransmitters.

Table 1.1 Major classes of neurotransmitters.

Type		Approximate levels in rat brain ($nmol \cdot g^{-1}$)
Type I: amino acids	Glutamate*	14000
	(Aspartate)*	4000
	γ-aminobutyrate	2500
	Glycine*	2000
Type II: amines and purines	ATP*	2500
	Acetylcholine	25
	Dopamine	6.5
	Noradrenaline	2.5
	Serotonin	2.5
	Adrenaline	1
	Histamine*	1
	Adenosine*	0.2
Type III: peptides	Cholecystokinin	0.5
	[Met]-enkephalin	0.35
	Substance P	0.1
	Vasoactive intestinal peptide	0.04
	Somatostatin	0.03
	Vasopressin	0.002

Data are approximate and do not reflect regional concentrations of transmitters. Data adapted from McGeer *et al.* (1987).

* A major proportion is present in non-transmitter pools.

have been divided into three categories (Table 1.1).

1 *Type I* neurotransmitters are simple amino acids such as *glutamate*, *γ-aminobutyrate (GABA)* and *glycine* which may account for transmission at up to 90% of all CNS synapses and are responsible for most fast information transfer. Even allowing for the presence of non-transmitter pools of the amino acids, these transmitters are present in micromoles per gram wet weight.

2 *Type II* neurotransmitters include the 'classical' transmitters *acetylcholine (ACh)*, the catecholamines (*dopamine*, *noradrenaline* and to a lesser extent *adrenaline*) and *5-hydroxytryptamine (5-HT)*. These are present in most areas of the brain at nanomoles per gram wet weight. Type II transmitters are predominantly slow acting and play a modulatory role in the CNS.

3 *Type III* neurotransmitters encompass a wide variety of *neuropeptides* which are characteristically present at very low concentrations (nanomoles to picamoles per gram of brain). Neuropeptides act predominantly as *neuromodulators* controlling the activity of diffuse receptors rather than at purely synaptic sites.

Many neurones contain more than one neurotransmitter, the most common combination being a Type II transmitter present in both SSVs and LDCVs and a Type III peptide transmitter present exclusively in the LDCVs.

1.2.2 Neurotransmitter receptors

Neurotransmitters and hormones both act by binding to receptors. Neurotransmitter receptors are exclusively located on the plasma membrane and may be of two types:

1 *Metabotropic receptors*: the binding of the ligand stimulates the production of a second messenger via a G-protein, closely analogous to the actions of hormones (*see* Fig. 5.3). Indeed a compound such as noradrenaline is used as a neurotransmitter in the brain and PNS and also as a hormone secreted into the bloodstream together with adrenaline by adrenal chromaffin cells. These *metabotropic receptors* are slow in operation and are frequently used to regulate the firing pattern of cells by controlling the phosphorylation of other receptors or by regulating ion channels on the cell body, dendrites or terminals.

The DNA sequences for most G-protein-coupled metabotropic receptors show common structural features and may be considered to belong to a *superfamily*. Multiple subtypes of virtually all the metabotropic receptors have been found; the isoforms may differ not only in their location, but also in their second messenger coupling. The second messengers generated by metabotropic receptors are as diverse as those produced by hormone receptors and include adenosine 3′,5′-cyclic phosphate (cAMP), arachidonate, inositol phosphates and diacylglycerol. *Presynaptic receptors* on or close to presynaptic terminals usually provide a negative feedback to decrease further release of transmitter. They are usually metabotropic and frequently act by modifying the ionic conductance of Ca^{2+}- and K^{+}-channels.

2 *Ionotropic receptors* contain integral ion channels which are activated by the binding of the transmitter, causing a very rapid change in the membrane potential. Ionotropic receptors are responsible for the fast transmission of information.

There are multiple receptors for most neurotransmitters; thus it is not possible to characterize a given neurotransmitter as having a particular action at all synapses, since this action depends on the nature of the postsynaptic receptor. As an illustration, the Type II transmitter ACh can interact with two types of ionotropic receptor ('nicotinic' receptors present at the nerve–muscle synapse and in the CNS) and no less than five subtypes of metabotropic receptors ('muscarinic' receptors present both in the CNS and the periphery). This diversity is becoming particularly apparent now that molecular genetic approaches can be applied to identifying the receptor genes. Thus the nature of the information transferred across a synapse by a given transmitter depends on the receptor which is present.

If the binding of a transmitter to a receptor increases the likelihood of the cell firing a nerve impulse, then the receptor is said to be *excitatory*. Conversely activation of an *inhibitory* receptor will decrease the likelihood of firing. This categorization does not depend upon the mechanism; for example, an ionotropic receptor which hyperpolarizes a neurone directly and a metabotropic receptor which hyperpolarizes by activating a K^{+} conductance are both classified as inhibitory.

1.2.3 Termination of neurotransmission

Some neurones are capable of firing several times per second. After each event the neurotransmitter must be cleared from the synaptic cleft to enable the postsynaptic receptor to return to its initial state in readiness for the next release. There are two ways in which this can occur — by *enzymatic degradation* or by *re-uptake*, either directly into the terminal or into surrounding glial cells (*see* Fig. 1.5).

Re-uptake is driven by the Na^+ gradient across the plasma membrane. Two or more Na^+ ions are cotransported with a molecule of transmitter and the energy available from this 'downhill' transport of the ions allows the transmitter to be concentrated many thousand-fold into the cell, lowering the extracellular concentration to submicromolar levels.

1.3 Fundamentals of neuroanatomy

A detailed description of neuroanatomy is outside the scope of this book. However, a basic knowledge is essential for the molecular neuroscientist in order to understand the literature and to select appropriate preparations for research. This section will discuss the major brain areas which provide sources of material and will refer to the standard laboratory animal — the rat (Fig. 1.7).

1.3.1 The spinal cord and peripheral nervous system

The spinal cord receives sensory information from the PNS mainly via pathways termed *dorsal roots* which are arranged segmentally down the spinal cord (Fig. 1.8). The cell bodies for these sensory neurones are grouped in *dorsal root ganglia*. The cell body of a sensory neurone (*see* Fig. 1.1) has a single process which divides to form a myelinated axon and dendrite.

The spinal cord controls movement in the limbs and trunk via *motor neurones* with cell bodies in the grey matter of the spinal cord and axons grouped in *ventral roots* (Fig. 1.8). *Spinal cord reflexes* involve a stimulation of motor neurones by incoming sensory input which is direct (perhaps via an interneurone) and does not involve brain.

Output from the spinal cord to smooth muscle, heart and secretory glands occurs via the *autonomic nervous system* (*ANS*) which is in turn classified into sympathetic and parasympathetic systems. The *sympathetic system* involves *preganglionic* cholinergic neurones with cell bodies in the spinal cord or brain stem (*see* Section 1.3.2) which synapse onto *postganglionic* noradrenergic neurones grouped in segmental ganglia, and which in turn innervate the target tissues preparing the body for the classic 'fight or flight' decision.

The *parasympathetic system* involves a different set of preganglionic cholinergic neurones which synapse at distinct ganglia onto postganglionic cholinergic neurones. These in turn innervate the same target tissues as the sympathetic system, but usually having an action antagonistic to the latter.

1.3.2 The brain stem

The brain stem is the region of brain between the spinal cord and the diencephalon (*see* Section 1.3.4) and contains three major regions: the *medulla, pons* and *midbrain* (Fig. 1.7). As well as containing all the nerve tracts responsible for transferring information between the spinal cord and brain, the brain stem controls respiration and heart beat and processes information for hearing, balance and taste.

Cell bodies in the brain stem are either clustered in *nuclei* or in a more diffuse *reticular formation*. Half the noradrenergic neurones in brain have their cell bodies in a nucleus within the brain stem called the *locus ceruleus*, and axons from these neurones diffusely innervate regions of the brain as distant as the olfactory bulb and the cerebral cortex.

Dopaminergic neurones outnumber noradrenergic neurones in brain. A major group of dopaminergic cell bodies in the brain stem is located in the *substantia nigra* and project their axons predominantly to the striatum. The system is involved in the control of voluntary movement and damage results in *Parkinson's disease*.

Serotoninergic (5-HT-secreting) cell bodies are present in large numbers in the brain stem, grouped in several distinct *raphe nuclei*. Their axons project down to the spinal cord where they control sensory and motor neurones, and also project to regions of brain including the cerebral cortex, hippocampus and striatum.

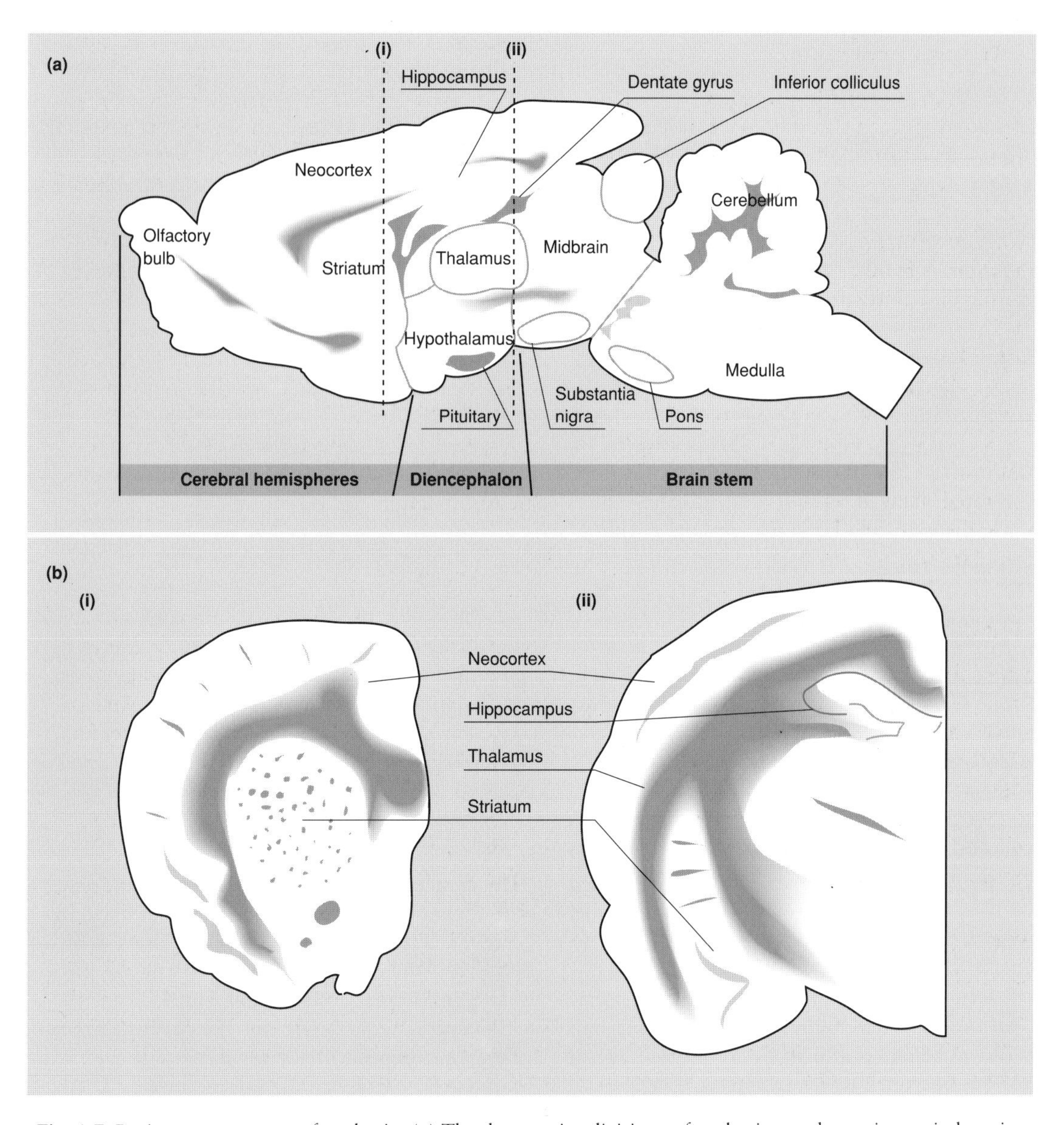

Fig. 1.7 Basic neuroanatomy of rat brain. (a) The three major divisions of rat brain are shown in saggital section. The brain stem includes the medulla, midbrain and cerebellum; the diencephalon comprises the thalamus, hypothalamus and pituitary, while the cerebral hemispheres include the striatum, olfactory bulb, neocortex, hippocampus and dentate gyrus. (b) Two half-coronal sections: (i) through the striatum, and (ii) through the hippocampus and thalamus.

1.3.3 The cerebellum

The cerebellum (Fig. 1.9), which is connected to the *pons* by tracts of nerve fibres termed *peduncles*, is involved in the control of movement and in the learning of motor skills.

The surface, or cortex, of the cerebellum contains a highly ordered array of neurones. The input into the cerebellar cortex occurs principally via the *mossy fibres* and also the *climbing fibres*. Mossy fibres originate from the brain stem and spinal cord and terminate in large, glutamatergic *glomeruli* which synapse onto the dendrites of local clusters of *granule cells*. Each mossy fibre branches to provide an input to about 400 granule cells. The small granule cells are present

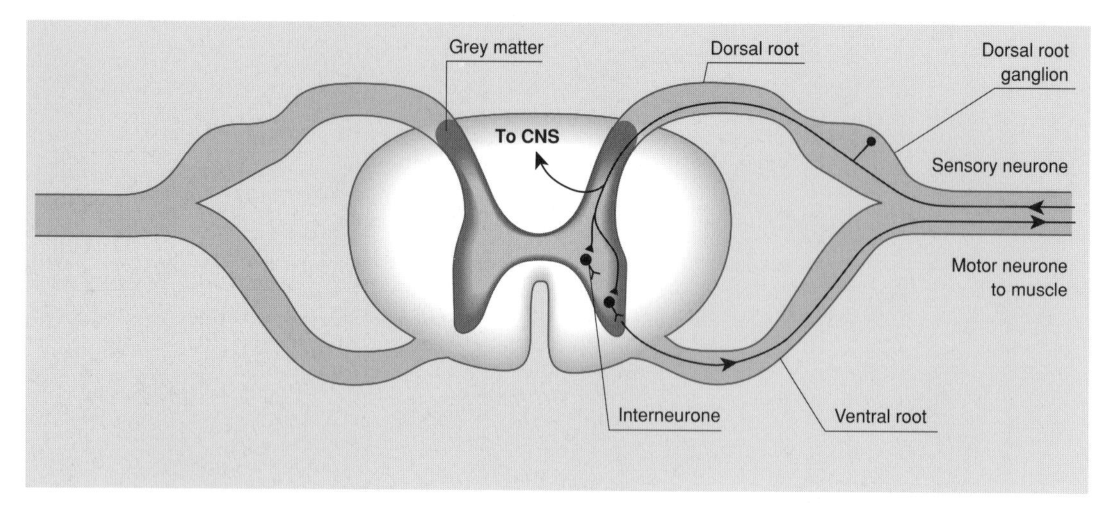

Fig. 1.8 The spinal cord, sensory and motor neurones. Sensory neurones, with cell bodies in the dorsal root ganglion, receive signals from sensory receptors and synapse onto motor neurones in the grey matter of the spinal cord, either directly or via short interneurones.

in vast numbers and their cell bodies pack the *granular layer* of the cerebellum. The axons from the granule cells ascend into the *molecular layer* where they bifurcate and give rise to *parallel fibres* whose glutamatergic boutons synapse onto the dendrites of a row of *Purkinje cells*. Purkinje neurones have large (50–80 μm) cell bodies, and extremely extensive dendritic trees, orientated at 90° to the parallel fibres. Each Purkinje cell receives an input from about 100 000 parallel fibres. *Climbing fibres* originate from the *inferior olive* in the cerebellar medulla and make synaptic contacts, which have been suggested to utilize aspartate as a transmitter, onto the soma and proximal dendrites (those close to the soma) of Purkinje neurones, which each receive a powerful excitatory input from a single climbing fibre. The only output from the cerebellar cortex occurs down the axon of the Purkinje cell. Three classes of inhibitory (GABAergic) interneurones are found in the cerebellar cortex (an *interneurone* is one whose processes are all contained within one anatomical region of brain, in contrast to *principal neurones* which *project* their axons to distant brain areas). Each type of GABAergic neurone is excited by parallel fibres; *Golgi cells* provide a negative feedback onto the granule cell dendrites at the glomeruli, *stellate cells* inhibit nearby Purkinje cells, while *basket cells* inhibit more distant Purkinje cells outside the 'beam' of excited parallel fibres.

During the development of the cerebellum, granule cell bodies migrate inward from the external surface layer along preformed cytoplasmic processes of a class of oligodendrocyte called *Bergmann glia*.

1.3.4 The diencephalon

The diencephalon (or 'between brain') comprises the thalamus which processes information to the cerebral cortex and the hypothalamus which regulates the PNS and, by controlling the pituitary, endocrine secretion (*see* Fig. 1.7).

Neurosecretory cells in the *hypothalamus* exocytose peptide *releasing factors* into portal capillaries which are directly connected to the *anterior pituitary* where they control hormone secretion into the bloodstream (*see* Fig. 12.2). Thus thyrotropin-releasing factor, corticotropin-releasing hormone, growth-hormone releasing hormone and luteinizing hormone-releasing hormone induce the release of the appropriate hormones from the anterior pituitary.

1.3.5 The cerebral hemispheres

The cerebral hemispheres comprise the cerebral cortex, basal ganglia, hippocampus and amygdala (*see* Fig. 1.7).

The *cerebral cortex* is a sheet of cells, smooth in the rat and folded in higher mammals, which

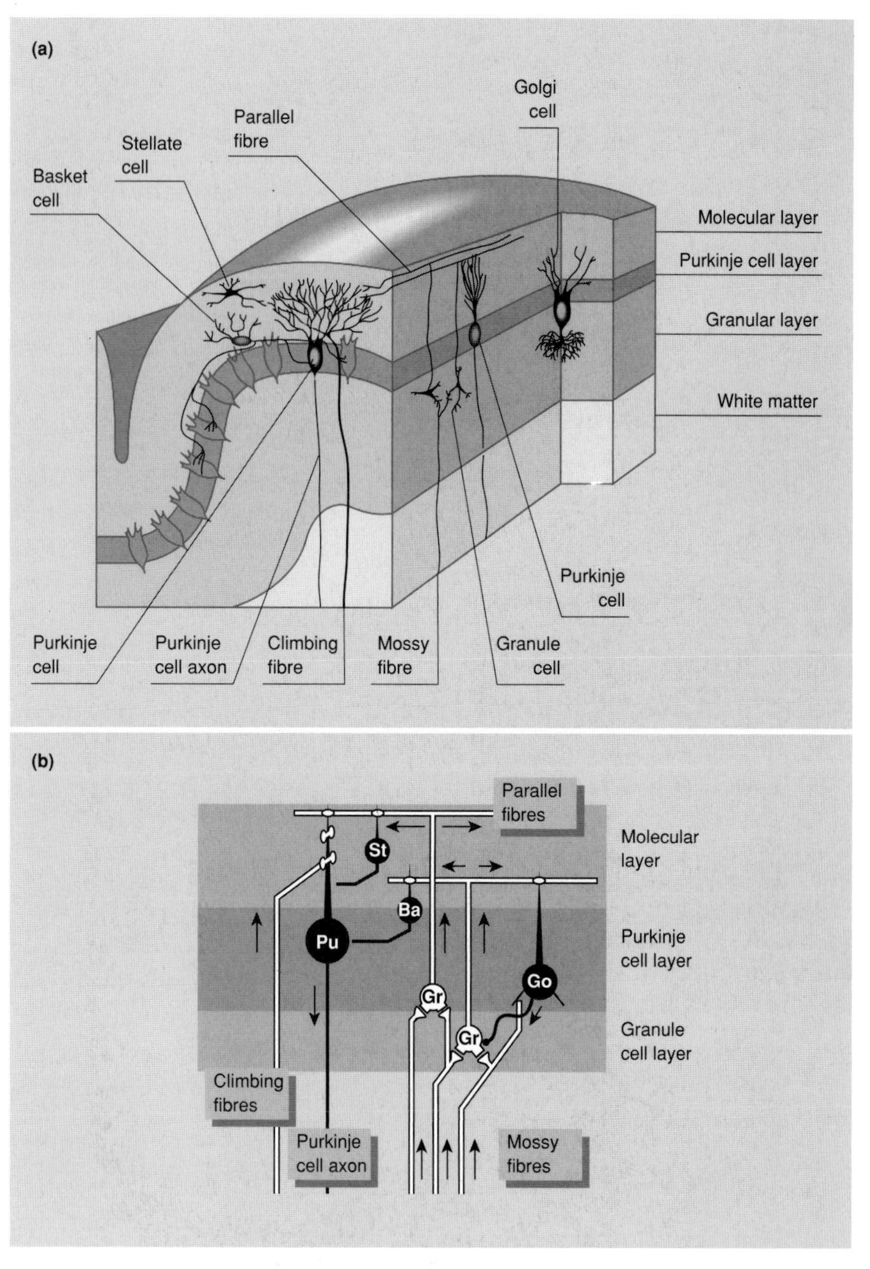

Fig. 1.9 The cerebellum. (a) Section through the cerebellar cortex. The main neurones are the numerous small granule cells which receive inputs from the mossy fibres and send out axons which bifurcate to form parallel fibres, and the Purkinje cells with large cell bodies and enormously extensive dendritic trees. In addition there are inhibitory basket cells, Golgi cells and stellate cells. The cerebellar cortex is organized into the molecular layer, containing the Purkinje dendrites and parallel fibres, the Purkinje cell layer containing the Purkinje cell bodies and the granular layer with the granule cell bodies. (b) Main circuits operative in the cerebellar cortex. Mossy fibres provide excitatory input from the brain stem, spinal cord, etc., via large glomeruli with the dendrites of the granule cells (Gr) which also receive inhibitory input from Golgi cells (Go). The granule cell parallel fibres form excitatory synapses onto the Purkinje cells (Pu) and also inhibitory Golgi, basket (Ba) and stellate (St) cells. These two latter cells synapse onto Purkinje cell soma and dendrites respectively. A Purkinje dendritic tree also receives a powerful excitatory input from a single climbing fibre. The only output from the cortex is via the Purkinje cell axon.

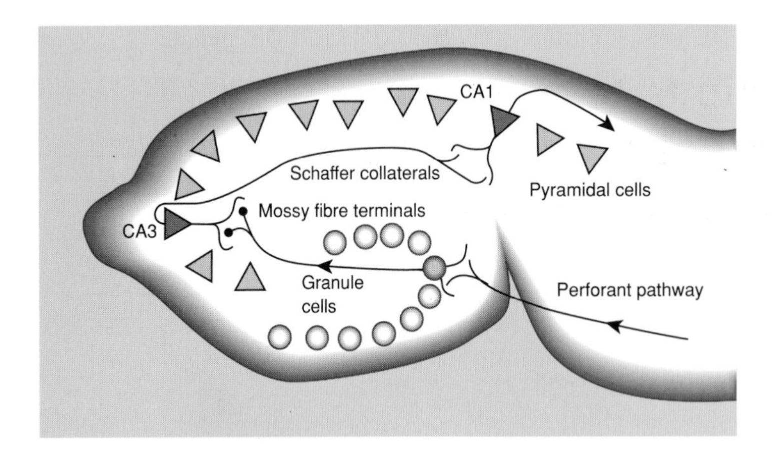

Fig. 1.10 The hippocampus. Schematic pathways in a transverse section of rat hippocampus. Input from the perforant pathway synapses onto granule cells whose axons in turn form synapses via large mossy fibre terminals with the dendrites of CA3 pyramidal neurones. CA3 neurones then synapse onto CA1 pyramidal cells in Schaffer collaterals.

covers the surface of the cerebral hemispheres and is responsible for the highest level of information processing. The *neocortex* is that portion of the cerebral cortex visible on the surface of brain, and has a six-layered structure, below which is the white matter made up of the mass of myelinated axons carrying information to and from the neocortex.

The *basal ganglia* largely comprise the *globus pallidus* and the *corpus striatum* (which is in turn divided anatomically into the *caudate nucleus* and the *putamen*). The *hippocampus* (Fig. 1.10), together with the associated *dentate gyrus* and *subiculum*, plays an essential role in the transformation of short-term memories into a permanent form for storage in the neocortex. Damage to the hippocampus results in an inability to form new memories, even though the recollection of events preceding the injury seems unimpaired. The *amygdala* controls the autonomic and endocrine systems.

In the following chapter we shall discuss the *in vitro* preparations from these diverse brain regions which can be used to give an insight into the molecular mechanisms of neurotransmission.

Further reading

Hall, J.W. (1992) *An Introduction to Molecular Neurobiology*. Sunderland, Mass.: Sinauer.

Isokawa, M., Levesque, M.F., Babb, T.L. & Engel, J., Jr (1993) Single mossy fiber axonal systems of human dentate granule cells studied in hippocampal slices from patients with temporal lobe epilepsy. *J. Neurosci.* **13**, 1511–1522.

Kandel, E.R., Schwartz, J.H. & Jessell, T.M. (1991) *Principles of Neural Science*. New York: Elsevier.

Levitan, I.B. & Kaczmarek, L.K. (1991) *The Neuron, Cell and Molecular Biology*. New York: Oxford University Press.

McGeer, P.L., Eccles, J.C. & McGeer, E.G. (1987) *Molecular Neurobiology of the Mammalian Brain*. New York: Plenum.

Siegel, G., Agranoff, B.W., Albers, R.W. & Molinoff, P. (1989) *Basic Neurochemistry*. New York: Raven Press.

Strange, P.G. (1992) *Brain Biochemistry and Brain Disorders*. Oxford: Oxford University Press.

2

Preparations for the study of synaptic transmission

The explosive development of molecular neuroscience in recent years has relied heavily on a number of preparations each of which have contributed to our present understanding. Before we deal with these advances in the remainder of the book we shall consider some of the major preparations, progressing from the complex to the simple; beginning with the brain slice and ending with the isolated membrane patch and systems for the expression of receptor and channel messenger ribonucleic acid (mRNA).

2.1 The brain slice

The brain slice is one of the oldest preparations for biochemical studies and has also been used for electrophysiology since the mid 1960s. Slices must be thin enough to allow adequate oxygenation of their core, but at the same time sufficiently thick to minimize the proportion of the slice which is damaged by the slicing itself, although they inevitably consist of a region of intact tissue sandwiched between two surface layers damaged by the cuts. In practice a compromise is usually reached at 0.3–0.4 mm. The orientation of the slice is important, particularly for the maintenance of local circuits for electrophysiology. The most widely utilized preparation for electrophysiology is the transverse hippocampal slice (shown schematically in Fig. 1.10) which retains a number of neuronal pathways and allows the study of processes by which transmission at individual synapses may undergo long-term regulation (*see* Section 13.5).

For neurochemical investigations an advantage of the slice over the isolated nerve terminal, or synaptosome, preparation (*see* Section 2.2) is that the slices may be electrically stimulated by field electrodes located on either side of the slice. In this way patterns of electrical stimulation may be applied which mimic the physiological input. However, while brain slices have been much employed for biochemical studies, it must be borne in mind that the preparation has three levels of heterogeneity: neurones vs glia, cell bodies vs nerve terminals and the inherent transmitter heterogeneity of the terminals themselves. Thus considerable effort must be devoted to unravelling the compartmentation of metabolism in this preparation.

2.2 Primary neuronal cultures

Primary neuronal cultures are those obtained directly from brain rather than from established cell lines (*see* Section 2.3). For biochemical purposes yield and homogeneity are the major criteria. While primary cultures may be made in limited yield and purity from a number of brain regions, one preparation which readily produces large numbers of neurones with >90% homogeneity is the cerebellar granule cell (Fig. 2.1). These cells give rise to the parallel fibres in the cerebellum which form glutamatergic synapses with the Purkinje cells (*see* Fig. 1.9). They are the most abundant single neuronal type in the

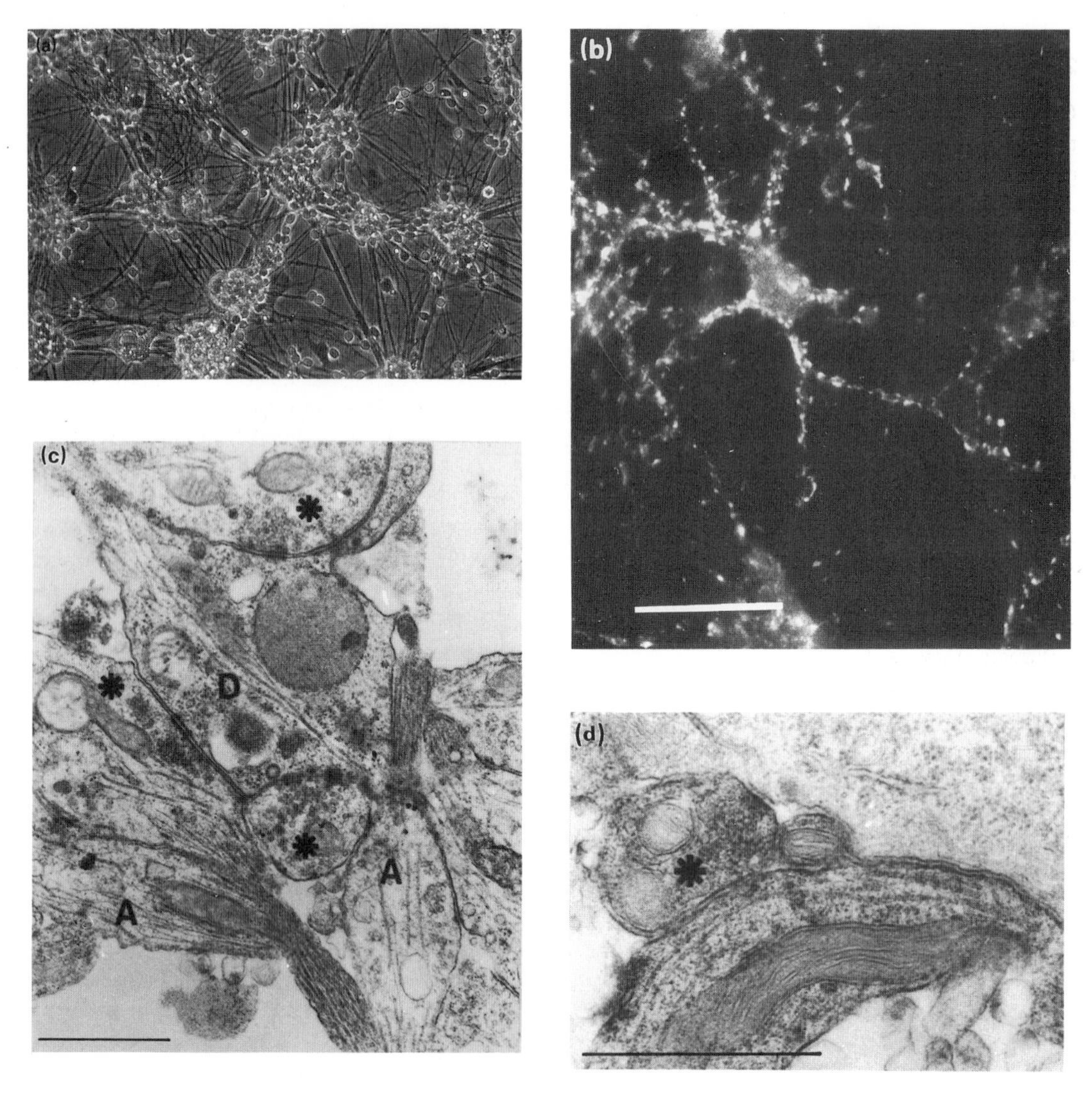

Fig. 2.1 Primary neuronal culture: the cerebellar granule cell. (a) Phase-contrast image of a primary culture of rat cerebellar granule cells maintained *in vitro* for 7 days. Note the clustering of neuronal cell bodies and the neurite bundles. (b) Immunofluorescence of a 14 DIV granule cell culture stained with antibodies to synaptophysin, a synaptic vesicle marker. Note the presence of presynaptic terminals on neurites and synapsing onto the cell soma. Bar = 10 μm. (c) Electron micrograph of a neurite region of a 12 DIV granule cell culture. Neurites are seen with axonal (A) and dendritic (D) characteristics. Synaptic vesicles are indicated by asterisks. Bar = 2.5 μm. (d) Higher magnification of an axo-somatic synapse. Bar = 1 μm. Data from Van-Vliet *et al.* (1989), by permission of Raven Press.

brain; the cerebellum of an 8-day rat containing about 50 million cells. The high purity of the culture relies on the exact developmental stage at which the preparation is made. At 8 days the Purkinje cells have differentiated and do not survive the dissociation and plating. In contrast the granule cells and glia are still largely undifferentiated allowing them to be plated. One day later the granule cells differentiate, at which stage a mitotic inhibitor is added to kill the still dividing glia.

2.3 Chromaffin cells

Chromaffin cells from the adrenal medulla share the same precursor cells as sympathetic neurones, possess a number of 'neurone-specific' proteins and can be considered as modified postganglionic sympathetic neurones. There are two distinct populations of cells releasing adrenaline and noradrenaline respectively, although these are rarely separated in practice. Acutely isolated bovine adrenal chromaffin cells (Fig. 2.2) can be obtained in high yield and purity by perfusion of bovine adrenal glands with collagenase-containing medium, following which the cells can be maintained in culture for several days. A chromaffin cell has a diameter of about 16 μm and contains some 30 000 large electron-dense secretory vesicles with a typical diameter of 280 nm. These vesicles appear to be closely related to neuronal LDCVs (*see* Section 8.2) although they are substantially larger (*see* Fig. 11.7). Depolarization- or receptor-evoked exocytosis from these cells has been extensively studied. It must be emphasized, however that exocytosis from these cells differs in several respects from small synaptic vesicle exocytosis in the CNS.

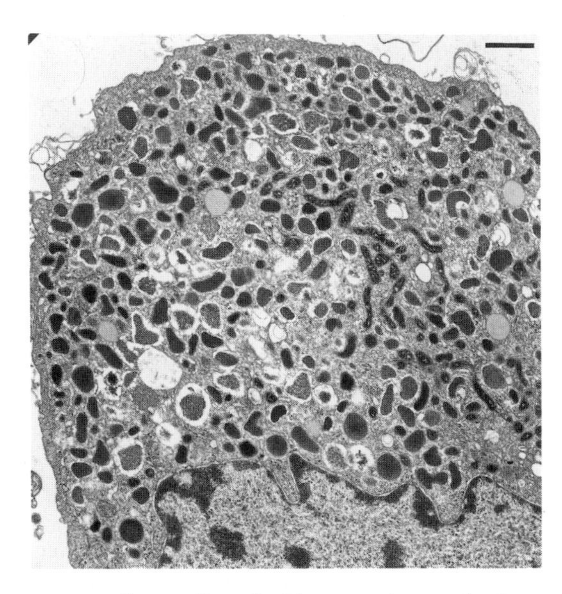

Fig. 2.2 Chromaffin cells. Electron micrograph of an isolated bovine adrenal chromaffin cell showing the packing of the cytoplasm with large dense-core (chromaffin) vesicles; bar = 1 μm. From Burgoyne (1991), by permission of Elsevier Science Publishers.

2.4 Neuronal- and glial-derived cell lines

The complex and heterogeneous organization of brain presents a particular challenge to the biochemist wishing to elucidate the coupling of metabotropic receptors to intraneuronal pathways of signal transduction. Unlike ionotropic receptors, which can be investigated electrophysiologically at a single neurone, investigation of metabotropic receptor coupling generally requires a homogeneous cell population with a defined, and preferably limited, receptor population such as an immortalized clonal cell line. Furthermore functional investigation of the staggering number of potential isoforms of ionotropic and metabotropic receptors, G-proteins and effectors revealed by molecular cloning can most clearly be performed by introducing the genes for specific isoforms into host cell lines normally lacking the endogenous gene product, or by specific mutagenesis to abolish expression of, for example, a receptor subunit. Cell lines are not without their problems: their relationship to *in vivo* neurones or glia may be rather distant, while the phenotype of cell lines may vary with culture conditions and between laboratories.

PC12 phaeochromocytoma cells, derived from chromaffin cell tumours can be maintained as a cell line demonstrating a non-neuronal phenotype. Under the influence of nerve growth factor the cells extend neurites (immature axons and dendrites) and differentiate into a neuronal cell type capable of releasing catecholamines such as dopamine (*see* Fig. 13.1).

The study of metabotropic signal transduction is, of course, not limited to neuronal and glial derived cell lines; thus a number of mutants of the *S49 lymphoma* cell line lacking the trimeric G-protein, G_s (*see* Section 5.3.1), possessing G-protein variants incapable of interacting with receptor or effector or alternatively lacking agonist-stimulated protein kinase A (PKA) have provided valuable information about the nature of G-protein coupling. *1321N1* astrocytoma cells which possess only the M_3 subclass of muscarinic acetylcholine receptors coupled via a pertussis toxin-insensitive G-protein to phospholipase C (*see* Section 5.4.2) allow this pathway to be studied without interference from other signal transduction pathways.

NG108–15 is a hybrid cell formed by the

fusion of separate neuroblastoma and glioma cell lines and related to sympathetic ganglion neurones. When the cells are dividing freely on a culture plate they show few processes (Fig. 2.3), however differentiation to a more neuronal morphology can be induced when the cells reach confluence and stop dividing, or by the addition of a growth factor or membrane-permeant cAMP analogue. The large (30 μm) soma facilitate patch-clamp analysis of ion currents (*see* Section 2.8) and voltage-dependent Na^+-, Ca^{2+}- and K^+-channels can be studied. Additionally the cells express receptors for bradykinin, enkephalins, noradrenaline, serotonin and ACh. In this last case the single M_4 subclass of muscarinic ACh receptors (*see* Section 10.4) negatively coupled to adenylyl kinase allows this isoform to be studied in isolation.

Cell lines can be used as vehicles to investigate the functional effects of deletions of sequence within receptors. A related approach is to construct chimeric receptors from the chromosomal DNA (cDNAs) or closely related but functionally distinguishable receptors in order to determine the function of specific regions within the receptor. Similarly the expression in cell lines of receptors which have been subjected to site-directed mutagenesis can enable critical residues to be identified, for example, in controlling ion channel specificity or kinetics (*see* Section 5.3).

2.5 Synaptoneurosomes

Synaptoneurosomes are isolated terminals with attached resealed postsynaptic membranes, and thus present a 'figure-of-eight' appearance (Fig.

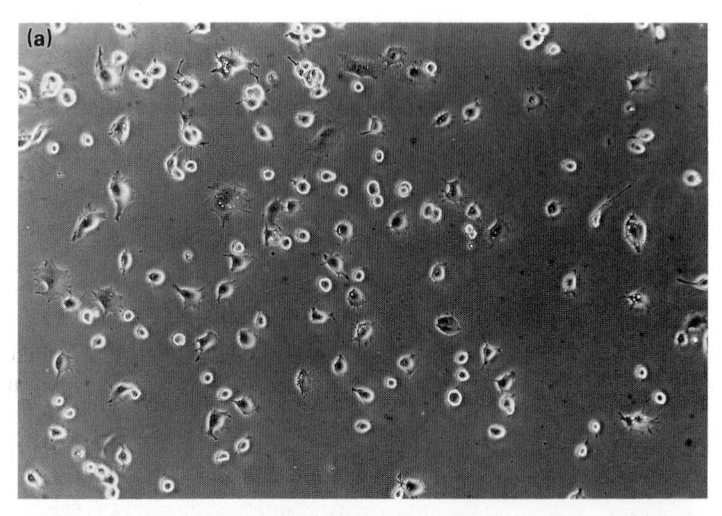

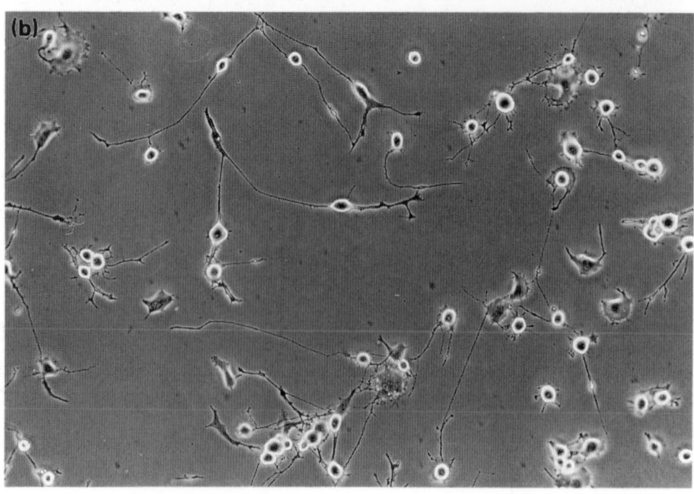

Fig. 2.3 NG108−15 neuroblastoma × glioma cells. (a) Cells viewed under phase-contrast show few processes prior to differentiation. (b) After differentiation with prostaglandin E1 and IBMX (a phosphodiesterase inhibitor) the cells differentiate and extend long neurites. From Docherty *et al.* (1991), by permission of Oxford University Press.

2.4). They are produced by homogenization of brain tissue in ionic media and forcing the resultant suspension through increasingly fine filters. While they are far from homogeneous, they have the advantage that postsynaptic as well as presynaptic events can be studied; for example, they have been employed to investigate second messenger responses to postsynaptic metabotropic receptors coupled to (1,4,5)-inositol triphosphate [(1,4,5)IP_3] formation (*see* Section 5.4.3).

2.6 Synaptosomes

Mitochondria are produced from a variety of tissues by differential centrifugation of tissue homogenates in sucrose. When a crude mitochondrial fraction from the brain is examined under the electron microscope it is found to contain plentiful particles which are evidently pinched-off nerve terminals. These *synaptosomes* as they were termed in 1964 have a characteristic appearance (Fig. 2.5) with a diameter of 0.5–1 μm, containing one or more small mitochondria frequently with an area of postsynaptic membrane adhering to it opposite a concentration of 50 nm diameter synaptic vesicles.

Mammalian CNS synaptosomes can be made most effectively from those brain regions such as the cerebral cortex and hippocampus which have clearly defined anatomical layers. This is because cleavage of these layers during homogenization shears off terminals from axons more readily than in regions lacking such a detailed layered structure.

A crude mitochondrial pellet, often referred to as a P2 pellet, containing predominantly a mixture of synaptosomes and free mitochondria from cell bodies and glia, is then further purified either by differential rate sedimentation through a percoll gradient in an angle rotor (Fig. 2.5), or alternatively by equilibrium gradient centrifugation of the P2 pellet into a ficoll gradient in a swing-out rotor.

Synaptosomes once prepared remain viable for several hours. This is perhaps not surprising since the *in situ* terminal must operate autonomously from the cell body except for the slow replacement of proteins and membrane components by the processes of axonal transport. They represent the simplest system which retains all the machinery for the uptake, synthesis, storage and exocytosis of neurotransmitters.

Virtually all neuronal preparations are heterogeneous in terms of transmitter content, since even the most closely defined anatomical region contains a variety of transmitters. Although it is possible to choose brain regions which optimize the concentration of the transmitter of interest, it is not possible to separate transmitter-specific subpopulations of synaptosomes on the basis of physical differences (size, density, etc.). Various immunological techniques have been tried to purify synaptosomal subpopulations from mammalian brain; one of these will be discussed in Section 10.2.

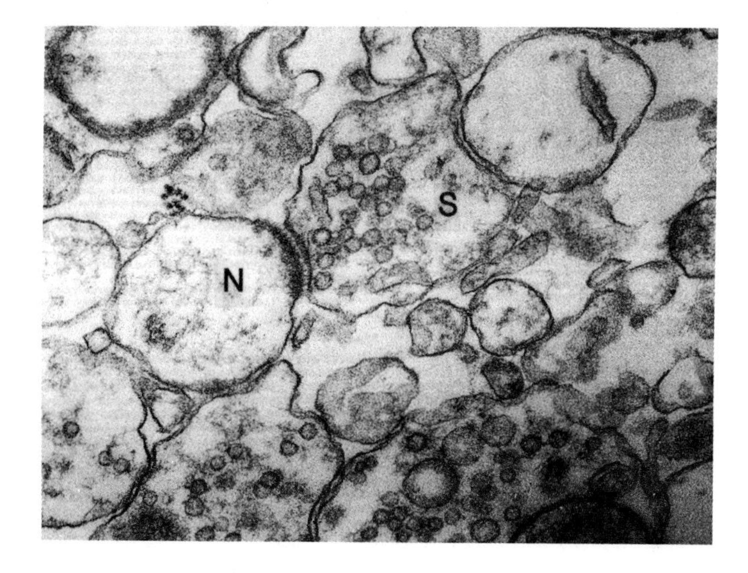

Fig. 2.4 The synaptoneurosome. Synaptoneurosomes are preparations containing resealed presynaptic (S) and postsynaptic (N) elements linked by a synapse to form a double organelle, in contrast to synaptosomes where any attached postsynaptic membrane is disrupted. Synaptoneurosomes are made by homogenization in a physiological ionic medium rather than sucrose, after which the homogenate is filtered through nylon mesh and ultimately a 5 μm pore filter. From Hollingsworth *et al.* (1985), by permission of the Society for Neuroscience.

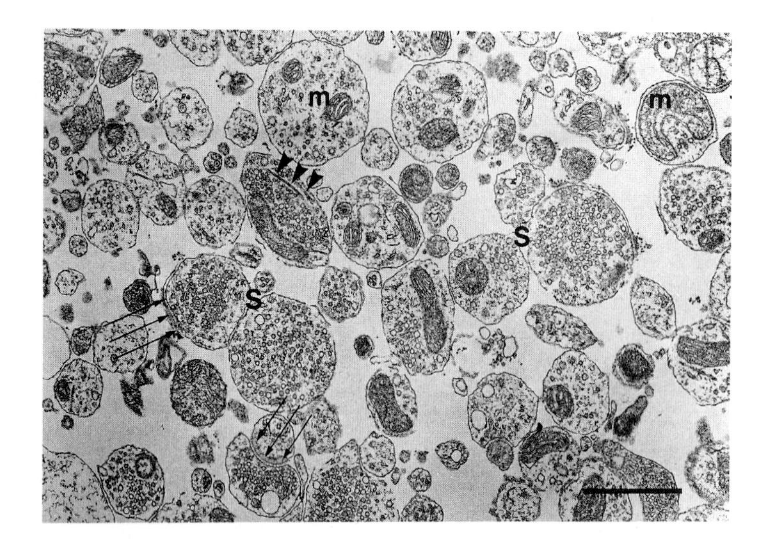

Fig. 2.5 The synaptosome. A field of synaptosomes from rat cerebral cortex purified on a Percoll gradient. Arrowheads indicate the presence of microtubules within the terminals while small arrows show postsynaptic densities attached to the terminals. S, synaptosomes; m, mitochondria. Bar = 1 μm. From Dunkley (1988), by permission of Elsevier Science Publishers.

2.6.1 Neurosecretosomes

The neurohypophysis or posterior pituitary (*see* Fig. 12.2) contains the nerve terminals of magnocellular neurones whose cell bodies are in the hypothalamus. Isolated nerve terminals can readily be obtained in high purity, particularly since the neurohypophysis contains no neuronal cell bodies. When depolarized with potassium chloride (KCl) they secrete the peptide neurohormones *oxytocin* and *vasopressin* which are contained within LDCVs with diameters between 200 and 400 nm. Most of the '*neurosecretosomes*' have a diameter of about 1 μm, but some are much larger, permitting patch clamping (*see* Section 2.8).

2.6.2 Permeabilized cell and synaptosome preparations

For studies requiring the introduction of impermeant compounds into secretory cells the permeability barrier which is presented by the plasma membrane must be overcome. At the single cell level this may be accomplished by microinjection, for example, via a patch electrode (*see* Section 2.8). For large populations of cells, where microinjection is impractical, agents capable of selective permeabilization of the plasma membrane are used.

The resistance of a vesicle to *osmotic lysis* is proportional to its radius of curvature. Consequently, hypotonic stress will lyse synaptosomes and larger cells, but not synaptic vesicles. Similarly the ice crystals produced during freezing will rupture the plasma membranes of cell and synaptosomes causing less damage to internal organelles. Thus *freeze-thawing* has been used to introduce impermeant proteins into synaptosomes, although the recovery of function is only partial. Another technique whose selectivity is based on relative diameter is *electroporation* whereby a suspension of cells is exposed to a transient high electrical field, typically by the discharge of a condenser across two plate electrodes; this burns a pair of 2–4 nm diameter holes in the plasma membrane at the points where the electric field is concentrated across the thin membrane (Fig. 2.6). Smaller organelles (e.g. secretory vesicles) survive as their smaller diameter results in a proportionately lower field across their membranes. Electroporation allows molecules up to 1 kDa to cross the membrane, and has been much used with chromaffin cells to obtain permeabilized preparations retaining the capacity for exocytosis. Synaptosomes are too small to allow this technique to be used.

Digitonin and the related *saponin* are plant alkaloids which form a complex with cholesterol in the plasma membrane and create holes large enough for proteins to pass in and out of the cell. Since plasma membranes contain the highest proportion of cholesterol they are most sensitive, although careful titration is necessary to avoid lysis of secretory vesicles. This technique has also been extensively employed to study chromaffin cell secretion which will be discussed in Chapter 8.

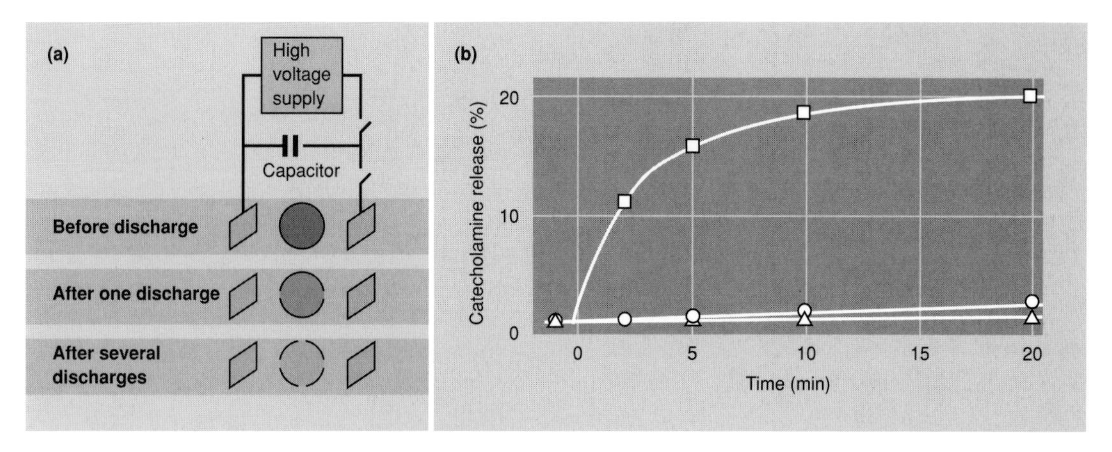

Fig. 2.6 Permeabilized preparations. (a) Electroporation: a suspension of cells (e.g. chromaffin cells) is exposed to a field of $2\,kV \cdot cm^{-1}$ ($200\,mV \cdot \mu m^{-1}$). Because the inside of a cell is at an isopotential, all the potential drop due to the field is concentrated across the plasma membrane. When this exceeds about 1.5 V two holes are 'burnt' in the membrane along the axis of the field. As the cell tumbles and the discharge is repeated several holes can be created in the membrane, and these remain open for an extended period. (b) Chromaffin cells suspended in media containing MgATP and $10\,\mu M$ Ca^{2+} were exposed to several discharges to render them leaky; some 20% of the total catecholamine is released over a period of 10 min (squares). No release occurred in the absence of Ca^{2+} (circles) or from non-permeabilized cells (triangles). No release of the cytoplasmic marker lactate dehydrogenase occurs, showing that the holes generated in the plasma membrane are too small to allow rapid loss of cytoplasmic proteins. However, dopamine β-hydroxylase which is co-stored in the vesicles is released in parallel with catecholamines, arguing for exocytosis rather than non-specific lysis. Data from Baker & Knight (1980).

Staphylococcal α-toxin (from *Staphyloccocus aureus*) is a soluble monomeric protein which oligomerizes in target membranes to form stable hexamers defining 2–4 nm diameter transmembrane pores. These allow the entry of molecules up to 1 kDa. The toxin is selective for the plasma membrane since the pores formed are too small for the soluble monomers to pass and gain entry to the cytoplasm. *Streptolysin-O* forms larger pores (approximately 10 nm diameter) in the plasma membrane sufficient to allow entry (and exit) of proteins up to the size of antibodies. This agent has been used in studies investigating the possible roles of PKC substrates in the control of noradrenaline (NA) exocytosis from synaptosomes (*see* Section 7.5.1).

2.7 Small synaptic vesicles

Synaptic vesicles were originally obtained as a pure fraction from rat brain synaptosomes in 1964 by the Whittaker group. Synaptosomes were exposed to hypotonic conditions in order to rupture the plasma membrane and release synaptic vesicles (which are resistant to hypotonic shock by virtue of their smaller size). The fraction was then applied to a sucrose density gradient and centrifuged to equilibrium. This has the disadvantage that lengthy centrifugation through sucrose leads to a loss of amino acids from the subpopulations of vesicles originally containing glutamate, GABA or glycine. This failure to find vesicular amino acids provided the main impetus for early theories of cytoplasmic release for the Group I amino acid neurotransmitters (*see* Section 9.5.5).

Further purification can be achieved by 'gel-permeation' on controlled pore glass bead columns, or chromatography on Sephacryl S-1000 columns. Small electron lucid synaptic vesicles appear under the electron microscope as a homogeneous population of 50–60 nm diameter vesicles. The distinctive nature of vesicle-associated proteins may also be exploited by immunoisolating vesicles with immobilized antibodies to vesicle-associated proteins such as synapsin I or synaptophysin (*see* Sections 7.4.3 and 7.4.4). In the latter technique, monoclonal antibodies against a cytoplasmic domain of synaptophysin were immobilized onto the surface of methacrylate 'immunobeads'. A 35 000 g supernatant of brain homogenate was equilibrated with the beads which were then sedimented, pulling down the purified synaptic vesicles. Only a limited yield of vesicles is obtained by these methods, frequently only 1 mg of vesicle protein from the brains of ten rats. This may be due to the attachment of synaptic vesicles to the cytoskeleton as only the fraction lacking such attachment would be liberated from hypotonically lysed synaptosomes. Additionally, hypotonic lysis may only temporarily rupture the plasma membrane before it reseals, allowing insufficient time for vesicles to dissociate from the cytoskeleton.

The electric organ of electric fish such as the ray *Torpedo* or the eel *Electrophorus* are a uniquely rich source of all the synaptic components of a cholinergic (ACh-secreting) synapse. In particular, *Torpedo* synaptic vesicles can be prepared by freezing the tissue in liquid nitrogen followed by mechanical crushing of the frozen tissue, and zonal centrifugation of a sucrose gradient.

2.8 Patch-clamp and the analysis of single ion channel conductances

The extraordinary high ion conductance of open channels and ionotropic receptors (some 10^4 ions can permeate per channel per millisecond) allied to the availability of stable high-gain amplifiers and the ability to *patch-clamp* minute areas of the plasma membrane has brought single channel ion conductances within the range of routine measurement. A glass micropipette with a fire-polished tip of diameter 1–5 μm is induced to form an extremely tight seal with the plasma membrane of a cell (Fig. 2.7). As long as the cell surface is free of extracellular matrix, slight suction can lead to the formation of a 'gigaohm seal' where the short-circuit resistance between the pipette interior and the extracellular fluid can be 1–100 GΩ, sufficient to allow detection of single-channel conductances. The technique can be used not only for isolated neurones grown in culture on coverslips, but also for cells in brain slices, where the debris on the surface of the slice is washed off by a stream of fluid before application of the patch.

In the *cell attached patch* mode the area of membrane thus isolated might contain only one or two channels. Since the typical resistance of a single channel is 10–100 GΩ (corresponding to 100–10 pS) suitable amplification can allow the recording of current passing through single channels as a function of applied voltage. The circuit is completed by current flowing across the much lower resistance of the remainder of the cell. Further suction will destroy this patch and allow *whole-cell recording* which measures the integrated current due to all the channels in the plasma membrane. The whole-cell mode also allows the cell to be dialysed with the contents of the pipette. Depending on the application this can be an advantage or a problem. For example, agents can be present at high concentration within the patch pipette in order to study their effect on channel behaviour; however channels frequently appear to inactivate merely because physiological cofactors in the cytoplasm are diluted away. A recent extension of the whole-cell mode allows changes in membrane capacitance to be measured; since this is proportional to surface area, capacitance changes due to vesicle exocytosis and endocytosis can be detected in, for example, chromaffin cells and mast cells (*see* Section 8.2).

In order to allow access to both sides of the plasma membrane, two conformations of isolated patch can be obtained. If the pipette in the cell attached mode is withdrawn, a 'bubble' of membrane forms which can be broken by exposure to air, leaving an *inside-out patch* with ready access to the original cytoplasmic face. Alternatively pipette withdrawal from a whole cell mode results in the formation of an *outside-out patch*. To decrease the disturbance to the cytoplasm in the whole cell mode, a *perforated patch* can be used. The polyene antibiotic *nystatin* present in the patch electrode permeabilizes the plasma membrane which contains cholesterol. The resulting pores allow the free passage of monovalent cations (and to a lesser extent Cl^-) but are impermeant to divalent cations and all larger molecules. The perforated patch thus allows electrical continuity with the cell, but prevents dialysis of the cell's contents.

2.9 *Xenopus* oocytes

Xenopus oocytes are widely employed for translating foreign mRNA. Not only are they capable of synthesizing, assembling and inserting a variety of functional receptors and voltage-gated ion channels after injection of exogenous mRNA, but they are also readily accessible to electro-

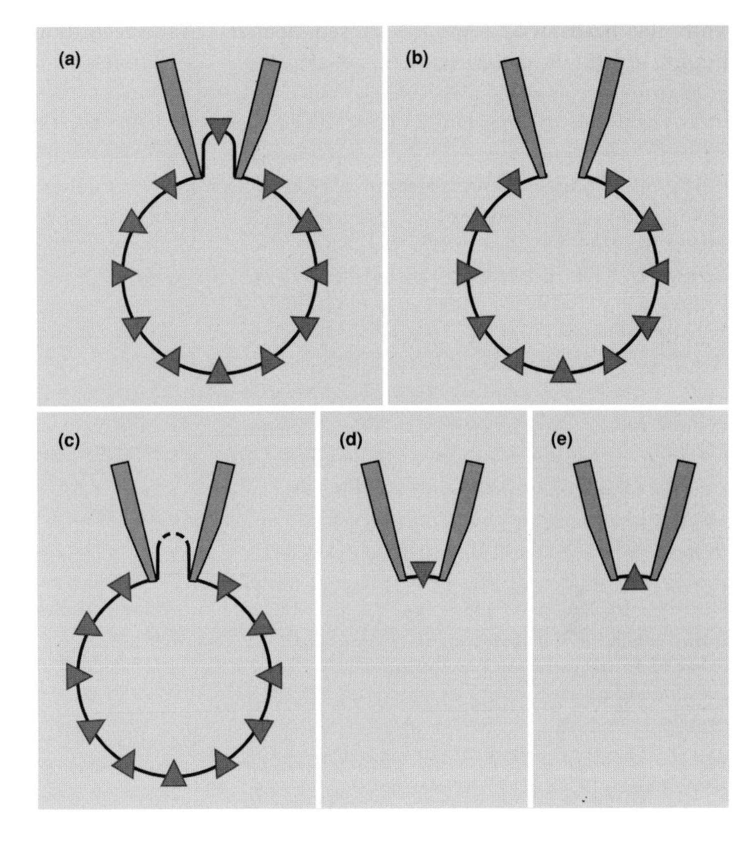

Fig. 2.7 Patch-clamp. (a) A carefully fire-polished electrode is induced to form a tight 'gigaohm' seal with a cell by slight suction, giving the *cell attached mode*. The conductance of single channels within the patch can be determined (the remainder of the plasma membrane providing a relatively low series resistance), but the internal environment cannot be manipulated. (b) Further suction breaks the 'bubble' of membrane within the electrode giving the *whole-cell mode* which measures the integrated current due to many channels. The cytoplasm is rapidly dialysed out by the contents of the patch electrode in this mode. (c) The nystatin patch permeabilizes the membrane to monovalent but not divalent ions and larger metabolites thus preventing dialysis of the cytoplasm. (d, e) Isolated patches can be right-side-out or inverted; single channel conductances can be measured and the environment on both sides of the membrane controlled.

physiology. Oocytes as dissected from the ovarian envelope are surrounded with cellular layers which can be removed manually or by treatment with collagenase in order to expose the plasma membrane surrounded only by a vitelline envelope. The denuded oocyte still possesses a number of voltage-gated conductances, but these are much smaller than those which can be induced by the injection of neuronal mRNA. Nanogram quantities of polyadenylated mRNA from a neuronal source is injected into individual oocytes. Alternately deoxyribonucleic acid (DNA) contained within a suitable vector can be injected directly into the nucleus. After allowing 8–36 h for protein synthesis, electrophysiology is used to detect successful synthesis and incorporation of ion conductances.

Either whole cell recordings or patch-clamp techniques may be used; the latter requires removal of the vitelline envelope in order to form a gigaohm seal. Individual ionotropic receptors which have been expressed in this system following injection of brain mRNA include those for glycine, GABA and glutamate (Chapter 9) while voltage-gated channels for Na^+, Ca^{2+} and K^+ have been detected (Chapter 4).

The oocytes possess an endogenous Ca^{2+}-activated Cl^- conductance; this can be exploited as an electrophysiological indicator of metabotropic receptors which act via phospholipase C and $(1,4,5)IP_3$ to release internal Ca^{2+} stores. Since injection of the purified mRNA for the receptor (obtained from the cDNA) is sufficient to evoke the final response, this shows that the G-protein and all the subsequent stages of the cascade are endogenous to the oocyte. Receptors which can be detected in this way include those for the neuropeptides neurotensin and substance P, the 5-HT-1C receptor, the M_1-muscarinic receptor and the mGluR1, metabotropic glutamate receptor.

The heterogeneity of expression inherent in the use of unfractionated mRNA can be decreased by using size-fractionated mRNA. This is useful for high molecular weight mRNAs which encode neuronal voltage-gated ion channels (*see* Fig. 9.14). If functional channels require more than one polypeptide, combinations of different mRNA fractions may be required to reconstitute the physiological activity.

After initial size-fractionation of mRNA by gel electrophoresis fractions are injected into a series of oocytes. Voltage-clamp electrophysiology is then used to screen the oocytes for any which possess the characteristics of the required receptor. mRNA from positive oocytes are then used to construct a cDNA library. Pools of cDNAs are screened by testing for the ability of antisense cDNA to hybridize to mouse brain mRNA. Density gradient centrifugation is then used to separate out mRNA : cDNA hybrids from unhybridized mRNA. The latter will be depleted in mRNA for the receptor and can be used as controls after injection into oocytes; while the mRNA capable of hybridizing with the cDNA should, after injection into cells, produce functional receptors. Rescreening the positive pool should result in the isolation of a single cDNA encoding a portion of the receptor.

There appears to be a selective expression of mRNA injected into oocytes; thus fewer receptor subtypes are expressed than are known to exist in brain. This may be due to selective degradation of certain messages on injection, to differences between the brain and *Xenopus* translational machinery, or to a failure of the oocyte to carry out all the posttranslational modifications (signal peptide cleavage, phosphorylation, glycosylation, etc.). Identification of metabotropic receptors relies on their ability to connect to the oocyte's endogenous G-proteins. There appears to be a difficulty in functional expression of cAMP-linked receptors in contrast to the phospholipase C-mediated ones detectable via activation of the endogenous Ca^{2+}-activated Cl^- channel.

2.10 *In situ* hybridization

In situ hybridization can identify, localize and to some extent quantify mRNA in cells in the nervous system. Isotopically or non-isotopically-labelled chromosomal RNA (cRNA) probes can be generated from the relevant cloned cDNA, or alternatively oligodeoxyribonucleotide probes can be produced on a DNA synthesizer avoiding the necessity for cloning. Synthetic probes are typically 30–50 bases in length and can be labelled isotopically (e.g. with ^{32}P or ^{35}S), or for better resolution with an antigenic group (such as biotin or digoxygenin) or enzyme (e.g. alkaline phosphatase) covalently attached to the 3′ tail. The tissue slice is fixed to prevent elution of the mRNA target, hybridized with the probe overnight and visualized by autoradiography, reaction with antibody or enzymatic development as appropriate. It should be emphasized that mRNA hybridization reveals the site of synthesis of a protein and not its final location in the neurone. This is particularly relevant in the case of presynaptically located receptors and channels.

Further reading

Primary neuronal cultures

Balazs, R., Gallo, V. & Kingsbury, A. (1988) Effect of depolarization on the maturation of cerebellar granule cells in culture. *Develop. Brain Res.* **40**, 269–276.

Burgoyne, R.D. & Cambray-Deakin, M.A. (1988) The cellular neurobiology of neuronal development: the cerebellar granule cell. *Brain Res.* **472**, 77–101.

Van-Vliet, B.J., Sebben, M., Dumuis, A., Gabrion, J., Bockaert, J. & Pin, J.P. (1989) Endogenous amino acid release from cultured cerebellar neuronal cells: effect of tetanus toxin on glutamate release. *J. Neurochem.* **52**, 1229–1239.

Chromaffin cells

Burgoyne, R.D. (1991) Control of exocytosis in adrenal chromaffin cells. *Biochim. Biophys. Acta* **1071**, 174–202.

Cell lines

Buckley, N.J., Hulme, E.C. & Birdsall, N.J.M. (1990) Use of clonal cell lines in the analysis of neurotransmitter receptor mechanisms and function. *Biochim. Biophys. Acta* **1055**, 43–53.

Docherty, R.J., Robbins, J. & Brown, D.A. (1991) NG 108–15 neuroblastoma × glioma hybrid cell line as a model neuronal system. In *Cellular Neurobiology — A Practical Approach*, J. Chad & H. Wheal (eds). Oxford: IRL Press, pp. 75–95.

Lendahl, U. & McKay, R.D.G. (1990) The use of cell lines in neurobiology. *Trends. Neurosci.* **13**, 132–137.

Synaptosomes

Dunkley, P.R. (1988) A rapid percoll gradient procedure for isolation of synaptosomes directly from an S1 fraction: homogeneity and morphology. *Brain Res.* **441**, 59–71.

Hollingsworth, E.B., McNeal, E.T., Burton, J.L., Williams, R.J., Daly, J.W. & Creveling, C.R. (1985) Biochemical characterization of a filtered synaptoneurosome preparation from guinea-pig cerebral cortex: cAMP-generating systems, receptors, and enzymes. *J. Neurosci.* **5**, 2240–2253.

Nordmann, J.J., Dayanithi, G. & Lemos, J.R. (1987) Isolated neurosecretory nerve endings as a tool for studying the mechanism of stimulus-secretion coupling. *Biosci. Rep.* 7, 411–425.

Cell permeabilization

Ahnert-Hilger, G. & Gratzl, M. (1988) Controlled manipulation of the cell interior by pore-forming proteins. *Trends Pharmacol. Sci.* **9**, 195–197.

Baker, P.F. & Knight, D.E. (1980) Gaining access to the site of exocytosis in bovine adrenal medullary cells. *J. Physiol. (Paris)* **76**, 497–504.

Bashford, C.L. & Pasternak, C.A. (1992) Ion modulation of membrane permeability. *J. Membrane Biol.* **103**, 79–94.

Hersey, S.J. & Perez, A. (1990) Permeable cell models in stimulus-secretion coupling. *Annu. Rev. Physiol.* **52**, 345–361.

Patch-clamp

Neher, E. (1992) Ion channels for communication between and within cells. *Science* **256**, 498–502.

Penner, R. & Neher, E. (1989) The patch-clamp technique in the study of secretion. *Trends. Neurosci.* **12**, 159–163.

Tang, J.M., Wang, J., Quant, F.N. & Eisenberg, R.S. (1990) Perfusing pipettes. *Pflug. Arch.* **416**, 347–350.

In situ hybrization

Emson, P.C. (1993) *In situ* hybridization as a methodological tool for the neuroscientist. *Trends Neurosci.* **16**, 9–16.

3

Ions and membrane potentials

3.1 Introduction

The foundation of modern neuroscience was laid in the middle of the present century by a generation of electrophysiologists who succeeded in establishing the ionic bases of three fundamental processes: (i) the resting membrane potential; (ii) the action potential; and (iii) fast synaptic transmission. Hodgkin, Huxley and Katz described how the contribution which an ion gradient makes to the membrane potential is proportional to the ion conductance. The important consequence was that this provided a mechanism for extremely rapid changes in membrane potential induced by changes in these conductances with relatively tiny net fluxes of ions. Since lipid bilayers are inherently impermeable to ions, it was necessary to propose the existence of discrete Na^+- or K^+-conducting *voltage-gated ion channels*, although some 30 years were to elapse before their molecular identification. This ionic hypothesis was extended to fast synaptic transmission itself by the finding that neurotransmitters bind to *ligand-activated ion channels*, providing excitatory or inhibitory inputs to the postsynaptic membrane.

3.2 The resting plasma membrane potential

Typical concentrations of major ions across the plasma membrane of a mammalian neurone are given in Table 3.1. As with other cells, K^+ is concentrated inside the cells while Na^+, Cl^- and particularly Ca^{2+} are at lower concentrations. A microelectrode connected to a high resistance potentiometer and inserted into a resting neurone will register a potential some 60–80 mV negative with respect to a reference electrode in the external medium (by convention this potential is given a negative sign). A small proportion of this *membrane potential* is due to the fact that adenosine triphosphate (ATP)-hydrolysing ion pumps in the plasma membrane are *electrogenic*, i.e. they cause the net expulsion of positive charge from the cells, either by the uncompensated *uniport* of Ca^{2+} in the case of the ATPase responsible for maintaining a low internal Ca^{2+}, or by the *antiport* of $3Na^+$ for $2K^+$ by the Na^+-pump, the $Na^+ + K^+$-ATPase. It must be emphasized at this stage that the term 'membrane potential' is misleading, since it is shorthand for 'the difference in electrical potential between two aqueous phases separated by the membrane' and must not be confused with any surface potential which the membrane itself might possess, for example, by virtue of charge on its phospholipids.

After allowing for the electrogenic component, the plasma membrane potential is due to the ion gradients across the membrane. Neurones possess K^+-channels which are open at the resting membrane potential. Since K^+ is present at a higher concentration within a cell than in the external medium it will attempt to flow out of the cell via the channel, carrying positive charge and leaving the interior negatively charged (i.e. generating a negative membrane potential). Only

Table 3.1 Typical ion concentrations across the plasma membranes and equilibrium potentials (E_x) for a typical mammalian neurone and the squid giant axon

	Mammalian neurone			Squid giant axon		
Ion	Intracellular (mM)	Extracellular (mM)	E_x (mV)	Intracellular (mM)	Extracellular (mM)	E_x (mV)
K^+	150	5.5	−90	400	10	−93
Na^+	15	150	+60	50	460	+56
Ca^{2+}	0.0001	1.5	+270	0.0001	10	+300
Cl^-	9	125	−70	100	540	−42

a minute ion translocation is required to generate the potential, after which further K^+-ion loss is prevented by an equilibrium between the concentration gradient tending to drive the ion out of the cell and the membrane potential tending to retain the positive ion in the negative interior.

The number of ions which must cross the membrane to establish a membrane potential is very small. The capacitance of a phospholipid bilayer membrane is close to $1\,\mu F \cdot cm^{-2}$, and it can be calculated that only about 10 000 K^+ ions must cross the membrane of a typical cell to generate a potential of −80 mV. Since K^+ would be present in the cytoplasm at about 100 mM, the cell might contain about 10^{10} ions, enormously in excess of the amount required to generate or restore the membrane potential.

3.2.1 The Nernst equilibrium potential

The relationship between membrane potential and the ion gradient can be placed on a quantitative basis by the application of simple thermodynamics. The capacity of an enzymatic reaction or an ion gradient to do work is a function of how far that process is away from equilibrium, and is given by the *Gibbs free energy* (ΔG). For an enzymatic reaction occurring in a single compartment, ΔG is given (in $kJ \cdot mol^{-1}$) by:

$$\Delta G = 2.3\,RT \log_{10} \frac{K}{\Gamma} \quad (3.1)$$

where K is the equilibrium constant for the reaction and Γ is the mass action ratio of the reaction which is actually observed in the experiment, R is the gas constant and T the absolute temperature. Putting values into this equation — for every order of magnitude a reaction is displaced from equilibrium, ΔG changes by $5.6\,kJ \cdot mol^{-1}$.

In the case of an ion gradient across a membrane the Gibbs free energy is again a function of the disequilibrium of the process; however this time two factors must be considered, the difference in concentration of the ion across the membrane and the difference in electrical potential between the two aqueous compartments (the *membrane potential*, V_m). The overall Gibbs free energy is a combination of both factors; for an ion X^{n+} distributed across a membrane:

$$\Delta G = nFV_m + 2.3\,RT \log_{10} \frac{(X^{n+}_{cyto})}{(X^{n+}_{out})} \quad (3.2)$$

This is the *electrochemical potential gradient* for the ion, and can be thought of either as the displacement of the ion from equilibrium or the measure of the capacity of the ion to do work.

When an ion gradient is at equilibrium, by definition $\Delta G = 0$. Thus the electrochemical potential equation above reduces to the following:

$$V_m = -\frac{2.3\,RT}{nF} \log_{10} \frac{(X^{n+}_{cyto})}{(X^{n+}_{out})} \quad (3.3)$$

This is the *Nernst equation* which gives the concentration gradient of the ion which would be at equilibrium with the observed membrane potential (V_m), or alternatively the particular membrane potential, or *equilibrium potential* (E_x), which would be generated from a given gradient if that were the only species permeable across the membrane.

3.2.2 Membrane potential for multiple permeant ions

Although a neurone has many ion gradients across its plasma membrane, so far we have only considered the effect of a dominant K^+ conductance generating a negative resting potential. If the same cell possessed only a Na^+-selective ion channel in the plasma membrane, Na^+ would tend to *enter* down its concentration gradient, carrying in positive charge and generating a *positive* membrane potential. The K^+ gradient would make no contribution as long as it has no means of permeating across the membrane. This illustrates a central principle: ion gradients only influence the membrane potential if the ion is permeable (Fig. 3.1).

Membrane potentials in neurones are changed with great rapidity by the opening and closing of ion channels gated either by neurotransmitters activating ionotropic receptors or by the membrane potential itself causing the voltage activation of channels. The conformational changes which open and close the channels will be discussed later in this book, here we shall restrict ourselves to the electrical consequences for the membrane potential.

Consider a neurone with ion gradients comparable to those shown in Table 3.1 and with K^+-channels which close at the same instant that Na^+-selective channels open. The equilibrium membrane potential will change from about −80 mV to about +60 mV. Even if the channel opening and closing is instantaneous, a finite time will be required for charge to flow through the Na^+-channel to establish the new membrane

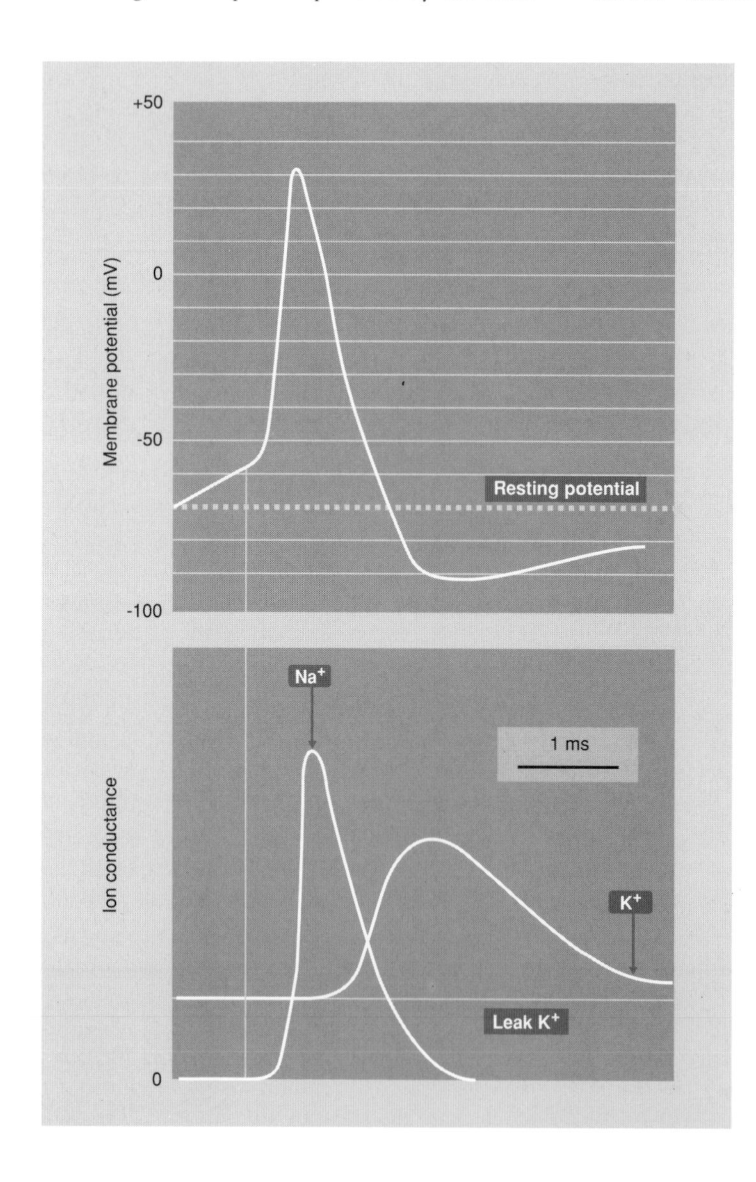

Fig. 3.1 The action potential. When a −70 mV resting membrane potential (maintained by a leak K^+ conductance) is reduced to about −55 mV voltage-activated Na^+-channels fire initiating an action potential which drives the membrane potential to a positive value. The Na^+-channels start to inactivate at the time when additional delayed K^+-channels activate, speeding the repolarization and even generating a transient after-hyperpolarization.

potential. The *time constant*, τ is the product of the membrane resistance and capacitance and will depend on the individual conductance and density of the channels.

In a real cell, more than one ion has a significant permeability at any one time. Under these conditions, the contribution of each ion to the membrane potential is proportional to its permeability, and the overall potential is given by the Goldman–Hodgkin–Katz equation:

$$V_m = \frac{2.3\,RT}{F}\left(\frac{P_K\,[K^+]_o + P_{Na}\,[Na^+]_o + P_{Cl}\,[Cl^-]_i}{P_K\,[K^+]_i + P_{Na}\,[Na^+]_i + P_{Cl}\,[Cl^-]_o}\right) \quad (3.4)$$

The ions which contribute significantly to the membrane potential are K^+, Na^+ and Cl^-. In a resting neurone where the K^+ conductance is dominant, the membrane potential across the neuronal membrane is about -80 mV, corresponding to a 30-fold K^+-gradient in/out. Both the K^+ and Cl^- gradients are close to thermodynamic equilibrium, and thus have little capacity to do work, whereas the Na^+ gradient is far away from equilibrium. The extreme case of disequilibrium is the Ca^{2+} distribution. $[Ca]_{in}$ is about 10 000 times lower than $[Ca]_{out}$.

Cl^- deserves some special mention. In resting cells the ion is close to electrochemical equilibrium, there being no Cl^--pump to maintain a disequilibrium in the face of a significant Cl^- conductance. However, a high Cl^- conductance will tend to oppose a plasma membrane depolarization which would otherwise be initiated by an increased Na^+ conductance. Inhibitory ionotropic receptors such as the glycine and GABA-A receptors (*see* Section 9.8) act by increasing Cl^- conductance and opposing the depolarization which would result from an increased Na^+ conductance.

3.2.3 The action potential

Neuronal membranes are *excitable*, which means that a depolarization of the membrane can be sustained and propagated along the membrane by the action of voltage-activated ion channels. The key ion channels in this process are voltage-activated Na^+- and K^+-channels. As discussed above, in a resting polarized neurone the ratio of Na^+ to K^+ conductance is very low, due to the ability of some classes of K^+-channels (but not Na^+-channels) to conduct under these conditions. If a neurone is depolarized to below -55 to -40 mV, for example by the injection of positive charge with a microelectrode (or more physiologically by the activation of excitatory ionotropic receptors increasing the membrane Na^+ conductance) voltage-activated Na^+-channels will open and amplify the initial imposed depolarization (*see* Fig. 3.1). For a brief period of 1–2 ms the increased Na^+ conductance will dominate the constant field equation, and the membrane potential will approach the positive Na^+ equilibrium potential. Voltage-activated Na^+-channels inactivate spontaneously after 1–2 ms, however and ΔE will begin to recover. The recovery of the initial polarized potential is hastened by the delayed opening of additional, voltage-activated, K^+-channels, which inactivate in a period of 5–10 ms and lead to a brief after-hyperpolarization. This cycle of depolarization and repolarization is an *action potential*.

The enhanced depolarization caused by opening of Na^+-channels has the effect of depolarizing the surrounding membrane. Any Na^+-channels sufficiently close to the origin will also fire, spreading the initial depolarization. This wave of depolarization is directional, since once Na^+-channels have opened and spontaneously closed, they remain *refractory* for 5–10 ms, and thus unable to respond to depolarization. If the initial depolarization is created at the axon hillock, there will be a unidirectional propagation of an action potential down the axon.

A neurone will fire an *action potential* down its axon if the summation of excitatory, inhibitory and spontaneous changes in membrane potential detected at the axon hillock are sufficient to decrease the resting membrane potential of some -80 mV to about -50 mV or below.

A naked axon has poor electrical properties and the action potential would die out rapidly if it were not for the continuous array of Na^+-channels down its length to regenerate the signal. However since each channel activation requires about 1 ms and if this must be repeated thousands of times down the axon, the propagation of the action potential is considerably slowed. Except for short *interneurones* which have axons only about 1 mm in length, mammalian axons overcome this problem of electrical leakage by insulating the axon with *myelin*. In the CNS myelin is synthesized by *oligodendroglia* (*see* Fig. 1.4). A glial cell can send out up to 60 processes each of which wraps around a short length (about 1 mm) of a different axon, greatly improving the electrical insulation of the axon. In the peripheral nervous system myelin is produced by *Schwann cells* which each insulate a single axon. In myelinated axons the Na^+-channels are clustered at the short gaps, or *nodes of Ranvier* between adjacent myelinated sections (*see* Fig. 1.2). Each segment of axonal myelination is about 1 mm in length. Propagation of the action potential occurs virtually instantaneously between the nodes which separate each segment. This is known as *saltatory*, or 'jumping' propagation. Na^+-channels are concentrated at the nodes, to such a high concentration that even K^+-channels are effectively excluded. An action potential can travel very rapidly between two nodes without the need for intermediate amplification and thus transmission is greatly speeded.

The speed of transmission down an unmyelinated axon can be raised by increasing the diameter of the axon. For example, some invertebrates such as the squid possess unmyelinated *giant axons* with diameters of 0.5 mm. The squid giant axon and its accompanying giant synapse are much exploited, since they allow experiments to be performed which are technically impossible with the tiny mammalian axons and terminals, such as direct electrophysiology and internal dialysis of the axoplasm.

The classic preparation in which the ionic basis of the action potential was described, the squid giant axon, possesses no mechanisms for modulating the size of an individual action potential, which is therefore an all-or-nothing affair. Thus the cell conveys information down the axon by modulating the frequency and pattern of discharges. Both somato/dendritic and nerve terminal action potentials *can* be modulated, frequently by the regulation of multiple K^+ conductances in these membranes, this is of primary importance in regulating neuronal excitability and transmitter release (*see* for example Section 9.7.1). Furthermore, somato/dendritic membranes are not totally inactive in the absence of receptor activation, but possess individual combinations of voltage-activated ion channels which can give a spontaneous and characteristic pattern of channel firing in these regions, which are modified rather than created by the synaptic input. A major role of the second messengers generated by the metabotropic receptors may be to produce a highly localized but relatively long-lasting modulation of the channels responsible for this inherent rhythm, frequently by phosphorylating ion channels.

3.3 Determination and manipulation of membrane potentials by non-electrophysiological techniques

The central theme of this book is to emphasize the biochemical techniques which have been brought to bear on the mechanisms of synaptic transmission. For many of the preparations reviewed in Chapter 2, direct electrophysiological techniques are inappropriate. This raises the problem of how to estimate and manipulate the plasma membrane potential in these preparations.

While it is possible to measure the gradient of K^+ across the plasma membrane by isotopic means (e.g. by using the isotope $^{86}Rb^+$ whose distribution mimics that of K^+) and thus to estimate the Nernst equilibrium potential for the ion as an estimate of resting membrane potential, this has been largely supplanted by continuous, non-destructive optical techniques which rely on the spectral shift or altered fluorescence yield which occurs when certain charged, membrane-permeant indicators accumulate within a cell or organelle in response to the membrane potential.

The indicators may be cationic or anionic. The cationic *carbocyanines* (Fig. 3.2) respond very rapidly to changes in membrane potential, however the presence of the high negative interior mitochondrial membrane potential means that the bulk of the dye will be accumulated within the mitochondrial matrix. It is sometimes stated that the cationic cyanine dyes selectively report the plasma membrane potential, but this is only because the high accumulations within the mitochondrion inhibit NADH-dehydrogenase and collapse the mitochondrial membrane potential, and the consequent toxic consequences of this must always be borne in mind.

The anionic *bisoxonols* (Fig. 3.2) avoid the problems due to mitochondrial accumulation, and are very sensitive with a large fluorescent yield. They however respond slowly to changes

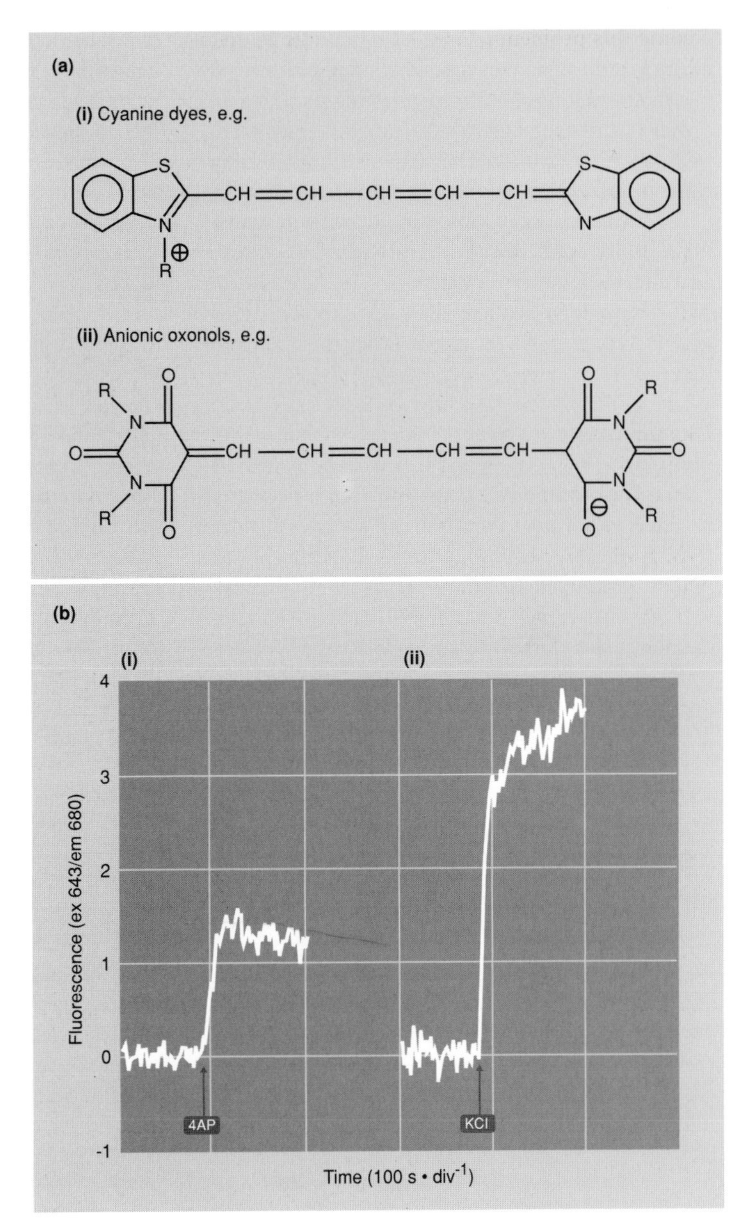

Fig. 3.2 Monitoring of plasma membrane potential in cell or organelle suspensions. (a) Two fluorescent indicators which are used for monitoring plasma membrane potential in organelle suspensions are (i) the cationic cyanine dyes, and (ii) the anionic oxonols. (b) Data showing the ability of the cyanine dye DiSC(5) to monitor the depolarization of a synaptosomal suspension exposed to (i) the K^+-channel inhibitor 4-aminopyridine (4AP), and (ii) 10 mM KCl. Data from Barrie *et al.* (1991).

in membrane potential, requiring up to 1 min to re-equilibrate after a transient. Despite their negative charge, the oxonols can still be used to monitor the negative membrane potential — the dye becoming less excluded from thë ell as depolarization proceeds. With both classes of indicator the membrane potential indicated is at best semi-quantitative. Calibration is obtained by defined additions of KCl to the medium and substituting these values, together with the internal $[K^+]$, into the Nernst equation for K^+ (Eqn 3.3).

3.3.1 Strategies for depolarizing neurochemical preparations

The price which must be paid for working with isolated neuronal preparations is that the physiological stimulus of the action potential is no longer present. In brain slice preparations single neurones may be stimulated by patch electrodes, axon bundles can be submitted to field stimulation simulating a train of physiological action potentials, or the entire slice can be exposed to an oscillatory field, while with synaptosomes an ionic means of depolarization must be employed.

The three modes of ionic depolarization (which are equally applied to synaptosome, cell culture and brain slices preparations) are KCl elevation, Na^+-channel activation and K^+-channel inhibition (Fig. 3.3); these will now be considered in turn.

Depolarization by elevated KCl. Because of the dominance of the K^+ conductance, the neuronal plasma membrane is a good approximation of a K^+-electrode and depolarizes when external $[K^+]$ is elevated. Although this is a standard and widely employed technique it does have some severe limitations. Since KCl causes a single, clamped, depolarization it will allow transient channels to fire only once before inactivating even though the plasma membrane remains depolarized. It is thus difficult to observe processes which rely upon the modulation of such channels since their effect on the plasma membrane potential will be negligible. Furthermore, the prolonged depolarization is non-physiological and can lead to the loss of cytoplasmic pools of transmitter by reversal of the plasma membrane transporters (*see* Section 9.5.5).

Depolarization by K^+-channel inhibition. A subset of K^+-channel, termed K_A^+ and sensitive to the mamba neurotoxin α-dendrotoxin, will be discussed in detail in Chapter 4. K_A^+-channels regulate neuronal excitability of synaptosomes and brain slices. Inhibition of the channel by either nanomolar dendrotoxin or millimolar 4-aminopyridine (4AP) causes the terminals to undergo spontaneous 'epileptic' action potentials, involving the repetitive firing of Na^+-channels, as though the terminals were still receiving action potentials down their severed axon. The effects of these inhibitors can be abolished by the Na^+-channel inhibitor tetrodotoxin.

Dendrotoxin and 4AP more closely mimic the physiological mechanism of terminal depolar-

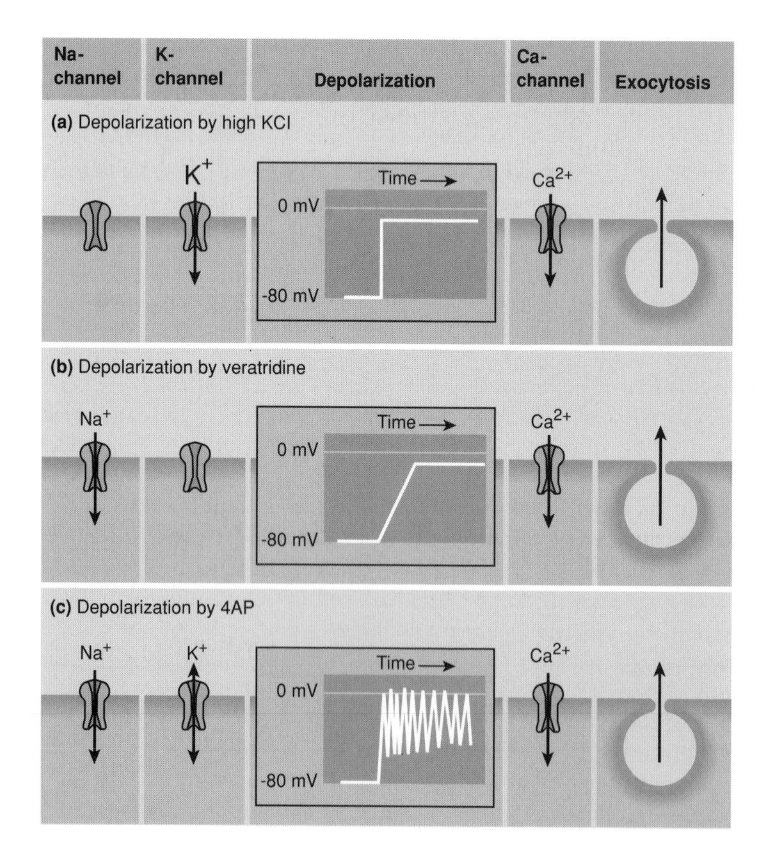

Fig. 3.3 *In vitro* depolarization strategies for neurochemical preparations. Three chemical means of depolarizing preparations such as synaptosomes, slices or cultured cells are shown. (a) Elevated KCl (10–50 mM) causes a clamped depolarization since non-inactivating K^+-channels provide the dominant conductance at the neuronal plasma membrane. (b) Veratridine causes a relatively slow depolarization since Na^+-channels must open spontaneously for the alkaloid to act by retaining them in the fully active state. (c) K_A^+-channel inhibitors such as 4-aminopyridine (4AP) and dendrotoxin can prolong action potentials in cell and slice preparations and induce spontaneous action potentials in synaptosomes. This method allows the effects of transiently active channels to be studied in this type of preparation.

ization than the alternative of high KCl. It is notable that a number of presynaptic regulatory events in synaptosomes and slices which rely on modulation of transient K^+-channels are only clearly apparent with 4AP depolarization rather than with KCl (*see* Section 9.7.1).

Depolarization by preventing Na^+-channel inactivation. Na^+-channels occasionally 'flicker' in polarized synaptosomes and cells. The rapid inactivation of individual channels can be inhibited by *veratridine*, and in a few seconds Na^+-channels accumulate in the open state, causing a long lasting increase in Na^+ conductance and associated depolarization. Veratridine must be used with caution. First there is an energy-dissipating Na^+-cycle across the plasma membrane due to a continuous influx of Na^+ which the Na^+/K^+-ATPase accelerates to counter, causing a lowering of ATP levels and a massive stimulation of mitochondrial respiration.

The second effect of veratridine is to collapse the Na^+-electrochemical potential across the plasma membrane: this implies not only that the membrane potential is depolarized (as occurs with KCl depolarization) but also that the Na^+ concentration gradient is dissipated (which does not occur with high KCl). This results in a dramatic reversal of a number of Na^+-coupled transport processes, such as Na^+ : amino acid cotransport (which will be discussed in detail in Chapter 9), resulting in an extensive Ca^{2+}-independent efflux of metabolic pools of amino acids such as glutamate, GABA and aspartate. Ca^{2+}-independent release of transmitter on addition of veratridine has been attributed to Na^+ influx triggering the release of mitochondrial Ca^{2+} stores, since the mitochondrial Ca^{2+} efflux pathway is activated by Na^+. However, this interpretation is generally incorrect since Ca^{2+} released from presynaptic mitochondria is not able to release transmitter as it is released into the bulk cytoplasm and not at the active zone which would be required for amino acid exocytosis (*see* Section 7.3).

3.4 Regulation of cytoplasmic pH

Cytoplasmic pH changes resulting from intervention in the metabolism of neurones are frequently overlooked. They are however crucial, since the internal pH must be maintained within a narrow range either side of pH 7.0 in order for metabolism to function effectively and to avoid the acidification of the cytoplasm resulting from the production of acidic metabolic end-products. If a typical cell was freely permeable to protons, a membrane potential of −60 to −90 mV would generate a pH_i of 1−1.5 pH acidic with respect to the medium, whereas the typical value which is observed is closer to 0.4 pH units.

Cells possess multiple pathways for the regulation of pH_i, the two best characterized being the Na^+/H^+ exchanger and the Cl^-/OH^- exchanger (Fig. 3.4). These play complementary roles; the former preventing excessive acidification while the latter corrects excess alkalinity. The Na^+/H^+ exchanger uses the sodium concentration gradient to expel protons. The carrier, which has been cloned, is electroneutral and would, if unregulated, be capable of increasing the cytoplasmic pH to about 8.4, utilizing a 10 : 1 $Na^+{}_{o/i}$ ratio. That the carrier does not do this is due to an allosteric proton binding site on the cytoplasmic face of the carrier which effectively inhibits the enzyme when pH_i rises above about 7.2. Amiloride, and more particularly the 5-*N* disubstituted alkylated analogues inhibit the carrier.

Another major pathway, the Cl^-/HCO_3^- exchanger, expels HCO_3^- in exchange for Cl^-. Since bicarbonate formed from carbonic acid within the cell leaves a proton behind, this transport, driven by the Cl^- concentration gradient, has the effect of acidifying the cytoplasm, compensating for an excessively high pH_i, although this is less common than excess acidification. Consistent with this role, the Cl^-/HCO_3^- exchanger shows a reciprocal pH dependency to that of the Na^+/H^+ exchanger, the pH regulation again being under the control of an allosteric H^+-binding site. The carrier has been cloned and at least four isoforms identified. The carrier is insensitive to amiloride, but is inhibited non-specifically by the stilbenes, SITS and DIDS.

A more complex carrier, which has not been identified at the molecular level, and which combines the stoichiometries of both of the previous carriers has been proposed to operate in the range of pH close to 7.2 where neither the Na^+/H^+ exchange nor the Cl^-/HCO_3^- exchange are kinetically functional.

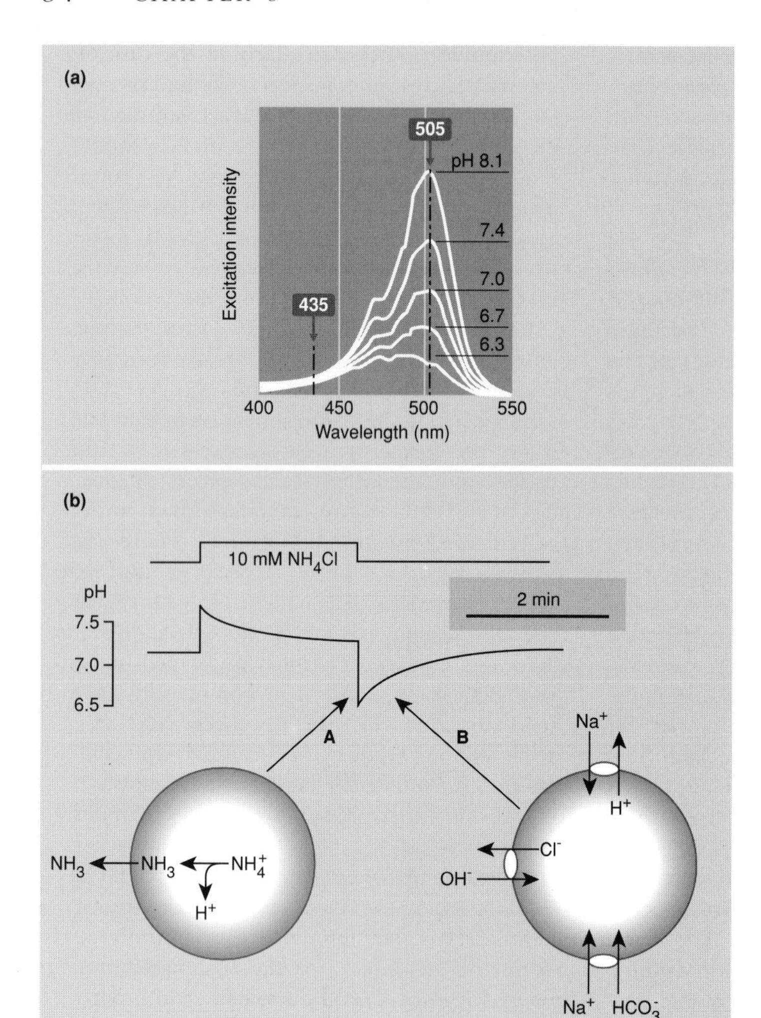

Fig. 3.4 Neuronal pH regulation. (a) Excitation spectrum of BCECF as a function of pH. The ratio of emission intensity alternately exciting at 435 nm and 505 nm allows pH to be calculated independently of dye loading. (b) Cytoplasmic pH transients can be induced *in vitro* by equilibrating cells with NH_4Cl. The non-ionized NH_3 enters the cell where it reprotonates increasing pH_i. Sudden removal of external NH_4Cl reverses this process (A) generating an acidic transient. The resting pH_i of about 7.1 is restored by a variety of transport processes (B) including Na^+/H^+ exchange, Cl^-/OH^- exchange and $Na^+ : HCO_3^-$ cotransport (which liberates OH^- as HCO_3^- dissociates to CO_2).

3.4.1 Measurement of cytoplasmic and intravesicular pH

Except in the case of extremely large cells or giant axons it is not possible to determine cytoplasmic pH directly by the use of micro pH-electrodes. Indirect techniques must be adopted which make use of fluorescent dyes whose accumulation or fluorescence is responsive to the cytoplasmic pH. Two techniques are described.

Fluorescein derivatives. Fluorescein is a phenolic weak acid which is very fluorescent in its unprotonated form and only slightly fluorescent when protonated. Since its pK is rather low (6.5) and it is relatively membrane permeable, it is not ideal for monitoring internal pH. Bis-carboxyethyl-carboxyfluorescein (BCECF) has a pK of 6.95 and by virtue of its carboxy groups is more hydrophilic and less membrane permeable than the native compound. The method of loading BCECF within cells relies on a technique pioneered for the Ca^{2+}-indicator fura-2 which is described in detail later (*see* Section 5.7.1). Briefly, the carboxyl groups are esterified as acetoxymethyl esters to give a membrane permeant form. Within the cell non-specific esterases cleave the ester groups, regenerating the free BCECF. The excitation spectrum has an isosbestic point at 450 nm and this can be used to provide a pH-independent calibration of the amount of accumulated dye, while the 505/435 nm ratio gives information on the pH

(Fig. 3.4). BCECF provides an accurate measure of the cytoplasmic pH and is little affected by the presence of acidic vesicles. First there is no extra accumulation of BCECF within these vesicles and second any BCECF which is formed within the highly acidic compartment tends to leak out slowly as the fully protonated form.

Acridine dyes. Acridine derivatives such as acridine orange and 9-aminoacridine are weak bases which accumulate within acidic compartments in the same way as isotopically-labelled weak bases. Their advantage lies in their ability to undergo concentration-dependent stacking of their planar molecules above a critical concentration. If the concentration of dye and organelles is adjusted so that stacking occurs within the acidic compartment then the acridines undergo a red-shift in their emission spectrum as stacked excimer complexes are formed. If one measures the fluorescence at the peak of the monomer excitation, the signal will decrease as the dyes are accumulated into the cells. It should be noted that this technique is best suited to the measurement of extensive acidification within, for example, secretory vesicles, rather than the typical 0.4 pH unit gradient across the plasma membrane.

The act of measuring a pH (or membrane potential) gradient decreases that gradient. For example, the accumulation of a weak base within an acidic compartment involves the protonation of the neutral base within the organelle hence using up some of the protons in the compartment. If the organelle does not possess a perfect homeostatic mechanism for restoring the gradient then the measured gradient will be unphysiologically low. This can be minimized by using the lowest concentrations of indicator that gives a detectable signal. This technique does not respect intracellular compartments, and a major limitation is that the presence of an acidic compartment such as secretory vesicles or lysosomes can lead to a greatly exaggerated accumulation of a weak base. A pH 5 interior of a secretory vesicle will result in a 100-fold accumulation of the base across the vesicle membrane; in cells possessing a significant number of secretory vesicles weak bases therefore give erroneous measurements of cytoplasmic pH. Conversely the alkaline interior of mitochondria can lead to an accumulation of weak acid within their matrices.

Alternatively, and particularly in the case of secretory vesicles the transport of large amounts of weak acid, weak base or charged membrane potential indicator can be exploited as a deliberate means to manipulate the transmembrane gradients. While the cytoplasmic pH is carefully regulated by the pathways discussed above, the internal pH of intracellular organelles containing H^+-pumps, such as mitochondria and synaptic vesicles, is largely a passive consequence of the amount of protons which can be pumped across the membrane before the membrane potential becomes large enough to oppose further transport. In the absence of other permeable ions a minute proton transport is sufficient to accumulate a high membrane potential ($\Delta\psi$) with negligible ΔpH. The availability of permeant anions (in the case of the synaptic vesicle) such as Cl^- or glutamate$^-$ (for glutamatergic synaptic vesicles) increases the electrical capacity of the vesicle by moving into the vesicle and neutralizing the membrane potential, allowing the pH gradient to increase at the expense of $\Delta\psi$. Conversely, the uptake of large amounts of weak base into the acidic compartment will decrease ΔpH, but since the vesicular H^+-translocating ATPase can continue to generate the same total proton electrochemical potential gradient, there will be a compensatory increase in the membrane potential. These manipulations can be exploited in order to determine whether the membrane potential or pH gradient provides the driving force for the accumulation of neurotransmitters into synaptic vesicles (*see* Fig. 9.7).

3.5 Neuronal bioenergetics

Neurones display no unique bioenergetic features. Nevertheless, an understanding of the major pathways for the synthesis and utilization of ATP is essential since virtually all the processes central to neuronal function (such as exocytosis, protein phosphorylation and membrane potential maintenance) directly or indirectly utilize ATP (Fig. 3.5). Additionally, one major field of study — brain anoxia or ischaemia — demands some understanding of the processes which occur when oxidative metabolism is disrupted.

Human brain accounts for 20% of the resting whole-body oxygen consumption. The extreme sensitivity of brain to any interruption in blood supply is in part a consequence of its extremely

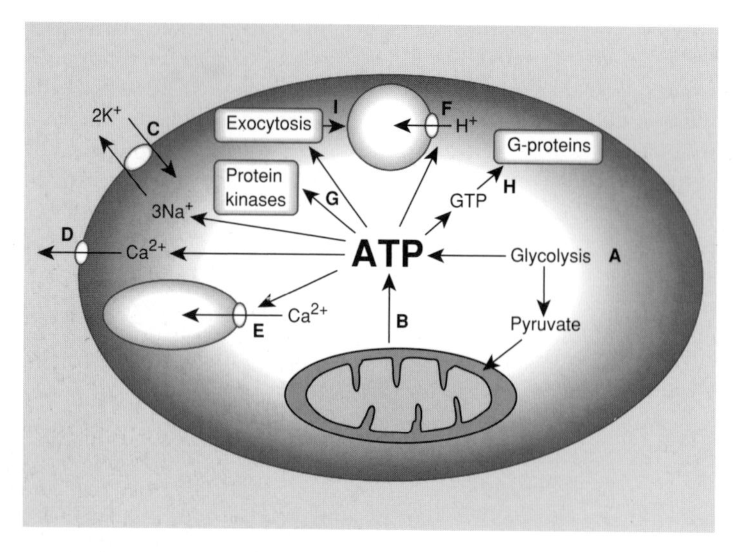

Fig. 3.5 ATP turnover in the nerve terminal. Terminals have no significant reserves of oxidizable substrates but can utilize glucose. Glycolysis (A) generates some ATP and also pyruvate which can be oxidized by the mitochondria generating the bulk of the terminal's ATP by oxidative phosphorylation (B). The greatest demand for ATP is by the Na^+/K^+-ATPase (C), while Ca^{2+}-ATPases are present on the plasma membrane (D) and internal membranes (E). A H^+-ATPase acidifies the lumen of synaptic vesicles (F). ATP is required as a substrate for protein kinases (G), for GTP synthesis and hence G-protein activity (H) and for undefined step(s) in the exocytotic process itself (I).

low glycogen stores, perhaps 1% of that in an equivalent mass of liver, limiting the possibility of temporary anaerobic metabolism. Neurones can utilize glucose or pyruvate as exogenous substrates (and ketone bodies such as β-hydroxybutyrate and acetoacetate under hypoglycaemic conditions) but appear unable to use the high concentrations of glutamate and aspartate within their cytoplasm.

The dehydrogenases in the mitochondrial matrix transfer electrons from substrates such as pyruvate to nicotinamide adenine dinucleotide

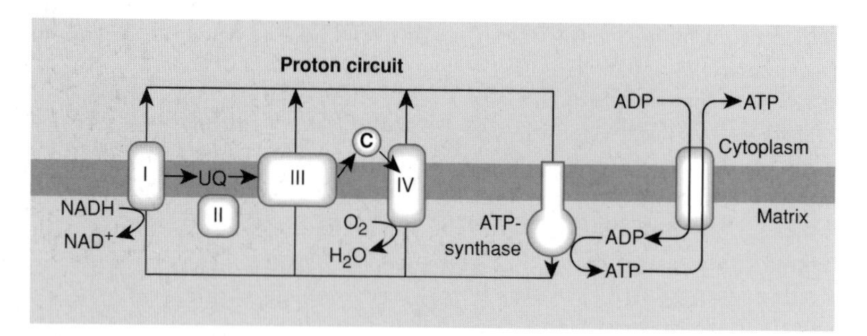

Fig. 3.6 The mitochondrial proton circuit. The mitochondrial respiratory chain consists of three proton pumps (Complexes I, III and IV (or cytochrome c oxidase)) in series with respect to the flow of electrons from NADH to O_2 and in parallel with respect to the proton circuit set between the respiratory chain and the ATP-synthase, which is an ATP-hydrolysing proton pump forced to run in reverse by the proton electrochemical gradient or proton-motive force (pmf) of some 240 mV set up by the proton pumping across the membrane. 180 mV of the pmf is in the form of a membrane potential (negative in the mitochondrial matrix) and this is the driving force for the accumulation of Ca^{2+} into the matrix under conditions of cellular Ca^{2+} overload. In ischaemia when there is neither oxygen nor substrate supply the pmf rapidly collapses; in anoxia when blood supply is retained in the absence of oxygen glycolysis accelerates enormously and may in the short-term maintain the pmf at a suboptimal level.

(NAD^+) forming NADH (Fig. 3.6). The mitochondrial respiratory chain consists of three proton pumps in series: Complex I, Complex III and Complex IV, or cytochrome c oxidase. The final electron sink is oxygen, which is reduced to water. The site of action of the Parkinsonism-inducing neurotoxin MPP^+ (*see* Section 11.6) appears to be Complex I. The protons which are pumped out by the respiratory chain generate a proton electrochemical gradient, or proton-motive force, of 240 mV (22 kJ $\cdot$ mol^{-1}) across the membrane, predominantly in the form of a membrane potential of some 180 mV. Protons re-enter the matrix via the ATP-synthase, which is an ATP-hydrolysing proton pump driven in reverse and utilizes three or four protons to synthesize one molecule of ATP from adenosine diphosphate (ADP). The proton 'current' flowing round the proton circuit, and hence the oxygen consumption, is directly proportional to the rate of ATP production by the mitochondria, which in turn equals the ATP demand of the cell. This is 'respiratory control'. A stringent criterion for the energetic integrity of a preparation such as the synaptosome is that it shows respiratory control, i.e. that the capacity for ATP production by their mitochondria is in excess of the ATP requirements of the terminal.

ATP is capable of doing work in the cell *not* because of any magic property inherent in the molecule itself (so-called 'high-energy bonds') but because the concentrations of ATP and its hydrolysis products ADP and P_i are held some 10 orders of magnitude away from equilibrium by mitochondrial ATP synthesis. Roughly this means cytoplasmic free nucleotide concentrations of 10 mM for ATP and 10 μm for ADP in the presence of 10 mM P_i. Each additional mole of ATP synthesized under these conditions requires a free energy input (ΔG) of some 57 kJ $\cdot$ mol^{-1}, and this amount of free energy is available, for example, to an ATP-driven ion pump for each mole of ATP hydrolysed under these conditions.

ΔG for ATP synthesis is difficult to measure experimentally in an intact cell or synaptosome: adenine nucleotides are compartmentalized between cytoplasm and mitochondrial matrix and much is bound and does not enter into the thermodynamic equation. The concentration of ATP itself, or even the gross ATP : ADP ratio, gives only a crude measure of the thermodynamic capacity of ATP hydrolysis to do useful work. The creatine phosphate : creatine ratio is better because it is in equilibrium with free cytoplasmic ATP and ADP via creatine kinase, which is active in neurones. On the other hand the so-called 'energy-charge' — a measure of the extent to which the β and γ-phosphate positions of ADP and ATP are saturated — has no thermodynamic meaning and should not be used.

A large number of agents commonly used in neuroscience will affect the ability of the mitochondrion to synthesize ATP. Some of these are summarized below:

1 *Protonophores.* The proton circuit discussed above can be artifactually short-circuited (uncoupled) by protonophores such as FCCP which make the mitochondrial membrane (and other membranes) permeable to protons. This allows protons to re-enter the matrix directly without going via the ATP-synthase. Protonophores cause a dramatic fall in ATP as ATP-synthase reverses and hydrolyses glycolytically-generated ATP. A protonophoric action should always be controlled for if a novel agent is found to block transmitter release since the latter requires high ATP levels.

2 *Ca^{2+}-ionophores.* $Ca^{2+}/2H^+$ exchange ionophores such as ionomycin are widely exploited to elevate neuronal $[Ca^{2+}]_c$. Unfortunately they also distribute into the mitochondrial inner membrane and set up a dissipative Ca^{2+} cycling which has an effect comparable to that of a protonophore. Ca^{2+}-ionophores must be very carefully titrated and any mitochondrial uncoupling action controlled for.

3 *Cyanide and other respiratory chain inhibitors.* When glucose is present in the medium the robust Pasteur effect in terminals and neurones allows glycolysis to accelerate greatly in the presence of respiratory chain inhibitors in an attempt to maintain ATP levels. However, at least with synaptosomes, this is only partially successful; ATP : ADP ratios fall sufficiently to limit exocytosis and, particularly in the presence of cyanide or under anoxic conditions, glycolysis soon fails.

4 *Deliberate or accidental hypoglycaemia.* Neurones and synaptosomes contain very limited stores of glucose. This can be demonstrated with the latter preparation which, after washing and resuspension in a glucose-free medium, becomes dependent on exogenous pyruvate as a mito-

chondrial substrate. The *malate–aspartate shuttle* (which is responsible for the mitochondrial reoxidation of NADH generated in glycolysis) is unfortunately inhibited by the same transaminase blockers which are used to block the metabolism of the amino acid GABA, e.g. aminoxyacetic acid. This generates a functionally 'hypoglycaemic' state since any pyruvate generated by glycolysis is reduced to lactate in order to reoxidize the NADH, rather than being made available to the mitochondrion.

Further reading

Black, J.A., Kocsis, J.D. & Waxman, S.G. (1990) Ion channel organization of the myelinated fiber. *Trends Neurosci.* **13**, 48–54.

Connors, B.W. & Gutnick, M.J. (1990) Intrinsic firing patterns of diverse neocortical neurons. *Trends Neurosci.* **13**, 99–104.

Action potentials and axonal propagation

Hodgkin, A.L. & Huxley, A.F. (1952) A quantitative description of membrane current and its application to conduction and excitation in nerves. *J. Physiol.* **117**, 500–544.

Membrane potential and pH determination and regulation

Barrie, A.P., Nicholls, D.G., Sanchez-Prieto, J. & Sihra, T.S. (1991) An ion channel locus for the protein kinase C potentiation of transmitter glutamate release from guinea pig cerebrocortical synaptosomes. *J. Neurochem.* 57, 1398–1404.

Grinstein, S., Cohen, S., Goetz-Smith, J.D. & Dixon, S.J. (1989) Measurements of cytoplasmic pH and of cellular volume for the detection of Na/H exchange in lymphocytes. *Methods Enzymol.* **173**, 777–790.

Madshus, I.H. (1988) Regulation of intracellular pH in eukaryotic cells. *Biochem. J.* **250**, 1–8.

Neuronal bioenergetics

Kauppinen, R.A. & Nicholls, D.G. (1986) Synaptosomal bioenergetics: the role of glycolysis, pyruvate oxidation and responses to hypoglycaemia. *Eur. J. Biochem.* **158**, 159–165.

McMahon, H.T. & Nicholls, D.G. (1991) The bioenergetics of neurotransmitter release. *Biochim. Biophys. Acta* **1059**, 243–264.

Nicholls, D.G. & Ferguson, S.J. (1992) *Bioenergetics*, Vol 2. London: Academic Press.

Njus, D., Kelley, P.M. & Harnadek, G.J. (1986) Bioenergetics of secretory vesicles. *Biochim. Biophys. Acta* **853**, 237–266.

Pressman, B.C. (1976) Ionophores. *Annu. Rev. Biochem.* **45**, 501–530.

Part 2
Proteins

4

The neuronal plasma membrane: transporters, pumps and channels

4.1 Introduction

The previous chapter emphasized that ion transport is central to an understanding of the synapse. In this chapter we shall consider the molecular structure of the transporters, ion pumps and channels which are present in synaptic membranes and are responsible for the ion conductance changes discussed in Chapter 3.

As in other excitable cells, three classes of transport process occur across neuronal membranes (Fig. 4.1):

1 *Transporters* reversibly bind the ion, ions or metabolites and undergo spontaneous conformational changes which expose the binding site(s) alternately to opposite sides of the membrane. Dissociation and association from the binding sites is governed by the local concentration of the ion and the affinity of the binding site, in the particular conformation. The conformational change of a transporter is *stoichiometric* with transport and must be repeated at each cycle. Transporters are much slower (perhaps 1000 ion $\cdot$ s^{-1}) than channels (which can conduct 10^7 ions $\cdot$ s^{-1}) although they generally display much greater specificity towards the transported species.

A transporter which carries one ion at a time (a *uniporter*) will transport the ion down its electrochemical gradient. Transporters which carry two species in the same direction (*symporter*) or in opposite directions (*antiporter*) can couple the gradient of one ion to drive the accumulation or expulsion of the second ion (Fig. 4.1).

2 *Ion pumps* are similar to transporters, except that the conformational changes are driven by metabolic energy (in the neuronal context almost invariably ATP hydrolysis). Synchronous with the conformational change is a change in the affinity of the binding sites, such that the binding site has a high affinity when it faces the compartment where the ion is present at low concentration and a low affinity when in contact with the high ion concentration. In this way the metabolic energy can be used to extract the ion from the low concentration compartment expelling into the high concentration compartment (*see* Fig. 4.5).

3 *Ion channels* undergo a conformational change from a non-conducting to a conducting state which may be triggered by voltage (voltage-activated channel) or the binding of a neurotransmitter (ligand-activated channel). Ion channels possess multiple transmembrane domains which may be part of a single polypeptide (Na^+- and

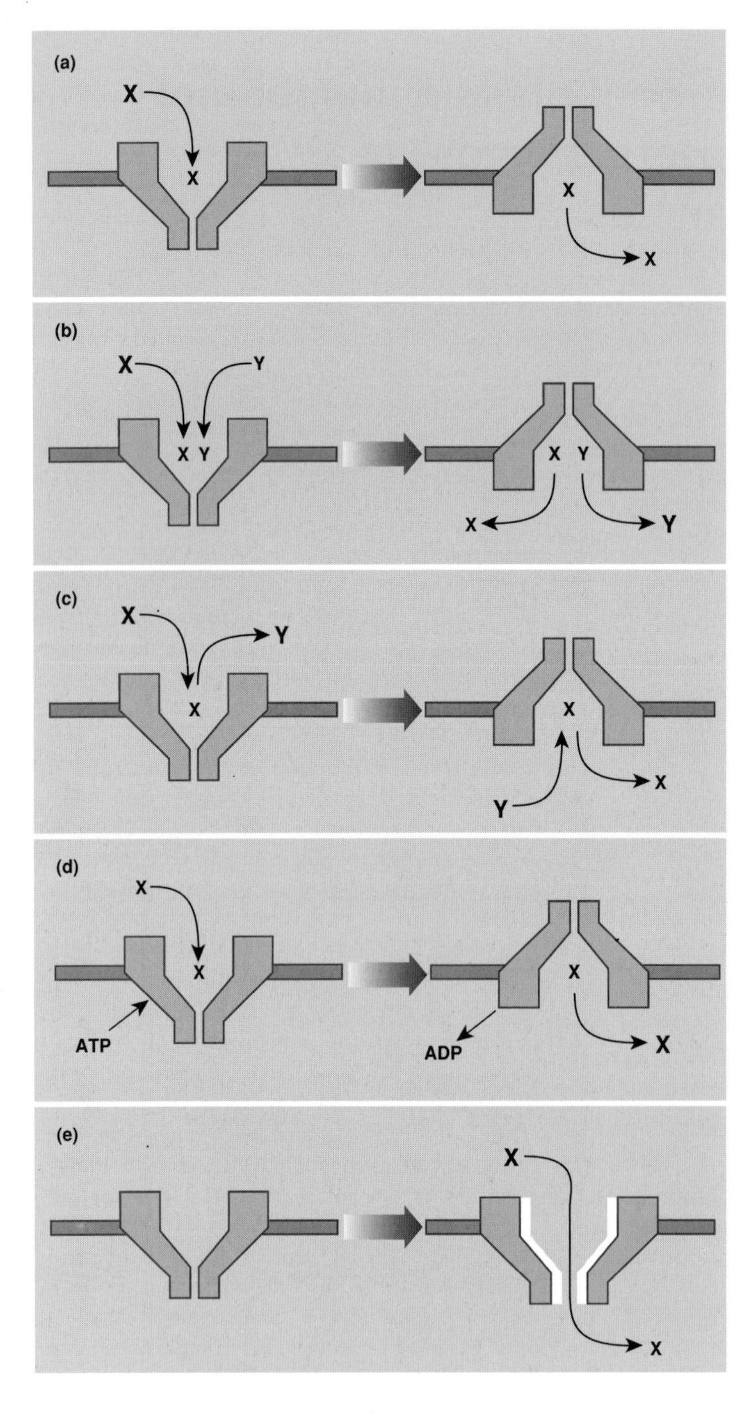

Fig. 4.1 Transporters, pumps and channels. A *transporter* undergoes a conformational change for each catalytic cycle, alternately exposing the binding site to the two aqueous phases. (a) A uniport mechanism involves the transport of a single species down its electrochemical gradient. (b) A symport involves the cotransport of two species (X, Y) in the same direction across the membrane. Transport of one species down its electrochemical gradient can drive the 'uphill' transport of the second species, e.g. a Na^+-coupled plasma membrane neurotransmitter transporter. (c) An antiport exchanges two species across the membrane. Downhill transport of one can drive the uphill transport of the second species in the opposite direction, e.g. the Na^+/Ca^{2+} exchanger in the plasma membrane. (d) An ATP-coupled *ion pump* can drive the uphill transport of an ion across a membrane. The free energy from ATP hydrolysis is used to change the affinity of the binding site from 'high' when picking up the ion from the low concentration compartment to 'low' when discharging the ion into the high concentration compartment, e.g. Ca^{2+}-ATPases. (e) An *ion channel* undergoes a conformational change which opens a central pore allowing the passage of typically 10 000 ion·ms^{-1}.

Ca^{2+}-channels) or a complex of multiple subunits (K^+-channels and ligand-activated channels). Activation of the channel causes a central pore within the channel to open and allow the passage of hydrated ions. The effect of a conformational change is essentially *catalytic*, allowing enormous numbers of ions to cross the membrane without stoichiometric oscillations in conformation until a further conformational change occurs which occludes the pore.

We shall now consider each class of transport mechanism in more detail, concentrating on generic features common to all synapses. Transmitter-specific aspects will be discussed in detail in the chapters dealing with individual transmitter classes.

4.2 Ion transporters

The predominant role of transporters in the neuronal plasma membrane is to utilize the free energy in the transmembrane Na^+ gradient (*see* Section 3.2) to drive either the accumulation of a species (by symport) or its expulsion (by antiport). Na^+-coupled symporters are used by neurones to accumulate neurotransmitters from the synapse as an aid to terminating transmission. Two classes of Na^+-coupled transmitter symporters have been described (Fig. 4.2). Protein purification of the glutamate, GABA and choline transporters in the late 1980s identified proteins of 65–80 kDa. Subsequent microsequencing allowed cDNA cloning of the GABA transporter and expression cloning and polymerase chain reaction (PCR) amplification of related sequences resulted in the isolation of cDNAs for the NA,

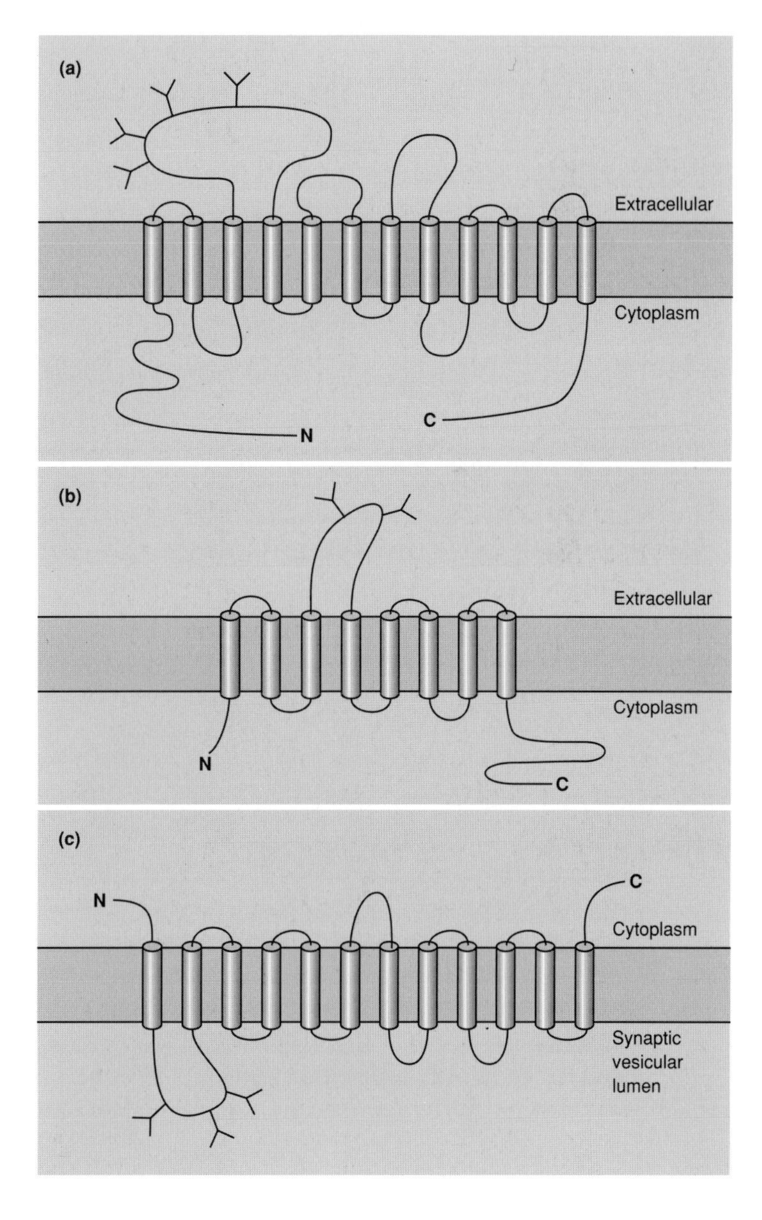

Fig. 4.2 Families of ion-coupled transmitter symporters. Putative transmembrane topology of (a) plasma membrane Na^+/Cl^--dependent neurotransmitter transporters (e.g. GABA); (b) plasma membrane Na^+/K^+-dependent glutamate transporters; (c) H^+-dependent vesicular monoamine transporters. Y, glycosylation sites.

5-HT, dopamine (DA) and choline transporters. The deduced polypeptides show 30–65% identity, with 12 putative transmembrane domains, cytoplasmic N- and C-termini and a large glycosylated extracellular loop between transmembrane domains III and IV. Interestingly the plasma membrane glutamate transporter (*see* Section 9.3) shows significant structural differences to the other carriers and may reflect a different 'family'.

The main Na^+-coupled antiporters in the plasma membrane are the Na^+/H^+ exchanger which has already been discussed in the context of cytoplasmic pH regulation (*see* Section 3.4) and the $3Na^+/Ca^{2+}$ exchanger which participates in lowering cytoplasmic Ca^{2+} after this has been increased by Ca^{2+}-channel activity.

4.2.1 The Na^+/Ca^{2+} exchanger

The $3Na^+/Ca^{2+}$ exchanger present in the plasma membrane of excitable cells has been sequenced (Fig. 4.3). The antiporter has 12 predicted transmembrane spans with a large cytoplasmic loop between spans 6 and 7. Under resting conditions the exchanger would operate close to thermodynamic equilibrium, since the Ca^{2+} electrochemical gradient across the polarized membrane quantified from the concentration gradient and the membrane potential is, at $35\,kJ \cdot mol^{-1}\,Ca^{2+}$, three times greater than the $12\,kJ \cdot mol^{-1}\,Na^+$ electrochemical gradient. It is frequently stated that this carrier can expel Ca^{2+} in parallel with the Ca^{2+}-ATPase; however the conditions under which this occurs may be rather specialized (*see*

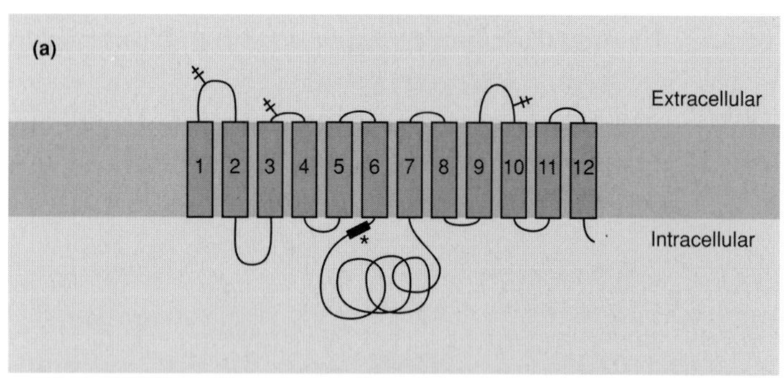

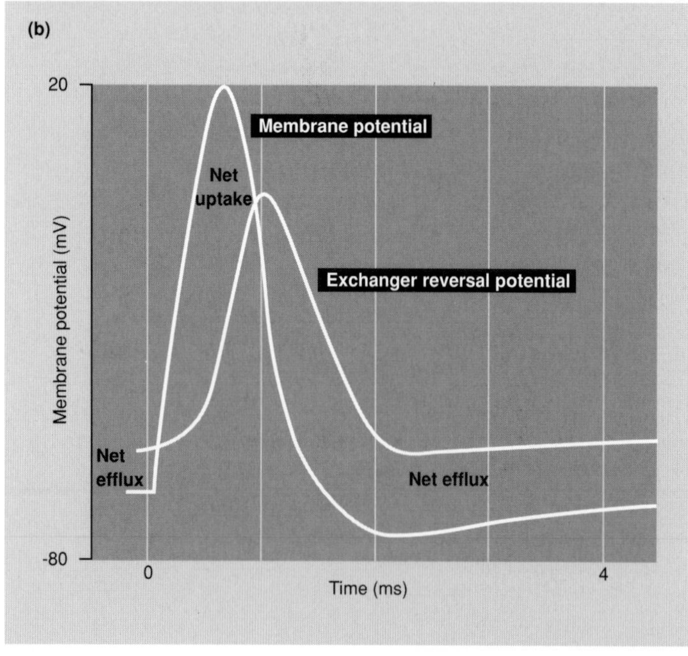

Fig. 4.3 Na^+/Ca^{2+} exchange. (a) Proposed transmembrane organization of the plasma membrane $3Na^+/Ca^{2+}$ exchanger. Potential extracellular glycosylation sites are shown; (*), region interacting with calmodulin. From Nicoll *et al.* (1990). (b) The electrogenic $3Na^+/Ca^{2+}$ exchanger operates close to equilibrium in neurones. In the polarized state there may be a net efflux of Ca^{2+}. During the rising phase of the action potential net uptake of Ca^{2+} is favoured thermodynamically, while net efflux would occur at the end of the repolarization and during any after hyperpolarization. Efflux in this last phase would be favoured by an activation of the transporter by elevated $[Ca^{2+}]_c$. Adapted from Blaustein (1988).

Fig. 4.3). The exchanger can be activated allosterically by internal Ca^{2+}; thus when $[Ca^{2+}]_c$ is in the region of 100 nM its activity is very low. The conditions under which the exchanger would be most likely to contribute to Ca^{2+} efflux would be when internal $[Ca^{2+}]_c$ is elevated after a train of action potentials and after the membrane potential has repolarized assisting this electrogenic carrier.

4.3 Neuronal ATP-driven cation pumps

The largest utilizers of ATP in the neurone are the ion pumps responsible for generating and maintaining the concentration gradients of Na^+, K^+ and Ca^{2+} across the plasma membrane and of H^+ across vesicular membranes. ATP-driven cation pumps (Fig. 4.4) can be classified into three types:

1 *F-type ATPases* are typified by the proton-translocating $F_1.F_0$-ATPase of the mitochondrial inner membrane. This ATPase is a multi-subunit complex, consisting of a transmembrane proton channel linked to a catalytic complex within the mitochondrial matrix. The reaction mechanism does not involve the formation of a transient, phosphorylated intermediate. The $F_1.F_0$-ATPase is restricted to the mitochondrial inner membrane and is dealt with in detail in a number of bioenergetic texts.

2 *V-ATPases* are proton-translocating ATPases responsible for the acidification of intracellular organelles such as lysosomes, endosomes and synaptic vesicles. The accumulation of Type I and Type II neurotransmitters into their respective synaptic vesicles is driven by the proton electrochemical gradient generated by the V-ATPase. This can amount to 140 mV: 1.5 pH units acidic within the vesicle lumen plus 50 mV $\Delta\psi$. Transmitter uptake into the vesicles can utilize either the membrane potential (e.g. in the case of glutamate), the ΔpH (for catecholamines) or both components (for GABA). The V-ATPase is a complex of at least eight different subunits, five of which to date have been cloned (Fig. 4.4). The V-ATPase shows some limited homology with the proton-translocating $F_1.F_0$-ATPase.

The V-ATPase is characterized by insensitivity

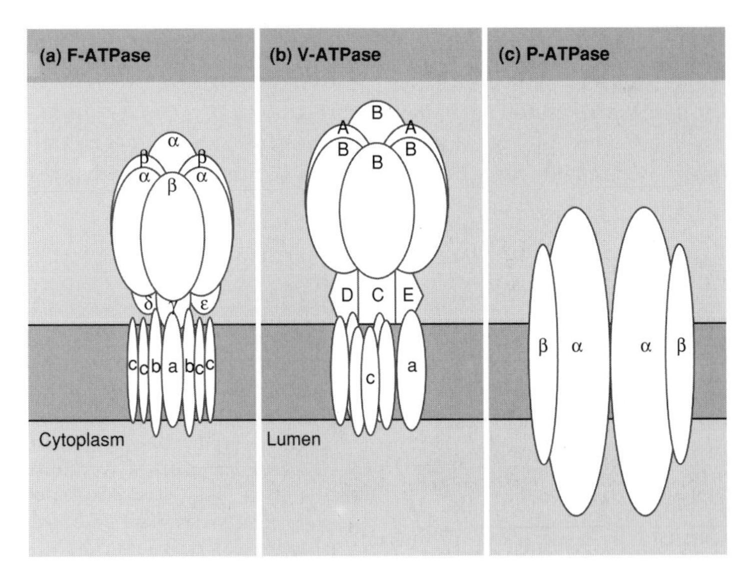

Fig. 4.4 Classes of ATP-hydrolysing proton pumps. Schematic model for the subunit structures of ion-translocating ATPases. (a) For the H^+-translocating F-ATPase of the mitochondrion the Greek symbols represent subunits comprising the soluble F_1 catalytic complex (with ATP binding site in the mitochondrial matrix) and the lower case letters the F_0 proton channel. (b) For the V-ATPase, the catalytic complex (capital letters) protrudes into the cytoplasm. The structure of the integral membrane complex (lower case letters) is known from electron diffraction studies. (c) The Na^+/K^+-ATPase, a typical P-ATPase, consists of a dimer of a single subunit with an accessory protein (β) which is required for insertion of the complex into the membrane. Adapted from Nelson (1991a).

to orthovanadate (which inhibits P-ATPases), relative insensitivity to the mitochondrial ATPase inhibitor dicyclohexylcarbodiimide (DCCD) but high sensitivity to *N*-ethylmaleimide and a specific inhibitor, *bafilomycin A1*. The protein accounts for about 20% of the total synaptic vesicle protein, and on average each vesicle would be expected to contain about four ATPase complexes.

3 *P-ATPases* comprise a family of cation pumps characterized by a reversible phosphorylation of an aspartyl residue during the catalytic cycle. The neuronal plasma membrane possesses two P-type ATPases: the Na^+/K^+-ATPase which maintains the concentration gradients of Na^+ and K^+ across the plasma membrane, and the plasma membrane Ca^{2+}-ATPase, which is largely responsible for maintaining the large electrochemical gradient of Ca^{2+} across the plasma membrane.

A common structural feature of P-type ATPases is a 100 kDa catalytic α-subunit which becomes temporarily phosphorylated during the hydrolysis of ATP (Fig. 4.5). Complete amino acid sequences have been obtained for most P-type ATPases, and they show clear homologies, and similar predicted structures based on hydropathy plots (Fig. 4.5). The ATP binding site is within the large cytosolic domain of the α-subunit. Covalent labelling of amino acid residues by agents which bind to the ATP site reveals that regions widely separated in primary sequence are brought close together in the tertiary structure to form the nucleotide binding site.

4.3.1 Na^+/K^+-ATPase

The Na^+/K^+-ATPase (Fig. 4.5) is the major utilizer of ATP in the neurone. $3Na^+$ are expelled and $2K^+$ accumulated for each ATP hydrolysed and because of this inbalance the transport is electrogenic. The maximal activity of the ATPase in neurones is very high, although normally the pump activity is restricted by feedback from the high Na^+ electrochemical gradient across the membrane. This in turn is a consequence of the low Na^+ permeability of the resting membrane (in the steady-state Na^+ cannot be pumped out of the cell faster than it can leak back in). If however this rate limitation is removed, for example, by adding the alkaloid veratridine to prevent Na^+-channel closure (Box 4.1) the respiration of a synaptosomal preparation can increase five-fold as the mitochondria compensate for the increased ATP demand.

The Na^+/K^+-ATPase consists of two subunits; a 112 kDa catalytic α-subunit with 1016 residues and a 38 kDa, 302 amino acid glycosylated β-subunit which is required for insertion of the complex into the membrane but which has no known catalytic role. The functional unit is probably $\alpha_2\beta_2$. The α-subunit has ten predicted transmembrane segments and in common with other P-type pumps undergoes an E_1-E_2 confor-

Box 4.1 Na^+-channel neurotoxins

Site 1 toxins: Act at the cytoplasmic face to inhibit channel opening. Toxins include tetrodotoxin from the Japanese puffer fish, saxitoxin and μ-conotoxins.

Site 2 toxins: Act in the hydrophobic core of the membrane and prevent channel closing. Toxins include the alkaloid veratridine, grayanotoxin and batrachotoxin

Site 3 toxins: Include the α-toxins from North African scorpions, and some sea anemone toxins. They inhibit channel inactivation

Site 4 toxins: Include American scorpion β-toxins. They alter the voltage-dependency of activation

Site 5 toxins: Include ciguatoxins which cause repetitive firing and persistent activation

Fig. 4.5 (*Opposite*) Cation-translocating ATPases and their catalytic cycles. (a) Hydropathy plots for the brain type 1 plasma membrane Ca^{2+}-ATPase (PMCA1), the sarcoplasmic reticulum Ca^{2+}-ATPase (SRCA) and the Na^+/K^+-ATPase α-subunit (NKA). N- and C-termini are cytoplasmic and there may be ten putative membrane spanning segments. From Shull & Greeb (1988), by permission of the American Society for Biochemistry and Molecular Biology. (b) Schematic catalytic cycle of the Na^+/K^+-ATPase: 1, In the E1 conformation the binding sites faces the cytoplasm and has a low affinity for K^+, which is replaced by Na^+; 2, ATP phosphorylates an aspartyl reside; 3, the phosphorylated E1 is unstable and undergoes a conformational change to E2; 4, the binding site of E2 faces outwards and has a low affinity for Na^+, which is exchanged for K^+; 5, K^+-E2 dephosphorylates; 6, dephospho-E2 is unstable and reverts to the E1 conformation.

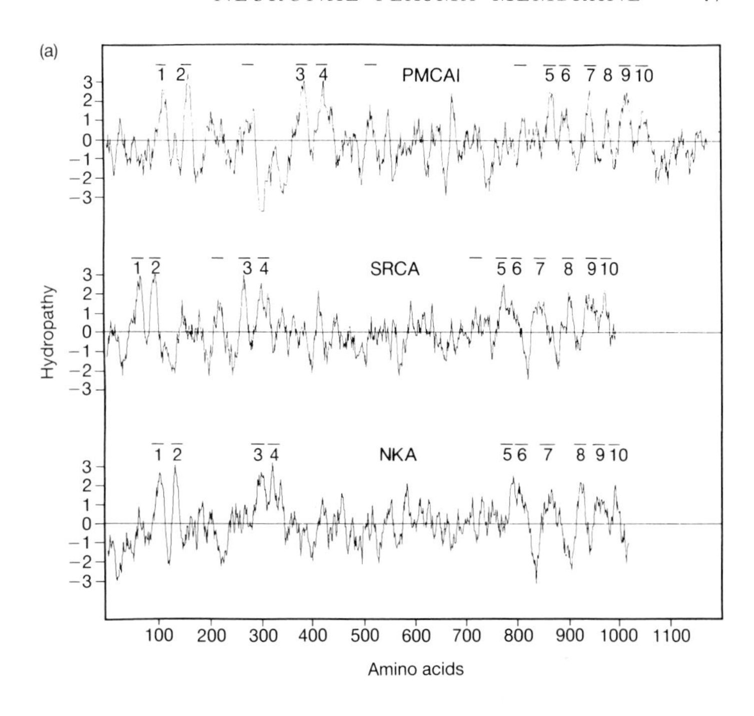
(a)
PMCAI
SRCA
NKA
Hydropathy
Amino acids

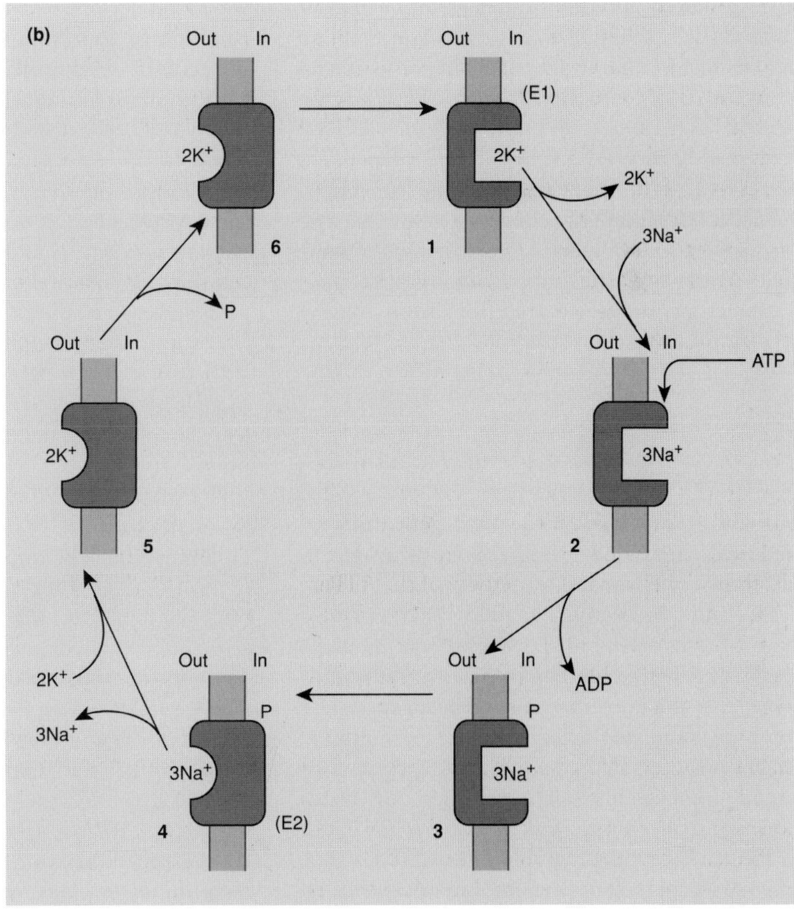
(b)
Out
In
(E1)
2K+
3Na+
ATP
ADP
P
(E2)
1
2
3
4
5
6

mational transition controlled by reversible phosphorylation of an aspartyl residue (Fig. 4.5).

4.3.2 Plasma membrane Ca^{2+}-ATPase

The Ca^{2+}-ATPase is a single polypeptide which associates with calmodulin and can therefore be purified by passing a detergent extract of neuronal plasma membranes down a calmodulin-affinity column. The isolated cDNA for the plasma membrane Ca^{2+}-ATPase indicates that the brain contains two isoforms of 130 and 133 kDa. The hydropathy profile and overall sequence is homologous with that of the Na^+/K^+-ATPase and the sarcoplasmic reticulum Ca^{2+}-ATPase (Fig. 4.5). The actual number of transmembrane-spanning regions is unclear, but may be as high as ten. The cytoplasmic C-terminal region possesses a region characteristic of calmodulin-binding domains of other proteins. At resting $[Ca^{2+}]_c$, calmodulin is dissociated from the enzyme; however when $[Ca^{2+}]_c$ rises and Ca^{2+} binds to calmodulin, this greatly increases calmodulin's affinity for the ATPase. The resultant Ca^{2+}-calmodulin–enzyme complex can transport Ca^{2+} with increased affinity. Thus the normal increase in activity of the Ca^{2+}-ATPase with $[Ca^{2+}]_c$ is amplified by the Ca^{2+}-dependent activation of the enzyme, allowing the Ca^{2+}-ATPase to respond very effectively to an increase in $[Ca^{2+}]_c$. There appears, however, be a short time-lag before this activation occurs. This hysteresis allows an elevated $[Ca^{2+}]_c$ to act as a second messenger for a limited time before the increased $[Ca^{2+}]_c$ is counteracted by the pump.

4.4 Ion channels

The ion channels involved in the generation of electrical signals are either *voltage-gated*, in which case they open in response to a depolarization from the resting membrane potential or *ligand-gated*, when their integral ion channels are opened by the binding of a neurotransmitter (Fig. 4.6); the latter are also termed *ionotropic receptors*. The two categories are not mutually exclusive since some channels may require both ligand binding and depolarization to open. Ligand-gated channels will be discussed in detail in the context of the individual transmitters, but since they possess sufficient homology to be classified into superfamilies, their common features will be discussed here.

4.4.1 Generic features of ionotropic receptors

The most detailed structural information is available for the nicotinic acetylcholine receptor, in fact the structure of other receptors is largely extrapolated from the uniquely precise information available for this ligand-gated channel. The muscle form of the nicotinic acetylcholine receptor (MnAChR) can be purified from *Torpedo* electric organ (*see* Section 10.3.1), whose cells, or electrocytes, are derived from muscle tissue. Purification of the receptor is aided by the existence of an 8 kDa peptide neurotoxin *α-bungarotoxin* from the venom of the cobra *Bungarus multicinctus*, which binds with extremely high affinity to the receptor and can be used in the radio-iodinated form to assay the receptor during its purification.

Under non-denaturing conditions the MnAChR from adult animals purifies as a complex of four subunits, α, β, γ and δ, with molecular weights 40–64 kDa. The stoichiometry of the subunits corresponds to $\alpha_2\beta\gamma\delta$ and this pentameric structure is supported by image reconstruction analysis carried out on electron micrographs of paracrystalline arrays of reconstituted MnAChR (Fig. 4.7). This indicates a central channel 7–9 nm in diameter and large extracellular domains for each subunit. Hydropathy plots for each of the subunits are consistent with four transmembrane α-helices (termed M1–M4) with extracellular N- and C-termini. Structural models predict that each subunit contributes one transmembrane α-helix to form the lining of the pore.

The structure of the MnAChR has been used as a generic model for other ionotropic receptors, although it must be emphasized that direct confirmation that the functional unit is a pentamer is only available for the MnAChR. The primary sequences and hydropathy plots of the individual subunits comprising the ionotropic MnAChR, 5-HT_3, GABA-A and glycine receptors are all sufficiently homologous to be considered members of the same superfamily, but the N-methyl-D-aspartate (NMDA) and AMPA/kainate (AMPA/KA) ionotropic glutamate receptors, while sharing the same putative transmembrane topology, are classed in a separate

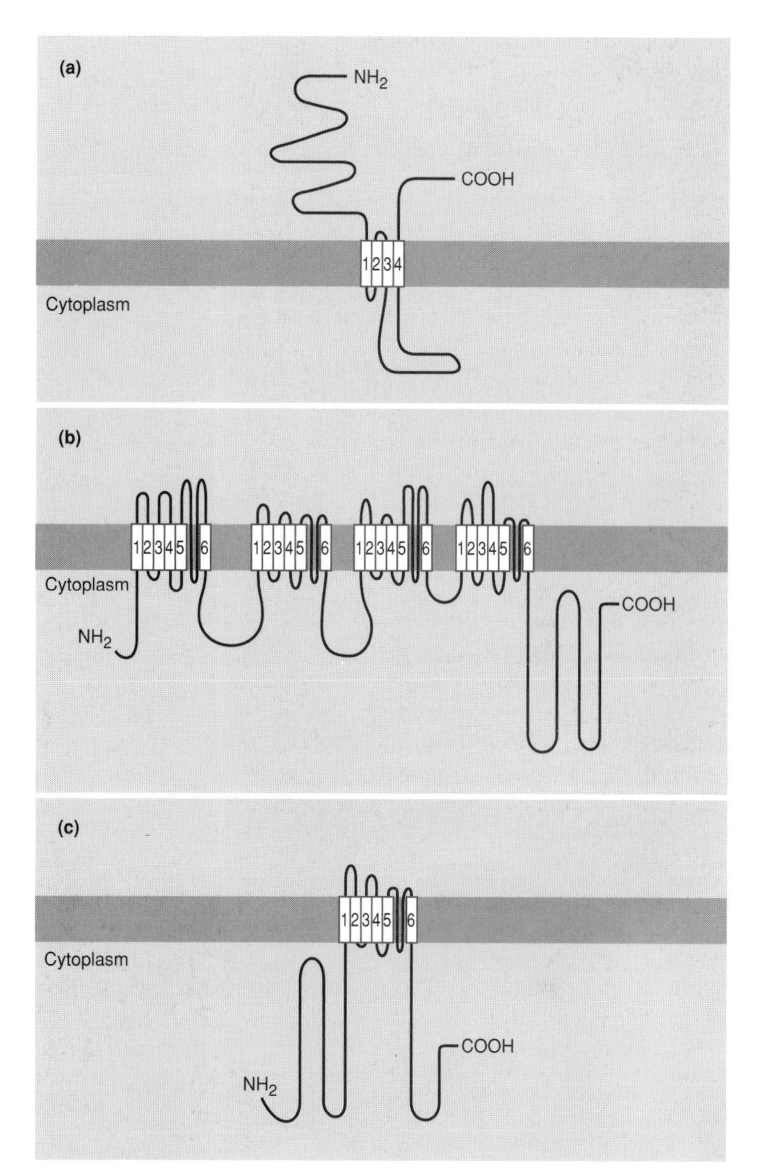

Fig. 4.6 Transmembrane organization of voltage-gated and ligand-gated ion channels. (a) One subunit of a typical ligand-gated ion channel (ionotropic receptor). Functional receptors are composed of five similar subunits. (b) The α-subunit of a Ca^{2+}-channel as an example of a voltage-gated ion channel. A single α-subunit can provide a functional Na^{+}- or Ca^{2+}-channel, although additional smaller subunits may assist with regulation and localization of the channel. (c) K^{+}-channels are assembled from four subunits to give structures closely homologous to the Na^{+}- or Ca^{2+}-channels.

superfamily (Fig. 4.8). In particular, immunocytochemistry suggests that the C-terminus of some ionotropic glutamate receptors may be cytoplasmic, suggesting that one of the apparent transmembrane helices may instead fold back as a hairpin (*see* Section 9.6.1).

For each of these ionotropic receptors multiple genes encoding homologous but distinct subunits have been identified. In the case of the GABA-A receptor, five classes of subunit have been identified and named (by analogy to the MnAChR) α, β, γ, δ and ρ (*see* Section 9.8). Closely homologous multiple genes have been found within these classes with (to date) six α, four β, three γ, one δ and two ρ isoforms. Alternatively spliced mRNA for some subunits has also been detected, where the same gene gives rise to two mRNAs. Individual receptors differ in expression systems depending on whether the native electrophysiology and pharmacology can be reconstituted with a single type of subunit (a homo-oligomer) or whether a combination of different subunits (a hetero-oligomer) is required. Techniques exist (discussed in Chapter 9) which allow the electrophysiological characteristics of a receptor in an individual neurone to be determined by patch

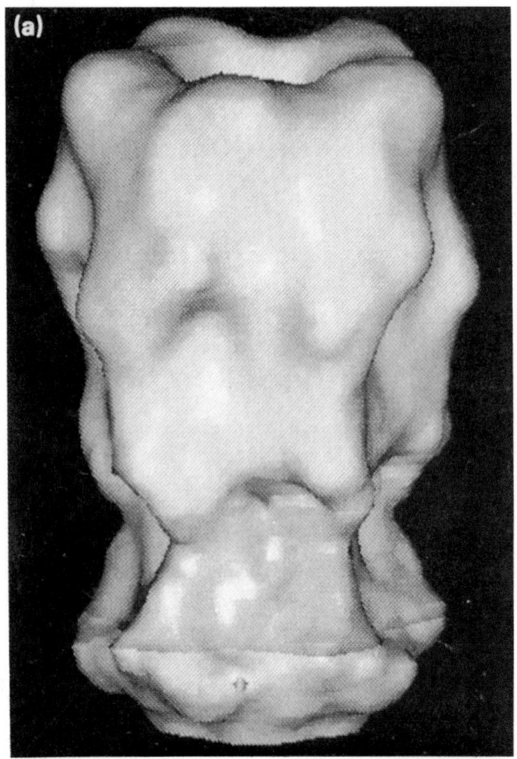

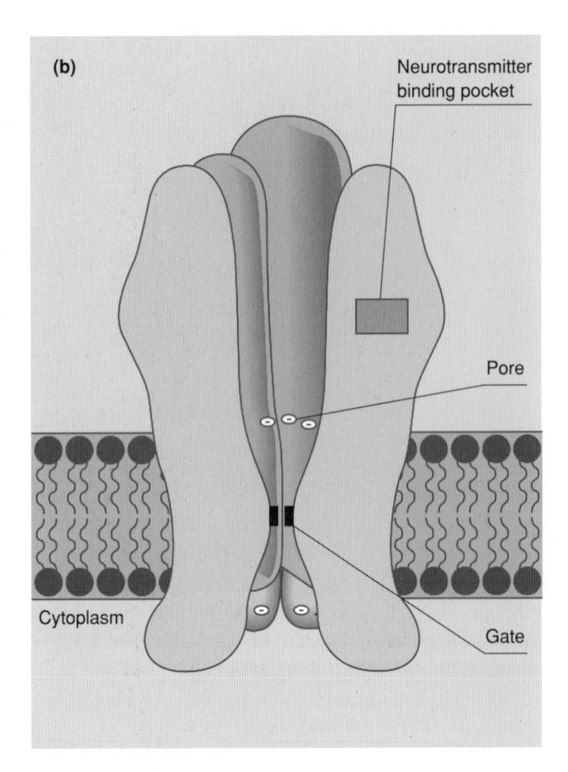

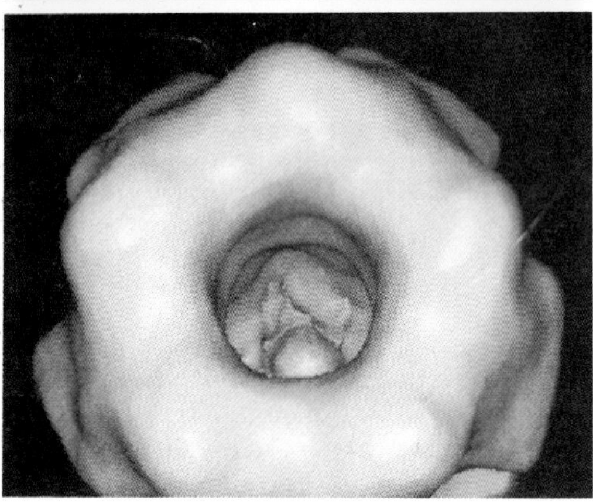

Fig. 4.7 The structure of the nicotinic ACh receptor — the archetypal ligand-gated receptor. (a) Side and top views of the channel obtained by image reconstruction of paracrystalline arrays of reconstituted receptor. The receptor is assembled from five subunits arranged around a central opening which narrows as it crosses the lipid bilayer. (b) Cross-sectional views of the channel — at the narrowest portion is the gate which is opened by conformational changes induced by the binding of transmitter to a relatively distant binding site. Charges on either side of the central pore determine the charge specificity of the transported ion. Adapted from Unwin (1993), reproduced with permission from *Neuron*. © Cell Press.

clamping, following which the cytoplasmic contents of the cell are sucked into the pipette and the mRNAs amplified by PCR to allow correlation of the electrophysiology with the pattern of subunit expression in that single cell. *In situ* hybridization reveals that the enormous number of potential permutations of subunits (at least 1000 for a GABA-A pentamer) appears to be somewhat restricted *in vivo* as, with some exceptions, most receptor isoforms appear to utilize a single subtype of each class of subunit rather than a mixture. Even so there may be 50–100 combinations of GABA-A receptor expressed in the brain, showing regional and developmental patterns.

In the case of the glutamate receptors, developmentally regulated single residue variants are produced by posttranscriptional mRNA base editing. Thus the deamination of a specific adenine residue in the mRNA of the AMPA/KA receptor causes an original glutamine codon to be interpreted as an arginine. This single change

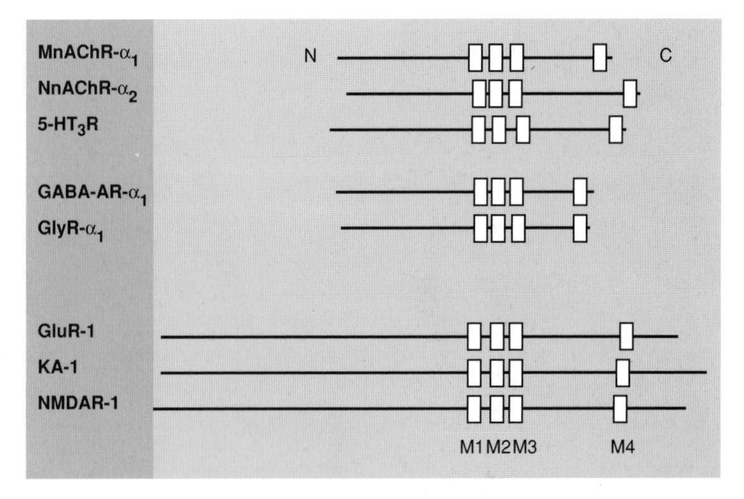

Fig. 4.8 Transmembrane organization of representative subunits of the ACh and glutamate ionotropic receptor families. Comparison of the polypeptide chains of muscle (MnAChR-α_1) and neuronal (NnA-ChR-α_2) nicotinic ACh receptor subunits, the 5-HT_3 receptor, α_1-subunits of the GABA-A and glycine receptors (all members of the ACh receptor family) and three members of the glutamate family, GluR-1, KA-1 and NMDAR-1. The blocks represent the transmembrane segments M1–M4. Note that the glutamate family has a much larger N-terminal extracellular extension. Adapted from Unwin (1993).

can dramatically alter the conductance of the channel (*see* Section 9.6.1).

4.4.2 Generic features of voltage-activated ion channels

The principal subunits of many of the isoforms of voltage-gated Na^+-, Ca^{2+}- and K^+-channels have been sequenced from cDNA clones. A common feature of each is a motif with six predicted membrane-spanning regions, S1–S6 (*see* Fig. 4.6). In addition to these classic transmembrane α-helices there has been recent evidence, which will be discussed in the context of K^+-channels below, that the extracellular loop between S5 and S6 may be folded back into the barrel of the channel to form a hairpin-like lining to the channel.

K^+-channel subunits possess just one such six-transmembrane motif (but probably function as tetramers), whereas for Na^+- and Ca^{2+}-channels the motif is repeated in four sequential domains in a single polypeptide with no less than 24 predicted membrane spanning regions in a polypeptide containing some 2000 amino acid residues (*see* Fig. 4.6). Together with an overall sequence homology, this structural homology allows all three classes of channel to be assigned to a superfamily. In addition to the α-subunit the channel complexes usually contain additional subunits whose function is regulatory or involved in cytoskeletal interaction.

A voltage-gated ion channel must have some means of detecting the electrical field across the membrane to enable it to respond to a change in membrane potential. Since the field strength across the hydrophobic core of a plasma membrane withstanding a membrane potential of $-80\,mV$ is some $150\,kV \cdot cm^{-1}$, it follows that a charged region of the primary sequence within the hydrophobic core will experience a force which will be a function of the membrane potential. If this force is sufficient to stabilize an otherwise unfavourable conformation, depolarization would lead to a relaxation of the channel structure to a lower energy conformation. In voltage-activated channels this relaxation allows the channel to conduct ions. Electrophysiologists can detect the charge movement associated with this conformational change as a brief *gating current* which can be separated from the much larger current due to ion flux through the channel. The steepness of the voltage activation of the channel requires that a minimum of 4–6 charges moves in the electrical field during activation.

The most likely candidate for this *voltage sensor* is contained within the fourth span (S4) in

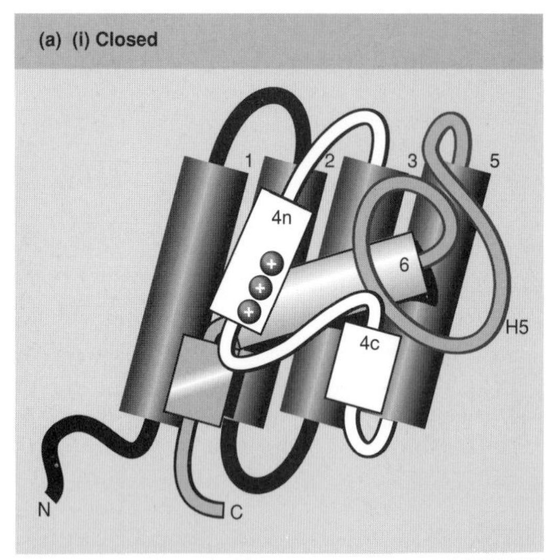

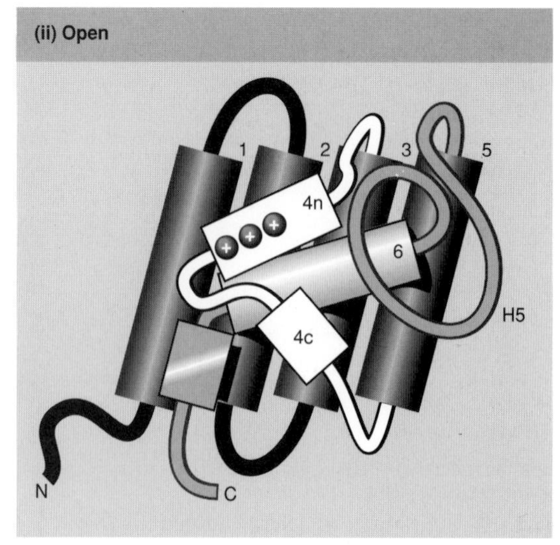

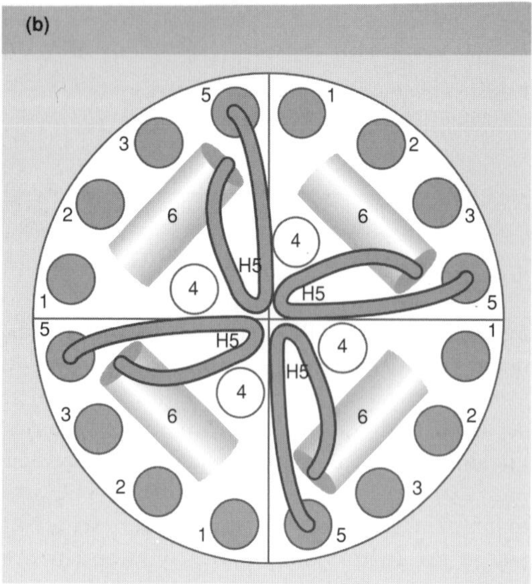

Fig. 4.9 A possible model for the structure of a voltage-activated ion channel. (a) (i) One domain is seen from the lumen of the channel. Transmembrane spans S1, S2, S3 and S5 (dark shading) may provide an outer ring for the channel. S6, which in the case of a K^+-channel subunit is distorted by a proline residue, may occupy an intermediate annulus, while the H5 loop between S5 and S6 may line the barrel of the channel as a β sheet. S4 appears to act as the voltage-sensor, carrying three or four positive arginine or lysine residues and pulled into the channel by the transmembrane field. (ii) Depolarization may cause the conformation to relax with an outward, twisting movement of S4 opening the channel. (b) Assembly of four such domains viewed from above. Adapted from Pongs (1992) and Guy & Conti (1990).

each domain (Fig. 4.9). A highly conserved pattern of positively charged arginine or lysine residues is found at every third residue in this α-helical region which form ionic interactions with regularly spaced anionic charges on adjacent helices. The depolarization-evoked conformational relaxation may involve a 'frame-shift' of this interaction to the next stable position triggering a conformational change in the whole channel. Support for this hypothesis has come from sequential site-directed mutagenesis to substitute the basic residues one by one. This results in a decrease in the steepness of the activation/potential relationship for the channel.

4.5 Voltage-activated K^+-channels

K^+-channels set the resting membrane potential, regulate the action potential and generally control excitability. Consistent with these multiple roles K^+-channels are remarkably diverse; for example, a single CA1 hippocampal neurone may have six or more distinct K^+ conductances which can influence the excitability of the cell. The formation of functional K^+-channels by the association of four 6-transmembrane-span subunits, rather than a single long 24 span polypeptide, and the presence of multiple genes for the subunits, theoretical allows for an enormous diversity of K^+-channel subtypes, comparable to that which we have just discussed for GABA-A receptors.

Three concerted approaches have been used to characterize K^+-channels: electrophysiology, pharmacology (including channel susceptibility to specific neurotoxins) and molecular genetics, although a complete synthesis between these three approaches has still to be made. A major obstacle to the understanding of neuronal K^+-channels, which stems from their assembly from four subunits, is that there is not necessarily a strict correlation between inhibitor sensitivity, which may be due to a sequence on a single subunit, and electrophysiological behaviour, which is a property of the complete tetrameric channel together with any additional regulatory subunits. While homomeric channels can be expressed and their electrophysiology characterized, there is still a very limited understanding of the properties of channels composed of defined heteromeric subunits.

We shall start by considering the electrophysiological behaviour of native K^+-channels.

4.5.1 K^+-channels defined by electrophysiology and pharmacology

K_A^+-channels (I_A) are closed at normal resting potentials and produce transient, rapidly inactivating, currents on depolarization. I_A currents may play a role in cells which fire repetitively in response to maintained depolarization, and control the excitability of the nerve terminal. In some neurones they may be activated by a variety of presynaptic G-protein linked receptors, thus hyperpolarizing the membrane and inhibiting transmission. A-channels can usually be inhibited by *dendrotoxin* and by 4-aminopyridine (Box 4.2).

Delayed rectifiers (I_K) are activated by depolarization after a short delay, and are slowly inactivating. The primary role of delayed rectifier K^+-channels is in the termination of the action potential (*see* Section 3.2.3). Channels with delayed rectifier characteristics are generally sensitive to tetraethylammonium (TEA).

Inward rectifiers have a greater tendency to allow K^+ to flow into the cell, for example during hyperpolarization, than out following depolarization. This can be caused either by a voltage-dependent block by internal Mg^{2+} or by channel opening during hyperpolarization. Recently cloned inward rectifier channels appear to have a dramatically different structure from the consensus models for voltage-gated channels discussed in Section 4.6.2.

At least two classes of *Ca^{2+}-activated K^+-channels* can be distinguished on the basis of ionic conductance and toxin sensitivity; these are the high conductance (100–250 pS) BK-channel and small-conductance (6–14 pS) SK-channel (1 pS corresponds to a conductance of $10^{-12}\,\Omega^{-1}$). The BK-channel is activated both by increased cytoplasmic Ca^{2+} and depolarization. It is very widely distributed, and provides a hyperpolarizing negative feedback when sufficient Ca^{2+} has entered to elevate the bulk $[Ca^{2+}]_c$ or in response to the liberation of Ca^{2+} from internal stores (*see* Section 5.8). The BK-channel is generally inhibitable by *charybdotoxin* (Box 4.2) although this toxin can also affect A-channels.

The small conductance (*SK*) Ca^{2+}-activated channels are responsible for the 'after-hyperpolarization' of neurones which follows a train of action potentials, being activated by the increase in neuronal $[Ca^{2+}]_c$. They are characterized by a small channel conductance (6–14 pS), lack of voltage dependency and half-maximal activation by 200–500 nM $[Ca^{2+}]_c$. They can be inhibited by apamin from bee venom (Box 4.2).

Hypoxia or glucose depletion frequently causes an increase in neuronal K^+ conductance and a resulting hyperpolarization. *ATP-sensitive K^+-channels*, which open when ATP decreases and close when ATP increases, are implicated in the glucose regulation of insulin secretion from pancreatic β-cells, and are also present in neurones, where they play a role in decreasing neuronal excitability under low energy conditions. *Sulphonylureas* inhibit the channel.

The *K_m^+-channel* is seen in a variety of neurones and is defined as a voltage and time-dependent K^+ current which is slightly active at resting potentials and is further activated by depolarization. It may thus act to stabilize the membrane potential. The conductance is suppressed by muscarinic receptor activation leading to depolarization and increased excitability. The inhibition is probably mediated by the increased $[Ca^{2+}]_c$ following (1,4,5)IP_3 liberation, since it is prevented by injecting Ca^{2+} chelators or by depletion of internal Ca^{2+} stores with ionomycin.

Box 4.2 K^+-channel neurotoxins and inhibitors

Dendrotoxins: α-, β-, γ- and δ-DTx (from the eastern green mamba *Dendroaspis angusticeps*) and DTxI (from the black mamba *Dendroaspis polyepis*) are 7 kDa single-chain polypeptide neurotoxins. Dendrotoxins bind to and inhibit a subset of K^+-channel subunits. At the presynaptic terminal DTx blocks a K^+-channel responsible for preventing spontaneous firing of the terminal and will thus induce repetitive firing and transmitter release. This may be seen both at the neuromuscular junction and with isolated synaptosomes.

β-bungarotoxin: βBTx (from the Taiwan krait *Bungarus multicinctus*) and related neurotoxins (including *taipoxin, notexin* and *crotoxin*) are polypeptide snake venom toxins which possess Ca^{2+}-dependent phospholipase A_2 (PLA_2) activity. The 22 kDa toxin has two chains linked by a disulphide bridge. The 13.5 kDa chain has PLA_2 activity, and is homologous with other PLA_2 enzymes, while the 7 kDa chain is homologous with dendrotoxin and hence targets the toxin to the membrane. The effect of βBTx is a complex combination of PLA_2 evoked disruption of membrane integrity (due to the formation of highly membrane-disruptive lysophosphatides) and direct inhibition of K^+-channels.

Mast cell degranulating peptide (from the venom of the honey bee *Apis mellifera*) is a very basic 22 residue peptide, originally found to trigger histamine release from mast cells, although this requires high concentrations. Nanomolar MCDP competes with DTx for binding to brain membranes and its action is generally similar to DTx.

Charybdotoxins (from the scorpion *Leiurus quinquestriatus*) inhibit Ca^{2+}-dependent K^+-channels but also compete with MCDP and DTx for binding to the DTx-sensitive channel. Finally, among synthetic K^+-channel antagonists, *4-aminopyridine* (4AP) inhibits DTx-sensitive K_A^+ channels at low concentrations; *tetraethylammonium* (TEA) is most sensitive for the delayed rectifier while Ba^{2+} has a wide specificity.

4.6 Channels defined by cloning and protein purification

Molecular cloning has successfully identified channels which behave, upon reconstitution, as fast-activating A-type channels, delayed rectifiers, and most recently as Ca^{2+}-activated channels.

Molecular investigations of K^+-channels was greatly aided by the analysis of a *Drosophila* mutation, *Shaker*, whose phenotype is an abnormal leg shaking during recovery from ether anaesthesia. Intracellular recording in *Drosophila* neurones showed that repolarization after an action potential was delayed, with a tendency to repetitive firing, strongly suggesting a defective K^+-channel subtype. The gene specifying this phenotype was localized to the X chromosome; the *Shaker* locus contains a 65-kb region of DNA with 14 exons, allowing for the possibility of multiple mRNAs by alternative splicing of the primary transcript (Fig. 4.10). When expressed in oocytes, A-type voltage-dependent K^+ currents are seen, although the inactivation kinetics can vary widely. For example *Shaker A* and *Shaker B* channels, which differ in their S6 region and C-termini as a result of alternative splicing, show respectively slow and rapid inactivation in response to prolonged depolarization.

Following *Shaker*, three further K^+-channel subfamilies have been cloned: designated *Shaw*, *Shal* and *Shab*. Mammalian homologues for each of these channels, as well as for *Shaker* have been identified by hybridization and the corresponding cDNAs cloned. However, whereas the *Drosophila* subtypes are largely derived by alternative splicing from a single gene, the different mammalian channels are largely the products of distinct genes.

An increasing number of rat nervous system K^+-channel clones have been characterized. The structure of the proteins deduced from the sequences correspond to molecular masses of 64–74 kDa (some 600 amino acids) with six potential transmembrane regions. This is one-quarter of the size of the major subunits of analogous Na^+- or Ca^{2+}-channels. By analogy to the Na^+- and Ca^{2+}-channels, which consist of one long polypeptide with 24 potential transmembrane segments, it is likely that functional K^+-channels are tetrameric. Two rat channels

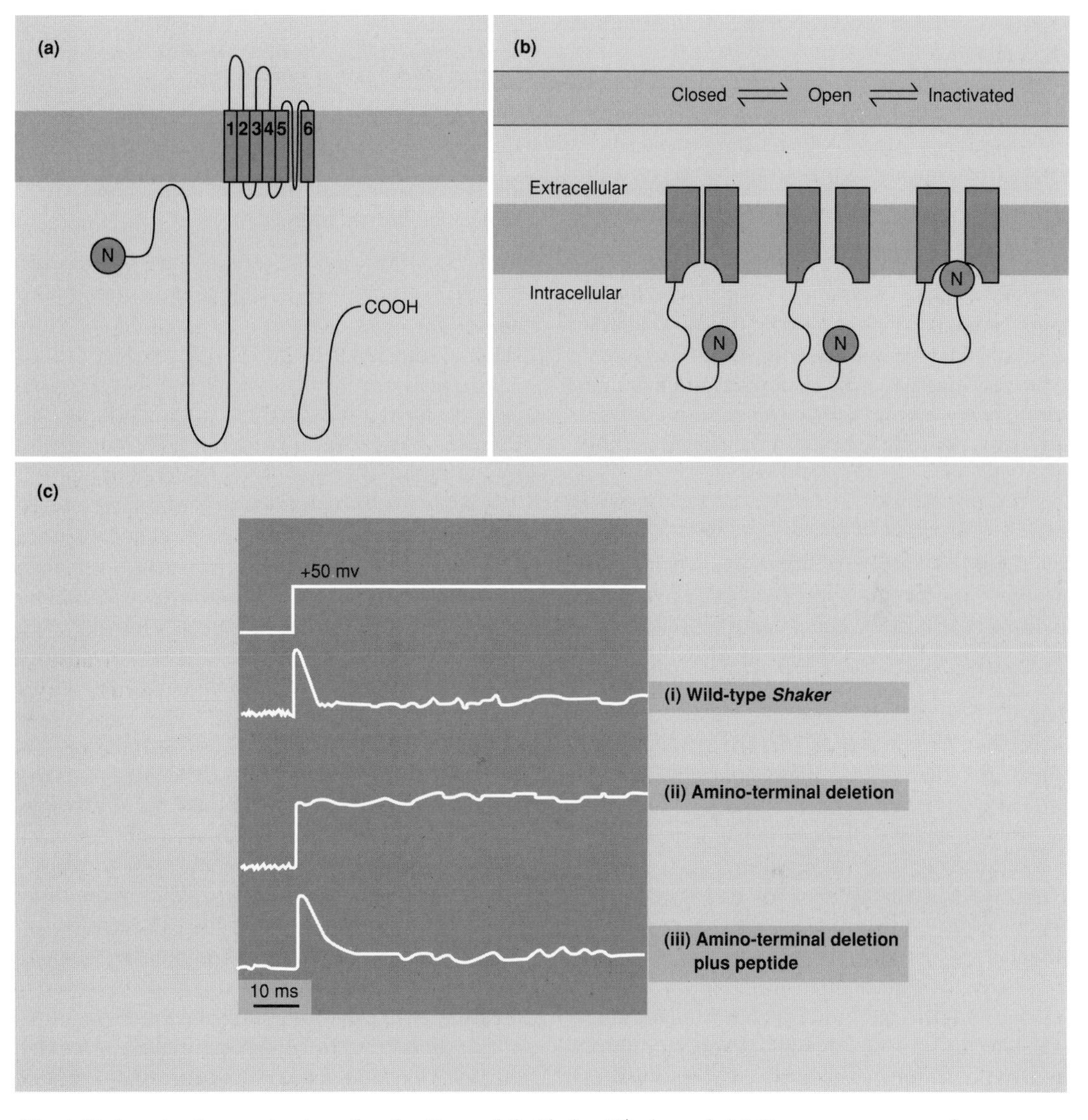

Fig. 4.10 Inactivation mechanisms for the *Drosophila Shaker* K^+-channel. (a) Consensus transmembrane organization of the *Shaker* K^+-channel showing the N-terminal region which may control the inactivation kinetics. (b) The 'ball and chain' model for the inactivation of the channel. When the channel is open it is proposed that it its susceptible to occlusion by the N-terminal cytoplasmic domain. (c) (i) *Shaker* channel subunits expressed in oocytes give a rapidly inactivating response to a step depolarization to +50 mV. (ii) Deletion of the first 20 amino acids of the N-terminus results in a non-activating channel. (iii) Readdition of a free 20-residue peptide corresponding to the N-terminal sequence to the cytoplasmic face of the patch restores inactivation. Adapted from Mackinnon (1991).

which show rapid, A-type, inactivation kinetics in *Xenopus* oocytes are Kv1.4 (RCK4 or RK3) and Kv4.2 (RK5). Immunocytochemical localization of these channels reveals that the two channels are segregated across the axon, with Kv1.4 being presynaptic and Kv4.2 being localized to somatodendritic membranes. Heteromeric multimers formed from Kv1.4 and Kv1.2 subunits may form the molecular basis of the presynaptic K_A^+-channel involved in the regulation of neurotransmitter release (*see* Section 3.3.1), while Kv1.1 and Kv1.2 have also been reported to form presynaptic heteromultimetric channels *in situ*.

cDNAs derived from the *slo* locus of *Drosophila* encode a series of closely related proteins of about 1200 residues derived by alternative splicing. While retaining the six putative transmembrane segments of other K^+-channels, there is little primary sequence homology to other cloned K^+-channels. The expressed channels require high micromolar cytoplasmic $[Ca^{2+}]_c$ for activity, although the conductance in wild-type *Drosophila* muscle which is defective in the *slo* mutant is much more sensitive to Ca^{2+}, suggesting that the *in situ* activity is not fully retained in the cloned channel.

4.6.1 Purification of toxin-binding proteins

In parallel to this genetic approach, conventional protein purification procedures have been performed, the assay being based on the binding of specific neurotoxin inhibitors of K^+ currents. Dendrotoxin (DTx) and β-bungarotoxin (βBTx, not to be confused with the α-bungarotoxin which binds to the MnAChR) compete for binding to K^+-channels, although the effect of the latter toxin is complicated by its phospholipase A_2 (PLA_2) activity. In the hippocampus, dendrotoxin inhibits fast inactivating (i.e. A-type) currents, whereas in sensory neurones slowly inactivating, delayed rectifier currents are affected.

DTx binds with high affinity ($K_d < 1$ nM) to a binding site in rat brain synaptic membranes. The binding sites for other K^+-channel inhibitory neurotoxins such as mast-cell degranulating peptide (MCDP), charybdotoxin and β-bungarotoxin copurify on a dendrotoxin-I affinity column, and a complex of peptides of molecular mass 80, 41 and 38 kDa is obtained. The 80 kDa peptide is glycosylated and has a peptide core of 65 kDa, close to that of the *Shaker* channel discussed above. The 80 kDa subunit can be phosphorylated by PKA, but not by PKC or CaMKII (Ca^{2+}-calmodulin-dependent protein kinase II) and corresponds to the peptide detected by molecular genetics (*see* Section 4.5.1). The roles of the smaller peptides in the purified complex are not clear, particularly since channel activity can be expressed with the 65 kDa *Drosophila* subunit alone. However, as will be discussed in the context of Na^+-channels, the small subunits may play a modulatory role, or anchor the channel in the correct location in the membrane.

4.6.2 Model building: the integration of electrophysiological, pharmacological and molecular information

Both the N- and C-termini of the cloned K^+-channels are cytoplasmic and the monomers are highly asymmetric in the membrane, with short extracellular loops (Fig. 4.10). Since there is currently no X-ray crystallographic data on the structure of the channels, all models must be to some extent speculative and the consensus model shown must not be considered definitive. The most productive approach to the structure/function relations within the K^+-channel subunits has examined the consequences of site-directed mutagenesis on the electrophysiological characteristics of the expressed channels; an approach which has also been particularly fruitful with the ligand-gated ionotropic receptors (*see* Section 9.6.1).

The 19-residue loop (H5) between S5 and S6 is highly conserved between different K^+-channels, and appears to be looped back into the channel to form a β-hairpin, four of which would be contributed by the individual subunits to constitute an eight-stranded 'barrel' lining the pore (*see* Fig. 4.9). Mutations in the H5 region profoundly affect the pharmacology of the channel. Residue 19 at the extracellular end of H5 controls the affinity of the channel for the inhibitor *tetraethylammonium* (*TEA*) when applied from the outside. Residue 11 is supposed to be at the intracellular bend of the hairpin since it determines the sensitivity for *internal* TEA. Consistent with the importance of the H5 barrel, chimeric K^+-channels in which H5 segments have been interchanged show single-channel conductances characteristic of the inserted H5 segment. Extracellular residues close to H5 also define the sensitivity of the channels to charybdotoxin.

The transmembrane segments S4 and S6 of the subunits may contribute a total of eight α-helices surrounding the β-barrel. The primary sequence of voltage-activated K^+-channels reveals the characteristic repeat in S4 of a positive Lys or Arg followed by two hydrophobic residues. This has been found in all voltage-gated ion channels, and corresponds to the voltage sensor discussed above. Finally segments S1, S2, S3 and S5 would contribute 16 α-helices to the outer layer of the channel (*see* Fig. 4.9).

Kinetic analysis of single-channel patch-clamp experiments suggests that voltage-gated K^+-channels undergo an ordered series of conformational states upon depolarization. At polarized potentials most channels are closed (state C). Depolarization causes a transition to a closed, but activated state (C*) presumably triggered by movement of the voltage sensor, following which there is a voltage-independent change to a second closed, activated state (C**) from which the channel is able to open (O). Opening and closing of the channel (i.e. the C** ↔ O transition) is reversible and voltage independent.

Repolarization depletes the two activated states due to the regeneration of the closed, resting state (C). However if depolarization is maintained the channel can inactivate. Two types of inactivation — *C-type* and *N-type* have been described, which can occur separately or together. N-type inactivation is believed to be governed by the N-terminal cytoplasmic sequence acting as an 'inactivation ball' capable of swinging into the channel and blocking it by binding to the S4–S5 intracellular loop (*see* Fig. 4.10). Evidence for this is found with N-terminal deleted mutants which inactivate very slowly. Channels which appear to close by this mechanism reactivate slowly. In contrast, channels which appear instead to inactivate slowly by a conformational change at the extracellular entrance of the pore (a C-type mechanism) reactivate rapidly.

Perhaps the most surprising recent finding from K^+-channel cloning studies is the identification of two very small inward rectifier K^+-channels by expression cloning: ROMK1 from rat kidney cells (not expressed in brain) and IRK1 from rat macrophages (strongly expressed in brain) and possessing respectively 391 and 428 amino acids. Although these channels possess regions homologous to the H5 channel-lining hairpin domain, only two additional membrane-spanning regions can be deduced from hydropathy plots, corresponding to S5 and S6 of the *Shaker* channels. It is possible that these novel channels still function as tetrameric complexes but have only a single annulus of transmembrane domains (*see* Fig. 4.9).

4.7 Voltage-activated Na^+-channels

Voltage-activated Na^+-channels are responsible for initiating action potentials. They are less diverse than K^+-channels or Ca^{2+}-channels but their key involvement in neuronal excitation is reflected in the existence of at least five different classes of neurotoxin which have evolved to interfere with their function (*see* Box 4.1).

Identification of the Na^+-channel polypeptides was first accomplished by protein purification from membranes which had been photoaffinity labelled with derivatized scorpion α-toxin. Even without covalent attachment, membranes solubilized with non-ionic detergents retain high-affinity binding of labelled tetrodotoxin or saxitoxin sufficiently to allow purification to be monitored. The main (α) subunit from mammalian brain has a molecular weight of 260 kDa and copurifies with two low molecular weight peptides, β_1 (36 kDa) and β_2 (33 kDa), the last being attached to the α-subunit by disulphide bonds. The skeletal muscle Na^+-channel lacks a β_2-subunit, while that from eel electric organ simply consists of the high molecular weight α-subunit.

All three subunits are heavily glycosylated, about 25% of the mass of the subunits being carbohydrate. Since only extracellular domains are glycosylated, this means that each subunit must have groups exposed on the extracellular surface. If N-linked glycosylation is inhibited with tunicamycin the Na^+-channel is neither assembled nor incorporated into the membrane.

Purified channels may be reconstituted into liposomes and then fused into planar lipid membranes which can be patched, allowing their electrical properties to be determined; batrachotoxin can be added to prevent inactivation of the channels. The subunits which appear to be part of the intact complex (α, β_1 and β_2 in the case of brain, α plus β for skeletal muscle or α alone for electroplax) are sufficient to allow full reconstitution of the physiological channel. Partial reconstitutions help to assign roles to the β-subunits. Brain β_1-subunit can be dissociated from the α-subunit by high ionic strength and results in a loss of functional activity. The subunit is no longer dissociable if tetrodotoxin is present. Loss of the β_2-subunit by treatment with disulphide-reducing agents has no discernable effect on the functions of the channel. Surprisingly, mRNA for neither β-subunit is required for functional expression of brain α-subunit mRNA in *Xenopus* oocytes although the inactivation kinetics of the expressed channel is much slower.

The availability of purified, functional Na^+-channel preparations has allowed cDNA clones to be isolated encoding the primary sequence of the α-subunit. Expression cDNA libraries from electroplax mRNA were screened both with antibodies against the purified subunit and with oligonucleotides complementary to segments of its mRNA sequence. The cDNA encodes a polypeptide of 1832 amino acids with four homologous repeated domains, each with six potential membrane-spanning regions homologous to a single K^+-channel subunit. In rat brain, three different mRNAs were isolated indicating the existence of three subtypes of channel, termed I, II and III. These isoforms each possess 200 amino acid inserts which are not present in the electroplax channel.

The α-subunit can be phosphorylated by cAMP-dependent protein kinase in intact synaptosomes and is therefore presumably a membrane-spanning subunit. A physiological significance for this phosphorylation is not established, although it may enhance channel inactivation. Protein kinase C will phosphorylate the subunit leading to a decreased maximal conductance but delayed inactivation. In addition to glycosylation, the α-subunit is further modified by fatty acylation and sulphation.

The close homology between Na^+- and Ca^{2+}-channels is demonstrated by the ability of point mutations of the Na^+-channel replacing Lys 1422 in repeat III and/or Ala 1714 in repeat IV by glutamate residues are sufficient to switch the channel specificity from Na^+ to Ca^{2+}.

4.8 Voltage-activated Ca^{2+}-channels

The major (α_1) subunits of the voltage-activated Ca^{2+}-channels (VACCs) which have been cloned to date show a strong homology with voltage-activated Na^+-channels, with four putative domains each possessing six likely transmembrane segments (Fig. 4.11).

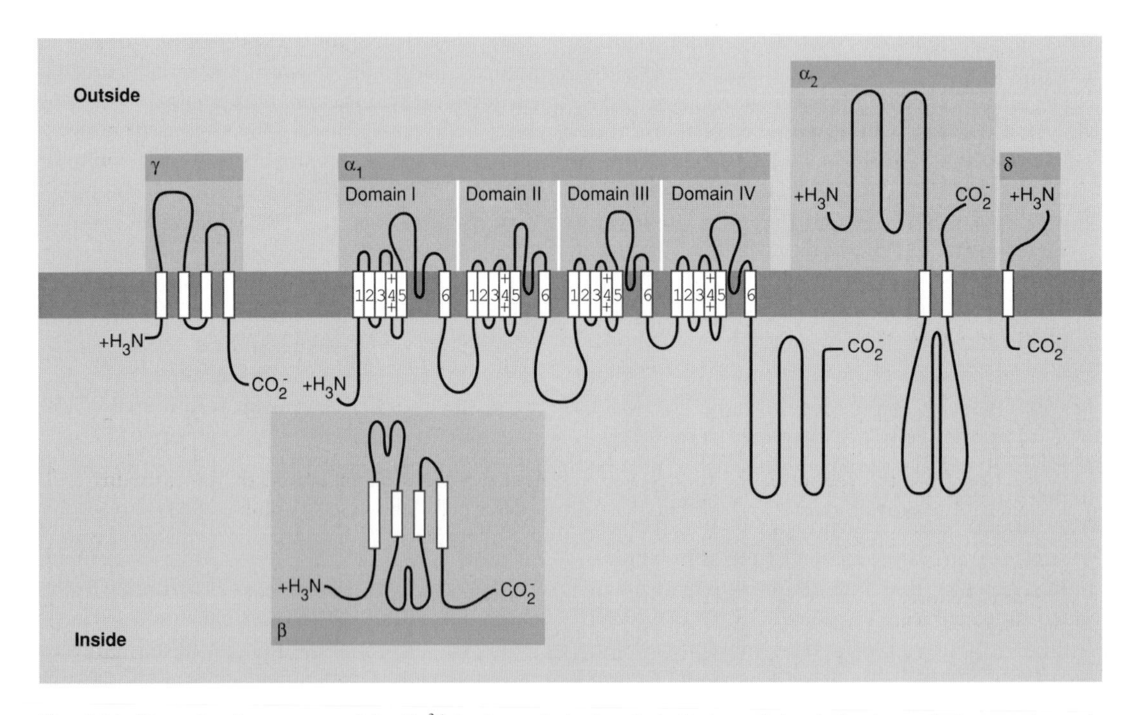

Fig. 4.11 The subunit structure of the Ca^{2+} L-channel. An 'exploded' view of the skeletal muscle L-type channel shows the predicted transmembrane folding of the subunits. The α_2-subunit is linked to α_1 by disulphide bonds, the δ-subunit is covalently linked to the α_2-subunit and is liberated from it on reduction. Adapted from Catterall & Striessnig (1992).

Box 4.3 Ca^{2+}-channel neurotoxins

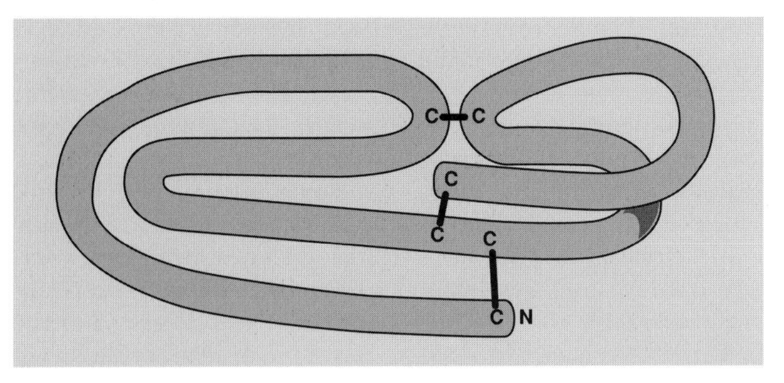

Schematic structure of a *Conus* toxin (ω-conotoxin-MVIIC). Adapted from sequence in Hillyard *et al.* (1992).

ω-conotoxin GVIA from the venom of the marine hunting snail *Conus geographicus* is a 27 amino acid polypeptide which potently inhibits a subclass of channel classified as N-type by electrophysiology. The distribution of conotoxin-sensitive channels is transmitter and species-specific: VACCs in chick synaptosomes are much more sensitive than corresponding mammalian channels. The toxin may inhibit mammalian Type II transmitter release (e.g. catecholamines) while being ineffective against amino acid release.

ω-conotoxin-MVIIC from *Conus magus* differs in selectivity from ω-conotoxin-GVIA. It was identified in a cDNA library constructed from venom ducts, using probes based on sequence of previous *Conus magus* toxins. The toxin inhibits some GVIA-resistant Ca^{2+} currents in hippocampal CA1 neurones and blocks Purkinje 'P' currents as well as 75% of ^{45}Ca uptake into KCl-depolarized synaptosomes. Its action is distinct from Aga-IVA (see below).

FTX is a partially characterized polyamine fraction from the venom of the funnel web spider, *Agelenopsis aperta*, and inhibits dendritic Ca^{2+} P-channels on cerebellar Purkinje neurones as well as release from the squid giant synapse.

ω-agatoxins: The *Agelenidae* family of funnel web spiders (not to be confused with the orb-web spiders, which produce toxins active against invertebrate glutamate receptors) produce a wide range of neurotoxins. α-Agatoxins are acylpolyamines which cause use-dependent block of insect postsynaptic glutamate receptors. μ-agatoxins are polypeptides which activate voltage-sensitive Na^+ channels. ω-agatoxins are polypeptide antagonists of voltage-sensitive Ca^{2+}-channels, particularly in insects — the natural prey of the spider. ω-Aga-IA and ω-Aga-IB are 7.5 kDa polypeptides which are active against dorsal root ganglion Ca^{2+}-channels but have little effect on presynaptic channels.

ω-Aga-IIA and ω-Aga-IIB are 11 and 8.5 kDa peptides respectively with a broad specificity against ω-conotoxin-GVIA sensitive Ca^{2+}-channels and to some extent L-channels. ω-Aga-IVA, is a 48 amino acid polypeptide which inhibits KCl-evoked ^{45}Ca entry into mammalian synaptosomes. Additionally it causes a partial block of somatic conotoxin-insensitive channels in hippocampal neurones.

Agatoxin-GI is distinct from the ω-agotoxins described above. It inhibits the non-inactivating Ca^{2+}-channels which are coupled to glutamate exocytosis in mammalian cortex and cerebellum.

A definitive classification of VACCs will only be possible when the individual peptides have each been purified, sequenced and expressed *in vitro*. While this is under way, the current classification relies on a combination of pharmacology (particularly exploiting the plethora of Ca^{2+}-channel antagonists and neurotoxins, Box 4.3), combined with electrophysiological criteria originally defined for channels of the cell body of the dorsal root ganglion by Richard Tsien and colleagues and classified as *L-type* (long-lasting); *T-type* (transient or tiny) and *N-type* (neuronal or neither).

The channels can be distinguished firstly by their single-channel conductances to 100 mM Ba^{2+} (used to optimize the current and to avoid inhibition which occurs when $[Ca^{2+}]$ increases at the original cytoplasmic face). L-channels have large conductances, N-channels are intermediate and T-channels have small conductances. The channels additionally differ in their voltage dependencies. The initial *holding potential* from which the patch must be depolarized in order to see the channel varies; at a partially depolarized holding potential of −30 mV N-channels are already inactivated and in contrast to L-channels, do not open with further depolarization. The extent of *depolarization* required to activate the channels also differs; T-channels are *low threshold*, requiring a smaller depolarization than do *high-threshold* L- and N-channels. L-channels are *non-inactivating* and continue to fire repeatedly when the patch is depolarized, whereas with N- and T-channels an initial burst of openings is seen after which the channels *inactivate*.

The L-, T- and N- classification, which was based on the Ca^{2+}-channels detectable by patching the cell soma, does not take account of any channels present on terminals or distal regions of dendrites which would not contribute to the signal at the soma. Additionally the proportion of high-threshold (i.e. non-T) voltage-activated Ca^{2+} current in dissociated neurones which is resistent to the combination of a dihydropyridine and ω-conotoxin varies from 10% in the superior cervical ganglion, to 50% in the cortex and hippocampus and to 90% for cerebellar Purkinje cells (Fig. 4.12). These non-L, non-N-channels

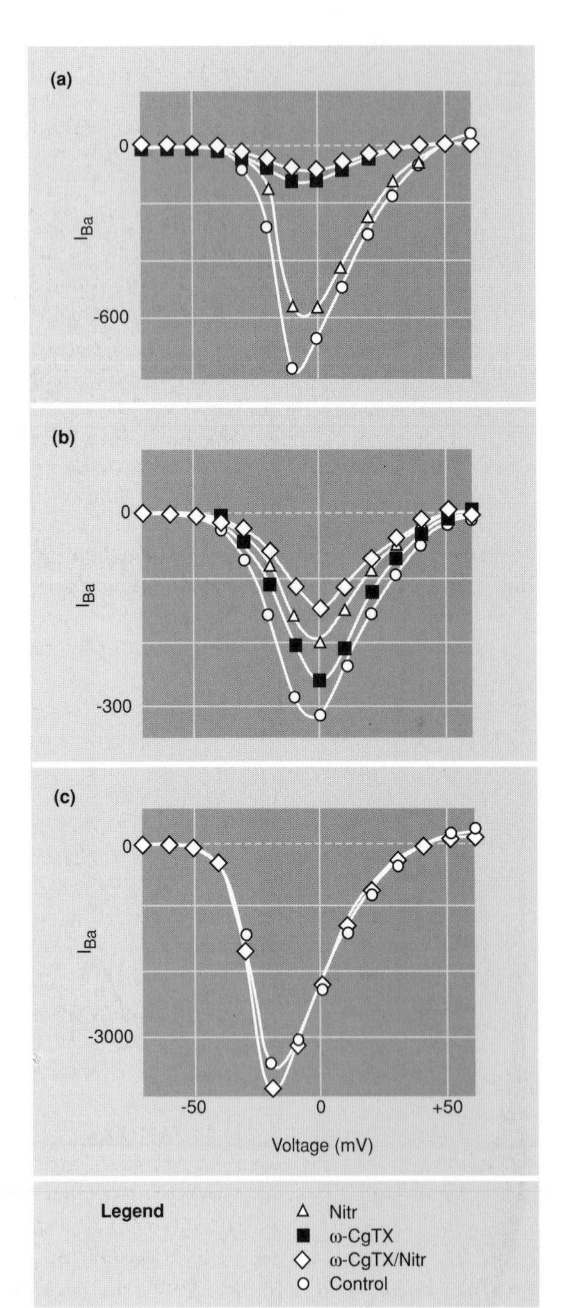

Fig. 4.12 Pharmacological and electrophysiological discrimination of L- and N-channels on the cell bodies of dorsal-root ganglion neurones. Whole-cell patch-clamp recordings of currents of the Ca^{2+} analogue Ba^{2+} in (a) sympathetic neurones, (b) hippocampal CA1 neurones, and (c) cerebellar Purkinje neurones were made at the end of a 24 ms depolarization from −70 mV to the voltages shown on the x axis. Experiments were repeated in the presence of the L-channel inhibitor nitrendipine (Nitr), the N-channel inhibitor ω-conotoxin-GVIA (ω-CgTX) and in the presence of both agents. The currents increase beyond the threshold at which the channels activate but decrease at positive potentials as the driving force for Ba^{2+} entry falls. In the sympathetic neurone most channels respond to ω-conotoxin-GVIA and few are resistant to both inhibitors whereas neither inhibitor has a significant effect on Purkinje cells, where 'P-channels' were originally decribed. From Regan *et al.* (1991), reproduced with permission from *Neuron*. © Cell Press.

are, despite being evidently heterogeneous sometimes grouped together as *P-channels*.

4.8.1 L-type Ca^{2+}-channels

L-channels are widely distributed, being present not only in brain but also in smooth muscle, cardiac muscle and at very high concentration in the transverse tubules of skeletal muscle (the infoldings of the plasma membrane in close proximity to the 'end-feet' of the sarcoplasmic reticulum and intimately involved in excitation-contraction coupling). In 100 mM Ba^{2+}, L-channels display a large unitary conductance of 25 pS and inactivate very slowly, if at all during prolonged depolarization.

Rat brain expresses at least four classes of Ca^{2+}-channel α_1-subunit, which have been designated in one nomenclature system rbA, rbB, rbC and rbD. rbC and rbD are most closely related to L-channels in non-neuronal tissue, suggesting that the brain expresses two distinct L-type channels. The neuronal L-channels are localized around neuronal cell bodies and at the base of major dendrites. There are few L-channels in presynaptic terminals, judged by synaptosomal insensitivity to L-channel inhibitors.

Most structural information is available for the skeletal muscle L-channel, which is closely homologous to the equivalent channel in the CNS. Lectin affinity chromatography of detergent solubilized transverse tubule membranes from rabbit skeletal muscle results in the isolation of an L-channel complex containing five polypeptides: α_1 (molecular weight 175 000), α_2 (150 000 in reduced form), β (52 000), γ (32 000) and δ (25 000) (Fig. 4.11). Each subunit has been cloned and sequenced, although the δ-subunit is encoded by the α_2 gene and is considered to be a peptide linked to the α_2-subunit.

The α_1-subunit binds both dihydropyridines and phenylalkylamines and is an *in vitro* substrate for a variety of protein kinases. It has a high degree of homology with the Na^+-channel with four internal repeats, each of which contains six possible membrane-spanning segments (Fig. 4.11). Proteolytic cleavage of an α_1-precursor peptide appears to be required to reduce a 212 kDa peptide predicted from the cDNA to the final 170 kDa subunit. The α_1-subunit or 'dihydropyridine receptor' yields a functionally active, dihydropyridine-sensitive Ca^{2+}-channel when expressed alone. Multiple isoforms of the α_1-subunit exist which can be grouped into four distinct subfamilies. Within each subfamily, the possibility of alternative splicing may exist, allowing an enormous scope for diversity.

The α_2-subunit is linked to the α_1-subunit through disulphide bonds. Both the α_2- and β-subunits are heavily glycosylated and do not have substantial hydrophobic domains (although α_2 may have two transmembrane segments). α_2 has a molecular mass of about 175 kDa in non-reducing gels, but loses a 25 kDa δ-subunit when its disulphide bonds are reduced. It has no homology with channel-forming peptides but when coexpressed with α_1 greatly increases the rate at which the latter is activated by depolarization, although the exact effect of the α_2-subunit on the kinetics is unclear at the time of writing. Rat brain expresses an alternatively spliced variant of the L-channel α_2-subunit. The β- and γ-subunits may play a regulatory role, the latter having homology with the Na^+-channel β_1-subunit.

L-channels respond to submicromolar concentrations of dihydropyridines which may either inhibit (e.g. nifedipine, nitrendipine) or activate (Bay-K 8644). Great care must be taken in intact cells before an effect of dihydropyridines is ascribed to a blockade of L-channels, since at sufficient concentrations the inhibitors can block virtually every transport process in the cell. To localize the region of binding of dihydropyridines, peptides from an α_1-subunit preparation which had been covalently labelled with a tritiated dihydropyridine were identified by immune precipitation with site-directed monoclonal antibodies (peptide mapping). A region in domain III at the extracellular end of transmembrane segment 6 was found to be the most likely site of interaction.

Phenylalkylamines such as verapamil are believed to act by blocking the central pore of the channel formed by the α_1-subunit, interacting with the cytoplasmic end of transmembrane segment 6 of domain IV. A third class of Ca^{2+}-channel blocker, the benzothiazepines such as diltiazem, act at sites which have not currently been characterized.

In neurones, L-type channels may be activated by phosphorylation of the α_1- and β-subunits by cAMP-dependent protein kinase and may be inactivated by the Ca^{2+}-dependent phosphatase calcineurin (*see* Section 5.9.3). In the absence of continued phosphorylation, for example, in the absence of ATP, the L-channels rapidly stop responding to membrane depolarization as they become dephosphorylated. L-channels in intact cells may undergo Ca^{2+}-dependent inactivation as the increase in $[Ca^{2+}]_c$ activates calcineurin and dephosphorylates the channel.

4.8.2 T-channels

T-channel currents occur in both excitable and non-excitable cells. They are characterized by a small unitary conductance of 8 pS (measured in 100 mM Ba^{2+}) and a brief burst of conductance followed by rapid inactivation. They are the only channels which can be cleanly separated in electrophysiological studies since they require only a slight depolarization to become active. This suggests that they may be involved in modulating the intrinsic excitability of a neurone. There are few studies at the biochemical level and no selective inhibitors currently available.

4.8.3 N-, P- and Q-channels

These provisional classifications will be considered together. N-type Ca^{2+}-channels determined electrophysiologically at the cell soma have unitary conductances of about 13 pS in 100 mM Ba^{2+}, and are characterized by a long burst of repetitive firing followed by incomplete inactivation. N-channels are high-threshold, requiring a strong depolarization to at least −20 mV for activation. The distribution of ω-conotoxin-GVIA-sensitive N-channels is species specific, in particular chick brain expresses a much higher proportion of N-channels than does mammalian brain.

Ca^{2+}-channels insensitive to both ω-conotoxin-GVIA and dihydropyridines are very common, particularly on non-somatic regions of the neuronal membrane. These have been termed *P-type* (Purkinje) channels, after the 90% of the Ca^{2+} current in Purkinje cells which is resistant to the combination of dihydropyridines and ω-conotoxin-GVIA but can be inhibited by FTX, a partially characterized polyamine fraction from the venom of the funnel web spider, *Agelenopsis aperta* (*see* Box 4.3). A channel sensitive to FTX mediates transmitter release at the squid giant synapse, while different polypeptide fractions from the same venom, including Aga-IVA and Aga-GI, as well as ω-conotoxin-MVIIC inhibit distinct subpopulations. This last toxin was identified in a cDNA library constructed from *Conus magus* venom ducts, using probes based on sequence of previous *Conus magus* toxins. ω-conotoxin-MVIIC and Aga-IVA allow further pharmacological dissection of the channels; while the Purkinje spine P-channel is sensitive to both toxins, the cloned BI (Class A) α_1-subunit combined with α_2 and the δ/β-subunits is ten-fold more sensitive to MVIIC and 100-fold less sensitive to Aga-IVA and has been termed a *Q-channel.*

As a generalization (with a number of exceptions) amino acid exocytosis in the mammalian brain seems to be coupled to dihydropyridine and ω-conotoxin-GVIA resistant P/Q-channels while Type II transmitters, e.g. catecholamines, tend to be coupled to ω-conotoxin-GVIA sensitive N-channels.

4.8.4 Cloning of non-L-channels

As in the case of the K^+-channels discussed in Section 4.6.2 the electrophysiological and pharmacological categorization of Ca^{2+}-channels will eventually be superceded by one based on molecular genetics. Four major classes of Ca^{2+}-channel α_1-subunit are expressed in rat brain (rbA, rbB, rbC and rbD) each possessing some 2000 residues and subunit molecular weights in the region of 250 kDa. Brain also expresses a 124 kDa alternatively spliced form of the skeletal muscle L-type α_2-subunit.

rbC and rbD gene products are closely related to L-type Ca^{2+}-channel α_1-subunits from other tissues, while rbA and rbB encode neural-specific isoforms. Three B-type channels have been cloned, sequenced and designated BI, BII and BIII. BI is brain specific and expressed at a high level in the cerebellum — predominantly in the Purkinje cell layer and also in CA3 hippocampal pyramidal neurones. In contrast to L-channels which are predominantly located on cell bodies, particularly in the region of the axon hillock, the BI-channel is located in dendritic regions (Fig. 4.13) although the distribution between postsynaptic and presynaptic sites is currently unclear. Although antibodies raised against the BI-subunit immunoprecipitate ω-conotoxin-

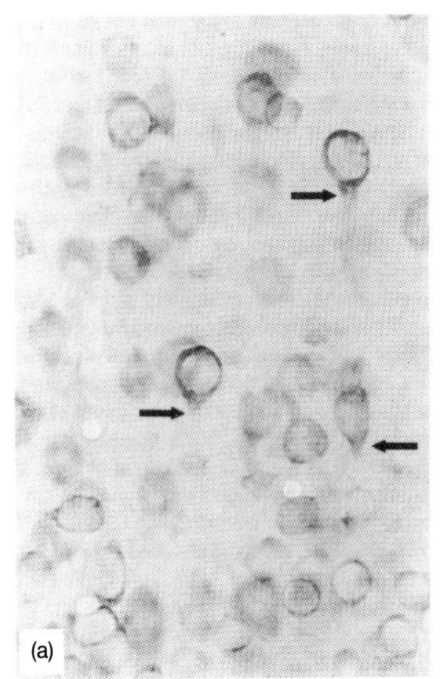

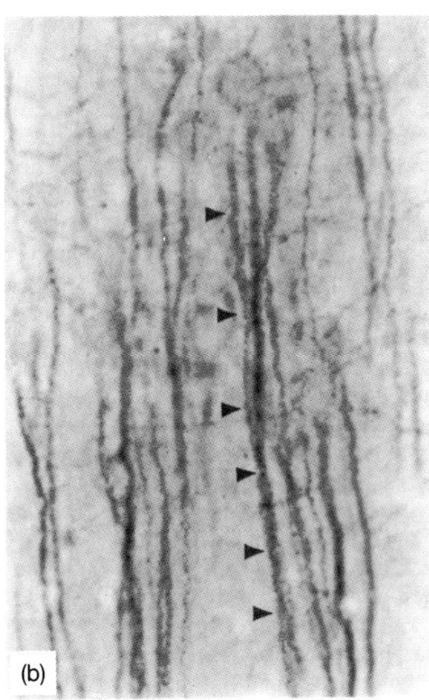

Fig. 4.13 Differential subcellular location of L-type and N-type Ca^{2+}-channels in rat cortical neurones. Sections of rat cerebral cortex were stained with monoclonal antibody against the α_2-subunits of neuronal L-type Ca^{2+}-channels (a) or with a polyclonal antibody raised against a sequence of the N-type cloned rbBI α_1-subunit (b). The L-channel antibody selectively labels cell soma and axon hillocks while the rbBI antibody mainly labels dendrites. Note that at this resolution it is not clear to what extent the latter is due to channels present on the dendrites themselves or on presynaptic terminals synapsing onto the dendrites. From Westenbroek *et al.* (1992), reproduced with permission from *Neuron*. © Cell Press.

GVIA-binding sites from rat brain, the channel expressed in oocytes is insensitive to both ω-conotoxin-GVIA and dihydropyridines and displays inactivation kinetics that do not correspond to electrophysiologically identified Ca^{2+}-channels.

A second non-L clone, BII, from rabbit brain is abundantly expressed in the cortex, hippocampus and corpus striatum and its expression correlates well with conotoxin-GVIA binding, suggesting that the cDNA might encode a further N-channel. BIII encodes a 2339 amino acid channel; transient expression of BIII cDNA results in a channel which can be totally inhibited by ω-conotoxin-GVIA and which displays kinetics closely corresponding to those characteristic of *in situ* N-channels. The three B-channels display sufficient sequence homology to justify their classification as a subfamily. Most variation occurs in the putative cytoplasmic loop between domains II and III (*see* Fig. 4.11) although the S5–S6 linker believed to line the channel (*see* Section 4.4.2) also shows some variability.

Further reading

Transporters

Amara, S.G. & Pachlczyk, T. (1991) Sodium-dependent neurotransmitter reuptake systems. *Curr. Op. Neurobiol.* **1**, 84–90.

Blaustein, M.P. (1988) Calcium transport and buffering in neurons. *Trends Neurosci.* **11**, 438–443.

Kanner, B.I. & Schuldiner, S. (1987) Mechanism of transport and storage of neurotransmitters. *Crit. Rev. Biochem.* **22**, 1–38.

Nicoll, D.A., Longoni, S. & Philipson, K.D. (1990) Molecular cloning and functional expression of the cardiac sarcolemmal Na/Ca exchanger. *Science* **250**, 562–565.

Pines, G., Danbolt, N.C., Bjorås, M., Zhang, Y., Bendahan, A., Eide, L., Koepsell, H., Storm-Mathisen, J., Seeberg, E. & Kanner, B.I. (1992) Cloning and expression of a rat brain L-glutamate transporter. *Nature* **360**, 464–467.

Schloss, P., Mayser, W. & Betz, H. (1992) Neurotransmitter transporters: a novel family of integral plasma membrane proteins. *FEBS Lett.* **307**, 76–80.

Uhl, G.R. (1992) Neurotransmitter transporters (plus): a promising new gene family. *Trends Neurosci.* **15**, 265–268.

Ion pumps

Bowman, E.J., Siebers, A. & Altendorf, K. (1988) Bafilomycins: a class of inhibitors of membrane ATPases from microorganisms, animals cells and plants cells. *Biochemistry* **85**, 7972–7976.

Forgac, M. (1989) Structure and function of vacuolar class of ATP-driven, proton pumps. *Physiol. Rev.* **69**, 765–796.

Nelson, N. (1991a) Structure and pharmacology of the proton-ATPases. *Trends Pharmacol. Sci.* **12**, 71–75.

Nelson, N. (1991b) Structure, molecular genetics and evolution of vacuolar H^+-ATPases. *J. Bioenerg. Biomembrane* **21**, 553–571.

Shull, G.E. & Greeb, J. (1988) Molecular cloning of two isoforms of the plasma membrane Ca-transporting ATPase from rat brain. *J. Biol. Chem.* **263**, 8646–8657.

Ligand-gated channel structure

Mitra, A.K., McCarthy, M.P. & Stroud, R.M. (1989) Three-dimensional structure of the nicotinic acetylcholine receptor, and location of the major associated 43 kD cytoskeletal protein, determined at 22 Å by low dose electron microscopy and X-ray diffraction to 12.5Å. *J. Cell Biol.* **109**, 755–774.

Unwin, N. (1993) Neurotransmitter action: Opening of ligand-gated ion channels. *Neuron* **10**(Suppl.), 31–41.

Voltage-gated channel structure

Catterall, W.A. (1992) Structure and function of voltage-gated sodium and calcium channels. *Curr. Op. Neurobiol.* **1**, 5–13.

Logothetis, D.E., Movahedi, S., Satler, C., Lindpaintner, K. & Nadal-Ginard, B. (1992) Incremental reductions of positive charge within the S4 region of a voltage-gated K^+ channel result in corresponding decreases in gating charge. *Neuron* **8**, 531–540.

Sakmann, B. (1992) Elementary steps in synaptic transmission revealed by currents through single ion channels. *Neuron* **8**, 613–629.

K^+-channels

Adelman, J.P., Shen, K.Z., Kavanaugh, M.P., Warren, R.A., Wu, Y.N., Lagrutta, A., Bond, C.T. & North, R.A. (1992) Calcium-activated potassium channels expressed from cloned cDNAs. *Neuron* **9**, 209–216.

Aldrich, R. (1993) Potassium channels: advent of a new family. *Nature* **362**, 107–108.

Amoroso, S., Schmid-Antomarchi, H., Fosset, M. & Lazdunski, M. (1990) Glucose; sulfonylureas and neurotransmitter release: Role of ATP-sensitive K^+ channels. *Science* **247**, 852–854.

Cook, N.S. (1988) The pharmacology of potassium channels and their therapeutic potential. *Trends Pharmacol. Sci.* **9**, 21–28.

Guy, H.R. & Conti, F. (1990) Pursuing the structure and function of voltage-gated channels. *Trends Neurosci.* **13**, 201–206.

Kavanaugh, M.P., Hurst, R.S., Yakel, J., Varnum, M.D., Adelman, J.P. & North, R.A. (1992) Multiple subunits of a voltage-dependent potassium channel contribute to the binding site for tetraethylammonium. *Neuron* **8**, 493–497.

Kirsch, G.E., Drewe, J.A., Hartmann, H.A., Taglialatela, M., De Biasi, M., Brown, A.M. & Joho, R.H. (1992) Differences between the deep pores of K^+ channels determined by an interacting pair of nonpolar amino acids. *Neuron* **8**, 499–505.

Latorre, R., Oberhauser, A., Labaraca, P. & Alvarez, O. (1989) Varieties of calcium-activated potassium channels. *Annu. Rev. Physiol.* **51**, 385–399.

Mackinnon, R. (1991) New insights into the structure and function of potassium channels. *Curr. Op. Neurobiol.* **1**, 14–19.

Moran, O., Schreibmayer, W., Wei, L., Dascal, N. & Lotan, I. (1992) Level of expression controls modes of gating of a K^+ channel. *FEBS Lett.* **302**, 21–25.

Pongs, O. (1992) Structural basis of voltage-gated K^+ channel pharmacology. *Trends Pharmacol. Sci.* **13**, 359–365.

Rehm, H. (1991) Neuronal voltage-dependent K-channels. *Eur. J. Biochem.* **202**, 701–713.

Salkoff, L., Baker, K., Butler, A., Covarrubias, M., Pak, M.D. & Wei, A. (1992) An essential 'set' of K^+ channels conserved in flies, mice and humans. *Trends Neurosci.* **15**, 161–166.

Sheng, M., Llao, Y.J., Jan, Y.N. & Jan, L.Y. (1993) Presynaptic A-current based on heteromultimeric K^+ channels detected *in vivo*. *Nature* **365**, 72–75.

Sheng, M., Tsaur, M.-L., Jan, Y.N. & Jan, L.Y. (1992) Subcellular segregation of two A-type K-channel proteins in rat central neurons. *Neuron* **9**, 271–284.

Smith, P.A., Chen, H., Kurenny, D.E., Selyanko, A.A. & Zidichouski, J.A. (1992) Regulation of the M current: transduction mechanism and role in ganglionic transmission. *Can. J. Physiol. Pharmacol* **70** (Suppl.), S12–S18.

Wang, H., Kunkel, D.D., Martin, T.M., Schwartzkroin, P.A. & Tempel, B.L. (1993) Heteromultimeric K^+ channels in terminal and juxtaparanodal regions of neurones. *Nature* **365**, 75–79.

K^+-channel neurotoxins

Strong, P.N. (1990) Potassium channel toxins. *Pharmacol. Ther.* **46**, 137–162.

Na^+-channels

Heinemann, S.H., Terlau, H., Stuhmer, W., Imoto, K. & Numa, S. (1992) Calcium channel characteristics conferred on the sodium channel by single mutations. *Nature* **356**, 441–443.

Ca^{2+}-channels

Armstrong, D.L. (1989) Calcium channel regulation by calcineurin, a Ca-activated phosphatase in mammalian brain. *Trends Neurosci.* **12**, 117–122.

Catterall, W.A. & Striessnig, J. (1992) Receptor sites for Ca^{2+} channel antagonists. *Trends Pharmacol. Sci.* **13**, 256–262.

Dubel, S.J., Starr, T.V.B., Hell, J., Ahlijanian, M.K., Enyeart, J.J., Catterall, W.A. & Snutch, T.P. (1992) Molecular cloning of the α_1 subunit of an omega-conotoxin-sensitive calcium channel. *Proc. Natl. Acad. Sci. USA* **89**, 5058–5062.

Hess, P. (1990) Calcium channels in vertebrate cells. *Annu. Rev. Neurosci.* **13**, 337–356.

Kim, H.-L., Kim, H., Lee, P., King, R.G. & Chin, H. (1992) Rat brain expresses an alternatively spliced form of the dihydropyridine-sensitive L-type calcium channel α_2 subunit. *Proc. Natl. Acad. Sci. USA* **89**, 3251–3255.

Niidome, T., Kim, M.-S., Friedrich, T. & Mori, Y. (1992) Molecular cloning and characterization of a novel calcium channel from rabbit brain. *FEBS Lett.* **308**, 7–13.

Regan, L.J., Sah, D.W.Y. & Bean, B.P. (1991) Ca channels in rat central and peripheral neurons: high-threshold current resistant to dihydropyridine blockers and omega-conotoxin. *Neuron* **6**, 269–280.

Sher, E., Biancardi, E., Passafaro, M. & Clementi, F. (1991) Physiopathology of neuronal voltage-activated calcium channels. *FASEB J.* **5**, 2677–2683.

Tsien, R.W., Ellinor, P.T. & Horne, W.A. (1991) Molecular diversity of voltage-dependent Ca channels. *Trends Neuronsci.* **11**, 431–438.

Tsien, R.W., Lipscombe, D., Madison, D.V., Bley, K.R. & Fox, A.P. (1988) Multiple types of neuronal calcium channels and their selective modulation. *Trends Pharmacol. Sci.* **12**, 349–354.

Westenbroek, R.E., Hell, J.W., Warner, C., Dubel, S.J., Snutch, T.P. & Catterall, W.A. (1992) Biochemical properties and subcellular distribution of an N-type calcium channel α_1 subunit. *Neuron* **9**, 1099–1115.

Williams, M.E., Brust, P.F., Feldman, D.H., Patthi, S., Simerson, S., Maroufi, A., McCue, A.F., Veliçeble, G., Ellis, S.B. & Harpold, M.M. (1992) Structure and functional expression of an omega-conotoxin-sensitive human N-type calcium channel. *Science* **257**, 389–395.

Ca^{2+}-channel neurotoxins

Gray, W.R., Olivera, B.M. & Cruz, L.J. (1988) Peptide toxins from venomous conus snails. *Annu. Rev. Biochem.* **57**, 665–700.

Hillyard, D.R., Monje, V.D., Mintz, I.M., Bean, B.P., Nadasdi, L., Ramachandran, J., Miljanich, G., Azimi-Zoonooz, A., McIntosh, J.M., Cruz, L.J., Imperial, J.S. & Olivera, B.M. (1992) A new conus peptide ligand for mammalian presynaptic Ca^{2+} channels. *Neuron* **9**, 69–77.

Jackson, H. & Parks, T.N. (1989) Spider toxins: recent applications in neurobiology. *Annu. Rev. Neurosci.* **12**, 405–415.

Mintz, I.M., Adams, M.E. & Bean, B.P. (1992) P-type calcium channels in rat central and peripheral neurones. *Neurone* **9**, 85–95.

Pocock, J.M. & Nicholls, D.G. (1992) A toxin (Aga-GI) from the venom of the spider. Agelenopsis aperta inhibits the mammalian presynaptic Ca channel coupled to glutamate exocytosis. *Eur. J. Pharmacol.* **226**, 343–350.

Uchitel, O.D., Protti, D.A., Sanchez, V., Cherksey, B.D., Sugimori, M. & Llinas, R. (1992) P-type voltage-dependent calcium channel mediates presynaptic calcium influx and transmitter release in mammalian synapses. *Proc. Natl. Acad. Sci. USA* **89**, 3330–3333.

Zernig, G. (1990) Widening potential of Ca antagonists: non L-type Ca channel interaction. *Trends Pharmacol. Sci.* **11**, 38–44.

5

Signal transduction within neurones

5.1 Introduction

A neurone *in situ* receives a continuous stream of input from what may amount to several thousand synapses onto its cell body and dendrites. As well as directly modulating the membrane potential, this input generates a complex interaction of second messengers, which include Ca^{2+}, cAMP, cyclic guanosine monophosphate (cGMP), inositol phosphates and diacylglycerol. These second messengers may modulate ion channel and receptor activity, initiate developmental changes in the cell, control the cytoskeleton and trigger and control exocytosis. For no cell is there a complete description of these interacting second messenger pathways, and there is of course no universal pattern of neuromodulation shared by all neurones. However, to provide an overview of a number of major pathways we shall consider the regulation of cytoplasmic Ca^{2+}.

In Fig. 5.1 a simplified flow chart is shown of some of the major interactions within a typical neurone which control the cytoplasmic free Ca^{2+} concentration $[Ca^{2+}]_c$. In the interests of clarity numerous cross-talk and feedback pathways have been omitted, as have pathways involving cGMP and arachidonic acid and its metabolites. The pathways depicted in Fig. 5.1 occur in many neurones and provide a framework for more detailed discussion. To provide some structure to the subsequent discussion I have arbitrarily divided the pathways in Fig. 5.1 into a number of stages which will be described in sequence.

5.2 Metabotropic and ionotropic receptors

The synaptic input onto a neurone is directed onto two classes of receptor (stage I of Fig. 5.1). Ionotropic receptors, which were discussed in

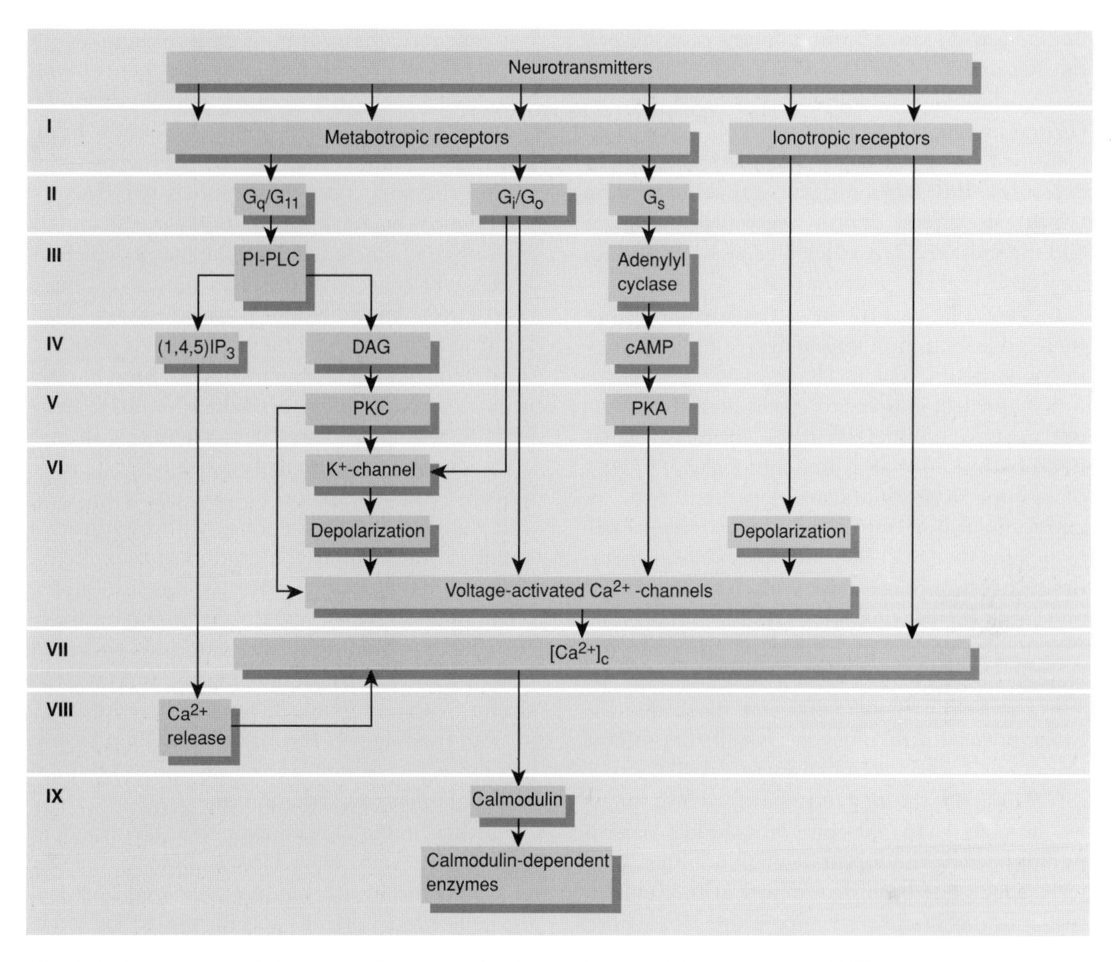

Fig. 5.1 An overview of the major signal transduction pathways within a neurone. (I) Neurotransmitters can activate metabotropic or ionotropic receptors. The former activate a range of G-proteins (II) which in turn regulate G-protein-coupled enzymes which control second messenger concentrations, such as phosphatidylinositol-specific phospholipase C (PI-PLC) and adenylyl cyclase (III). Diacylglycerol (DAG) and cAMP generated by these enzymes (IV) activate proteins kinases C (PKC) and A (PKA) respectively (V). In some neurones PKC can inhibit K^+-channels (VI), modulating depolarization and hence Ca^{2+} voltage-activated channel activity and cytoplasmic free Ca^{2+} concentrations (VII). Some K^+-channels can also be directly activated by G_i via a membrane-delimited mechanism. In some cells PKA can phosphorylate and activate Ca^{2+}-channels. The other second messenger generated by PKC, namely $(1,4,5)IP_3$, can release Ca^{2+} from internal stores (VIII) and contribute to the regulation of cytoplasmic Ca^{2+}. Ca^{2+} can activate calmodulin and hence a range of calmodulin-dependent protein kinases and phosphatases (IX).

the previous chapter, have integral ion channels and act in a submillisecond time scale. These receptors can modulate cytoplasmic $[Ca^{2+}]_c$ in two ways. Firstly those with predominantly monovalent cation permeabilities will depolarize the membrane as Na^+ enters the cell. This may be sufficient to activate Ca^{2+}-channels in the vicinity of the receptor, in which case Ca^{2+} will enter the cell. In addition to this indirect effect, some ionotropic receptors can directly conduct Ca^{2+} and elevate $[Ca^{2+}]_c$ independently of voltage-activated channel activity; these include the glutamatergic NMDA receptor and some isoforms of the AMPA/KA receptor which will be discussed in Chapter 9.

In contrast to the submillisecond responses of ionotropic receptors, *metabotropic* or *G-protein-coupled* receptors are slower acting but can initiate a wide range of modulatory responses in the neurone. In the last years there has been an enormous proliferation in the number of G-protein-coupled receptors which have been identified. Currently the number is well in excess of 100 including the 7-transmembrane G-protein-coupled photon receptor rhodopsin, and the most recently elucidated class, the olfactory receptors (*see* Section 6.3).

In view of this diversity, it is fortunate that there appears to be a strong degree of homology in the structure and mechanism of these receptors. G-protein-coupled receptors possess a single polypeptide with 450–1200 residues. Their hydropathy plots and immunological mapping of epitope accessibility are consistent with an extracellular N-terminus, which can vary greatly in length, seven 20–25 amino acid transmembrane hydrophobic spans with two small and one large cytoplasmic loops and a cytoplasmic C-terminus (shown in a linear two-dimensional view in Fig. 5.2a).

In common with virtually every integral membrane protein, difficulties in crystallization have prevented X-ray determination of detailed structure. However, the likely structure can be inferred by homology to the photon-coupled proton pump bacteriorhodopsin from halophilic bacteria which has been determined at 0.35–1 nm resolution from electron microscopy of ordered two-dimensional arrays of the protein. Seven-transmembrane segments were found in a slightly tilted cylindrical array with a deeply buried retinal cofactor. Although there is little detailed sequence homology between bacteriorhodopsin and G-protein-coupled receptors, this structure has been adopted as a working model. By comparing the primary sequences of some 200 metabotropic receptors, defining the minimum length of the loops connecting adjacent helices and examining the distribution of hydrophilic (inward facing) and hydrophobic (possible lipid-exposed) residues, a consensus arrangement of the α-helices has been proposed (Fig. 5.2b).

The most detailed information for transmitter binding is available for the β-adrenergic receptor which will now be taken as a typical example (Fig. 5.2a). An individual point mutation of an aspartic acid (Asp) residue at position 113 in the third hydrophobic domain, which is present only in metabotropic receptors which bind biogenic amines, results in a large decrease in the affinity with which agonists and antagonists bind. Since biogenic amines are protonated, it is possible that this Asp residue interacts with the positive amine group. Similarly, mutation of two closely apposed serine (Ser) residues in the fifth transmembrane domain decrease the binding affinity of ligands containing the catechol hydroxyl groups characteristic of the biogenic amines, consistent with an involvement of these residues in the binding of the agonists. Once again, metabotropic receptors which do not bind catecholamines lack these serines.

Further interactions have been identified in other transmembrane helices, leading to a model in which the ligand binding is shared between the helices, anchored by specific molecular interactions and located quite deeply within the structure of the receptor (Fig. 5.2c). This is supported by studies with fluorescent agonists whose emission spectra indicate a hydrophobic environment for the ligand protected from aqueous quenching agents. It is also consistent with the location of the retinal cofactor in bacteriorhodopsin.

Fig. 5.2 (*Opposite*) 7-Transmembrane G-protein-coupled receptors. (a) A linear representation of the transmembrane organization of a β-adrenoceptor. Note the extracellular N-terminus and cytoplasmic C-terminus. The transmembrane α-helices are numbered 1–7. Amino acid residues in square boxes can be deleted without affecting ligand binding or protein folding. The acidic Asp 113 (D, highlighted in TM3) is conserved in all receptors with small cationic ligands, consistent with binding of the positively charged group, while two serines at positions 204 and 207 in TM5 are present in receptors binding catecholamines and would allow hydrogen binding of the adjacent hydroxyl groups of the ligand. The filled squares at the N-terminus represent glycosylation sites. From Tota *et al.* (1991). (b) Consensus arrangement of α-helices looking down on the membrane from the external phase. (c) Binding of small cationic ligands between the transmembrane helices. D is the aspartate residue on TM3. Adapted from Tota *et al.* (1991).

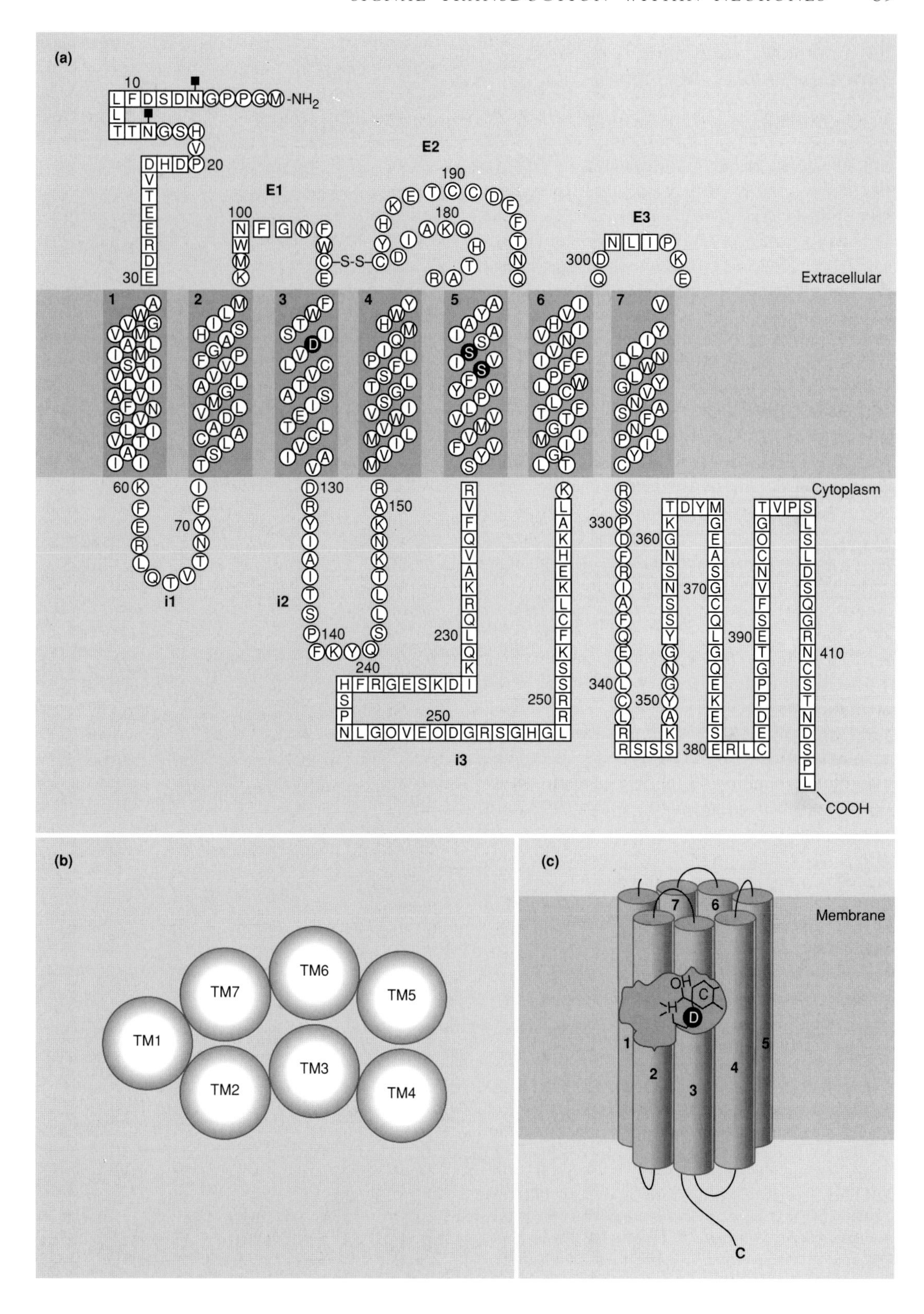
(a)
-NH2
E1
E2
E3
-S-S-
Extracellular
Cytoplasm
i1
i2
i3
COOH
(b)
TM1
TM2
TM3
TM4
TM5
TM6
TM7
(c)
Membrane
C

5.3 Interaction of trimeric G-proteins with metabotropic receptors

In the inactive state G-proteins are heterotrimers with a molecule of guanosine diphosphate (GDP) non-covalently bound to the α-subunit (Fig. 5.3). Trimeric G-proteins (which must be distinguished from the small monomeric G-proteins discussed in Section 7.4.5) contain an α- (39–46 kDa), β- (37 kDa) and γ- (8 kDa) subunit. Although none of the subunits possess obvious hydrophobic regions, the αβγ trimer is associated with the cytoplasmic face of the plasma membrane via isoprenylene or myristoyl anchors.

The conformational change in the receptor which results from the binding of the neurotransmitter allows an appropriate G-protein to dock with intracellular domains of the receptor (stage II of Fig. 5.1). Site-directed mutagenesis experiments implicate the first 20 amino acids of the large cytoplasmic loop (termed i3) of the receptor in the coupling of the receptor to the G-protein.

The conformational change in the G-protein when it binds to the activated receptor allows cytoplasmic guanosine triphosphate (GTP) to displace bound GDP from the α-subunit (Fig. 5.3). This in turn allows the conformation of the receptor to relax, which can be detected as a decrease in the affinity with which the receptor binds its agonist, facilitating its dissociation. The GTP-bound α-subunit now dissociates from the βγ-subunits which remain closely associated as a dimer. The G-protein activation is very rapid and many molecules of G-protein can be activated by a single activated receptor.

The Gα-subunits are considered to be highly mobile along the inner surface of the membrane, shuttling between receptor and effector systems. These may be enzymes such as adenylyl cyclase or ion channels such as a voltage-activated Ca^{2+}-channel. At any one moment there may be more than one signal transduced in the same cell and the role of distinct G-proteins in preventing these

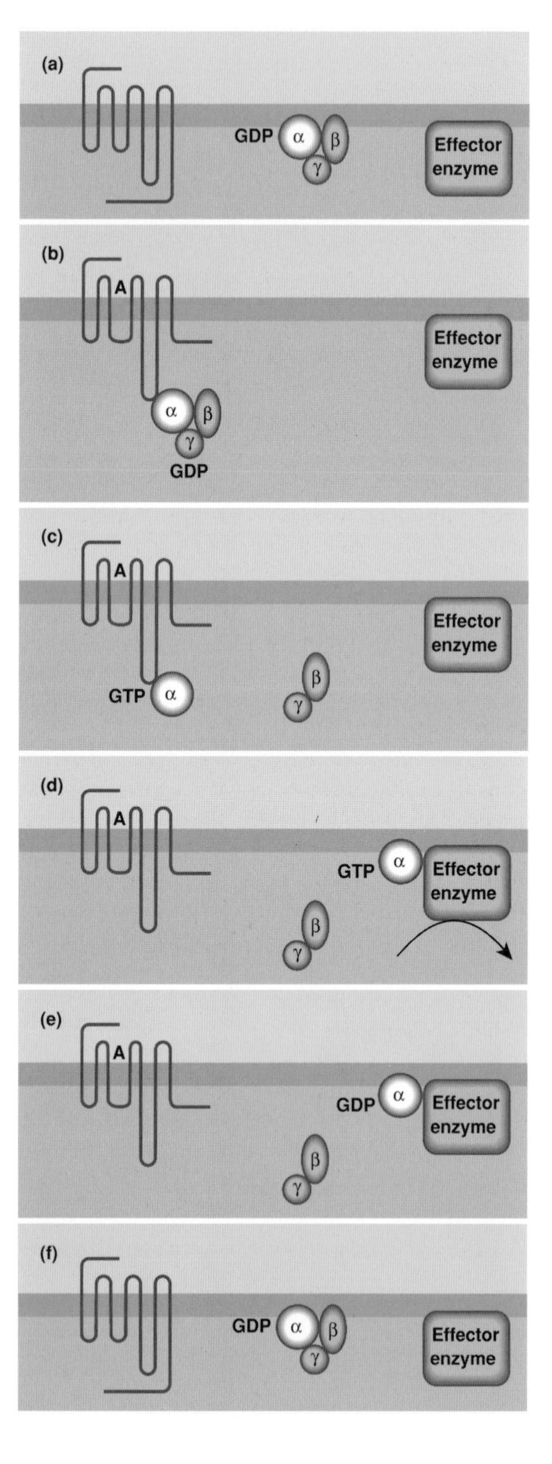

Fig. 5.3 The G-protein catalytic cycle. (a) In the inactive form the G-protein is a trimer with GDP bound to the α-subunit. (b) Agonist binding (A) to the receptor causes a conformational change in the cytoplasmic domain allowing the G-protein to bind, GTP to exchange for GDP, and (c) the βγ-subunits to dissociate. (d) The active GTP-α diffuses to the effector enzyme causing activation. Multiple effectors can be activated by the α-subunit until its endogenous GTPase activity hydrolyses GTP to GDP (e). If the receptor is still activated by agonist the cycle can be repeated, otherwise (f) the cycle reverts to the inactive state.

signals from interfering with each other is only beginning to be resolved (*see* Section 5.3.1).

G-protein α-subunits remain catalytically active until the bound GTP is hydrolysed to GDP by their low intrinsic GTPase activity. There may thus be time for a single α-subunit to activate many effector proteins. Hydrolysis can be hastened by *GAP* or GTPase-activating proteins (which must not be confused with the growth associated protein GAP-43 — *see* Section 7.5.1). *In vitro* the non-hydrolysable analogue *GTP-γ-S* can indefinitely prolong the active state while conversely *GDP-β-S* can bind firmly to the nucleotide site and prevent G-protein activation. Fluoride can activate G-proteins in the presence of aluminium ions by forming AlF_4^- which can bind to Gα-subunits at the site normally occupied by the terminal phosphate of GTP and thus stabilize its active conformation.

There are currently cDNAs for at least 21 distinct α-subunits (grouped into G_s, G_i, G_q and G_o subfamilies). In addition at least four β- and six γ-subunits have been cloned, allowing for an immense number of theoretical permutations. The α-subunits define the effector selectivity of the G-protein (Table 5.1) whereas the $\beta\gamma$ dimer may define the receptor specificity, since inhibiting the expression of specific β isoforms can cause selective blockade of different G-protein-coupled functions in the same cell.

G_s activates adenylyl cyclase. *Cholera toxin* constitutively activates $G\alpha_s$ by adenosine diphosphate (ADP) ribosylation at a specific Arg residue which inhibits its intrinsic GTPase activity and hence its inactivation. G_q and G_{11} are both members of the pertussis- and cholera-toxin insensitive G_q subfamily, whose major function is to activate the β class of the phosphatidylinositol-specific phospholipase C (PI-PLC) family, generating the second messengers diacylglycerol and (1,4,5)IP_3. Both $G_q\alpha$ and $G_{11}\alpha$ are present in brain, although the former is more abundant.

The G_i subfamily include three isoforms of $G\alpha_i$, two splice variants of $G\alpha_o$ as well as the rod outer segment G-protein transducin, G_t. G_i and G_o are present in brain in much higher concentrations than G_s or G_q. G_i inhibits adenylyl cyclase and also can regulate certain K^+- and Ca^{2+}-channels by a 'membrane delimited' mechanism probably involving a direct interaction between G-protein and ion channel (*see* Section 5.6.5). *Pertussis toxin*, from the whooping cough bacterium *Bordatella pertussis* inactives $G\alpha_o$ and $G\alpha_i$ (which possess a specific Cys residue near their carboxyl terminals) by an ADP ribosylation which prevents their activation. Although the α-subunit of transducin resembles the G_i subfamily of G-proteins, it is, unlike the other members of this group, constitutively activated by cholera toxin in the same way as G_s.

5.3.1 G-proteins and the control of convergent and divergent signals

G-protein-linked signal transduction is a complex network of divergent and convergent signalling. A given neurotransmitter may bind to multiple classes and isoforms of metabotropic receptor. On the other hand a wide range of transmitters may all converge upon a single effector system — the most extreme example being the range of transmitters which are capable of regulating neuronal K^+- and Ca^{2+}-channels by the 'mem-

Table 5.1 G-proteins and effectors.

Receptors	G-protein	Typical action of activated α-subunit
Diverse	G_s	Activation of adenylyl cyclase Ca^{2+}-channel activation
Olfactory	G_{olf}	Activation of adenylyl cyclase
Rhodopsin	G_t (transducin)	Activation of cGMP phosphodiesterase
Diverse	G_o, G_{i1}, G_{i2}, G_{i3}	Inhibition of adenylyl cyclase Ca^{2+}-channel inhibition K^+-channel activation
Diverse	G_{11}, G_q	Activation of phospholipase C

brane-delimited' mechanism (*see* Section 5.6.5). One neurone in which convergent and divergent pathways controlling ion channels has been examined in detail is the superior cervical ganglion cell (Fig. 5.4), where no less than nine separate agonists can converge to inhibit the same Ca^{2+} conductance.

Some specificity is gained through the recognition of specific G-proteins by receptors; however receptors may be promiscuous in their choice of G-proteins. Thus recombinant muscarinic M_2 receptors can couple *in vitro* to G_o, G_{i1}, and G_{i3}, while M_1 and M_3 receptors in certain cells can stimulate inositol phosphate production via both pertussis toxin sensitive and insensitive G-proteins. Even β-adrenergic receptors appear under some circumstances to activate G-proteins other than G_s. At the effector end of the pathway, individual K^+-channels can be stimulated *in vitro* by a variety of recombinant activated Gα-subunits.

Since Gα-subunits are mobile along the cytoplasmic face of the membrane, this redundancy would seem to lead inevitably to extensive cross-talk and interference. However, considerable specificity is frequently still retained. For example, D_2-dopamine receptors in anterior pituitary cells concomitantly reduce Ca^{2+} currents and increase K^+ currents. Polyclonal antibodies against specific G-proteins were introduced into the cells via a whole cell patch. Anti-$G\alpha_o$ was found to reduce Ca^{2+}-channel block while $G\alpha_{i3}$ attenuated the activation of K^+ conductance. The ability of distinct G-proteins to affect K^+- and Ca^{2+}-channels is supported by a study in which a point mutation of an aspartate residue within an expressed α_2-adrenoceptor abolished the ability of the receptor to activate K^+ currents while retaining the ability to inhibit Ca^{2+}-channels, indicating that the mutation prevented the receptor from activating a specific subpopulation of G-protein targeted to the K^+-channel.

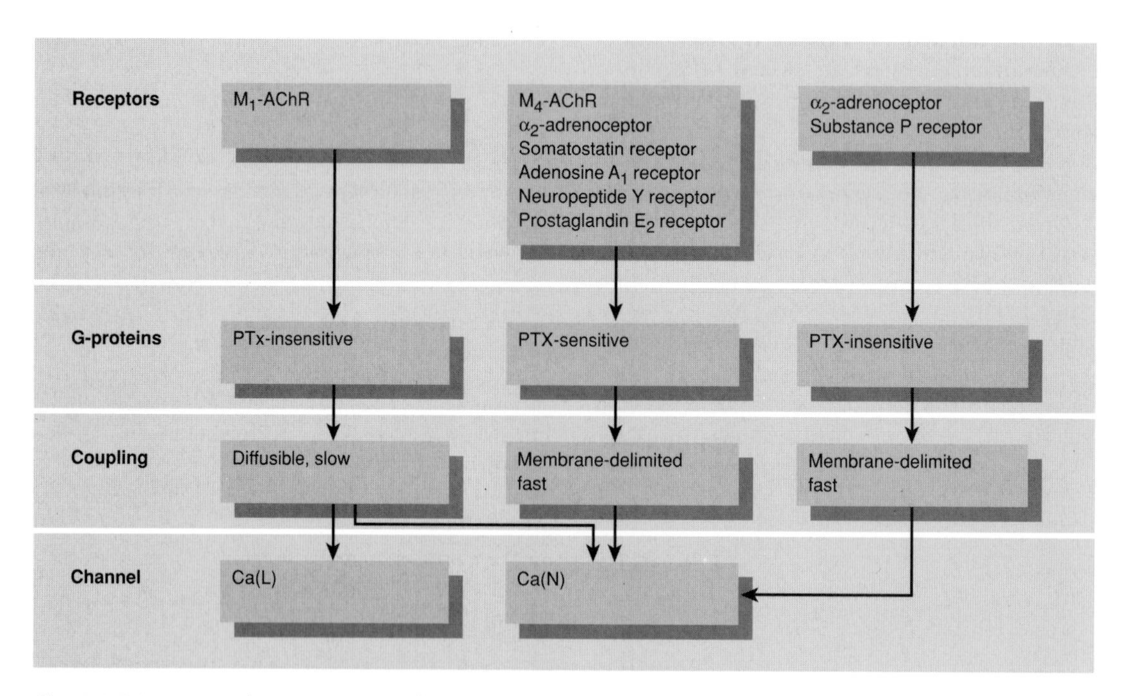

Fig. 5.4 Divergent and convergent signalling in G-protein-mediated signal transduction. In superior cervical ganglion cells, an N-type Ca^{2+}-channel is inhibited by six distinct receptors acting via a pertussis toxin (PTx)-sensitive G-protein interacting with the channel via a fast membrane-delimited mechanism and altering the channel's voltage dependency and opening kinetics. Muscarinic M_1 receptors acting via a PTx-insensitive G-protein, generate an unidentified cytoplasmic second messenger which reduces the probability of channel opening and has similar effects on L-type Ca^{2+}-channels in the cells. Finally α_2-adrenergic agonists can inhibit the channel via a PTx-insensitive membrane-delimited pathway. Adapted from Hille (1992). AChR, acetylcholine receptor.

5.4 Effector enzymes activated by G-proteins

The activated α-subunits of G-proteins can diffuse and collide with enzymes which synthesize or degrade second messengers. We shall now consider the major effector enzymes (stage III of Fig. 5.1), although later direct membrane-delimited interactions with ion channels will be discussed.

5.4.1 Adenylyl cyclase and cAMP

Adenylyl cyclase is a integral 120 kDa plasma membrane glycoprotein, whose hydropathy plots indicate the presence of two sets of six membrane-spanning segments linked by a 240 residue cytoplasmic loop containing a catalytic site which is duplicated at the C-terminus (Fig. 5.5). The structure thus resembles that of some membrane transport proteins (*see* Section 4.2) even though the cyclase has no transport activity. Adenylyl cyclase is active as a cyclase-$G_s\alpha$-ATP complex and is also activated independently by Ca^{2+}/calmodulin. This one of numerous examples of cross-talk between Ca^{2+} and other second messengers.

Adenylyl cyclase can be inhibited by another Gα-subunit, $G\alpha_i$. The mechanism of this inhibition is still not well understood, since a direct effect of $G\alpha_i$ on purified cyclase cannot be demonstrated. It is possible that liberated βγ dimers may directly inhibit certain forms of adenylyl cyclase, alternatively βγ dimers may compete with the cyclase for liberated $G\alpha_s$-subunits reducing the concentration available for the cyclase. The plant terpinoid *forskolin* can directly activate adenylyl cyclase and is a tool to investigate cAMP-mediated events in intact neurones, although it is of limited specificity.

The steady-state concentration of cAMP in the cell is a result of the balance between adenylyl cyclase and phosphodiesterase activities. The only target for cAMP is protein kinase A (*see* Section 5.5.1). It should be noted in passing that cAMP plays a somewhat limited role in mammalian nerve terminals.

5.4.2 Phosphatidylinositol-specific phospholipase C

Phosphatidylinositol 4,5 bisphosphate (PIP_2) is a minor phospholipid generated from phosphatidylinositol (PI) by specific kinases and found mainly in the inner leaflet of the plasma membrane (Fig. 5.6). Three forms of PI kinase have been described. 4-Kinases include integral proteins, potently inhibited by adenosine in non-neuronal tissue, or soluble proteins relatively insensitive to adenosine. 5-Kinases are mostly cytosolic. The third kinase, phosphatidylinositol-3-kinase, phosphorylates PI, phosphatidylinositol 4 phosphate (PIP) and PIP_2 at the 3-position. Although this prevents the phos-

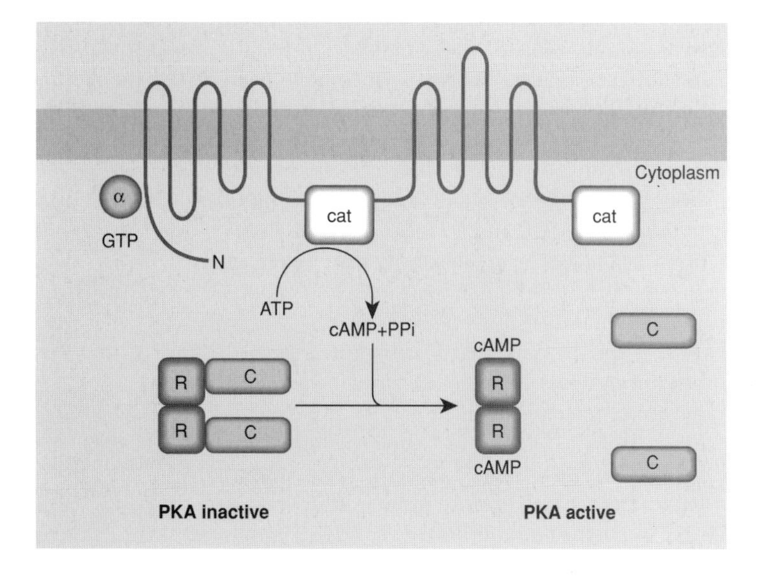

Fig. 5.5 Adenylyl cyclase and protein kinase A (PKA). Adenylyl cyclase is an integral 120 kDa plasma membrane glycoprotein with duplicated catalytic domains (cat) and activated by Gs_α and calmodulin (some forms). cAMP activates PKA causing dissociation of constitutively active catalytic subunits (C) from regulatory domains (R) which otherwise inhibit activity.

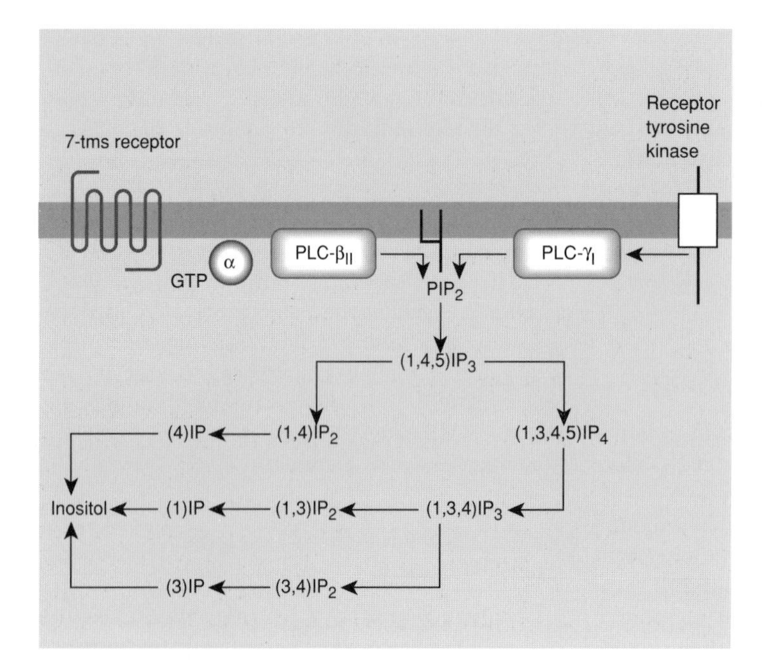

Fig. 5.6 PI-PLC and inositol phosphate metabolism. Phosphatidylinositol bisphosphate (PIP_2) hydrolysis can be induced by activation of a 7-transmembrane receptor via a G-protein and PLC-β_{II} or by a tyrosine receptor kinase activation of PLC-γ_I. The first product is (1,4,5)IP_3 which releases Ca^{2+} from intracellular stores. (1,4,5)IP_3 can be phosphorylated to (1,3,4,5)IP_4 whose role is controversial. Hydrolysis of these inositol phosphates generates inositol which can be reincorporated into phosphatidylinositols.

pholipids from being substrates for PI-PLC, the novel phosphoinositide *phosphatidylinositol 3,4,5 trisphosphate* (PIP_3) which can be generated may be involved in tyrosine-receptor kinase mediated activation of a specific protein kinase C (PKC) isoform, PKC-ζ (*see* Section 13.2.2).

A class of *phosphoinositide-specific phospholipase* C enzymes associate peripherally with the inner face of the plasma membrane and can hydrolyse PIP_2 generating two second messengers; a membrane-phase diacylglycerol (DAG) and a soluble inositol 1,4,5-trisphosphate. To date four classes of PLC isoenzymes have been described, designated β, γ, δ and ε. An earlier described α isoform may be a proteolytic degradation product of δ. The PLC classes are the products of separate genes and show limited homology. The β, γ and δ isoenzymes predominate in brain: PLC-γ is uniformly distributed in neurones, PLC-β is high in the cortex and hippocampus, while PLC-δ is high in glia.

PLC-γ can be activated by single transmembrane *receptor tyrosine kinases* by a G-protein-insensitive mechanism which will be discussed in Section 13.2.2. Relative to G-protein linked receptors, tyrosine receptor kinases generate inositol phosphates much more slowly and with a longer latency, and additionally activate a range of effectors including *phosphatidylinositol-3-kinase* which generates the putative lipid-phase messenger PIP_3.

PLC-β_1 can be activated synergistically by elevated $[Ca^{2+}]_c$ (in the range 0.1–1 μM) and activated G-protein. It is important however to distinguish between this direct metabotropic receptor-mediated activation of PI-PLC and an indirect effect due to an ionotropic or depolarization-evoked elevation of $[Ca^{2+}]_c$ which may in turn activate the phospholipase.

Although a wide range of receptors are coupled via G-proteins to PI-PLC (Table 5.2) the nature of the G-protein coupling is difficult to study since coupling between the receptor and PLC in

Table 5.2 Some neurotransmitter receptors coupled to PIP_2 hydrolysis.

Transmitter	Receptor	Subtype
Noradrenaline	Adrenoceptor	α_1
ACh	Muscarinic	M_1, M_3, M_5
5-HT		5-HT1C, 5-HT2
Histamine		H1
Glutamate	mGluR	1,5
Bradykinin		B2
Vasopressin		V1
Substance P		NK1, NK2, NK3

isolated membranes is very labile. Recently the pertussis toxin-insensitive G_q family has been characterized and shown to specifically activate β_1-PLC, although not all G-protein coupling to PLC is pertussis toxin insensitive. A further complication is that PLC-β_2 may be activatable by the liberated $\beta\gamma$ dimer rather than the α-subunit of the G-protein, and this may account for the pertussis toxin sensitivity. A negative feedback mechanism exists for the PI-PLC pathway whereby PKC activated by PLC-liberated DAG can phosphorylate and inactivate PLC-β at a specific serine residue.

While PI-PLC is by far the best characterized receptor-activated phospholipase, there are indications that other phospholipases may be activated during signal transduction, although the directness of their coupling to receptor activation is currently debated. In many cells a discrepancy is observed between the time courses of (1,4,5)IP_3 and DAG production, the latter being more prolonged. This sustained DAG formation has been ascribed to the activation of a *phospholipase D*, generating a free base such as choline or ethanolamine together with phosphatidic acid which could then be dephosphorylated to DAG (Fig. 5.7). The signal transduction pathway responsible for activating PLD is currently contentious. It has also been suggested that the mismatch between DAG and inositol phosphates might be due to a phosphatidylcholine-specific PLC. However the significance of this pathway in the CNS is unclear. It should also be noted that PLC can hydrolyse PI and PIP as well as PIP_2, bypassing the production of (1,4,5)IP_3.

Phospholipase A_2 (PLA_2) predominantly liberates polyunsaturated fatty acids (often arachidonate) from the 2-position of phospholipids, leaving a lysophospholipid. A cytosolic 110 kDa PLA_2 has been cloned which possesses sequence homology to Ca^{2+}-dependent PKC isoforms in the Ca^{2+}/phospholipid binding domain, suggesting that it may undergo a translocation from the cytoplasm to the membrane in a similar way to activated PKC (*see* Section 5.5.2) Arachidonic acid (AA) is a major fatty acid at the 2-position of many phospholipids and is liberated by PLA_2 action. AA can be further metabolized via the lipoxygenase and cyclo-oxygenase pathways (Fig. 5.8).

5.4.3 Pathways of inositol phosphate metabolism

Although an enormous number of inositol phosphate isomers are possible (many of which have been detected *in situ*), an undisputed second messenger role has only been established for (1,4,5)IP_3 (Fig. 5.6). (1,4,5)IP_3 binds to an intracellular ion channel releasing internal Ca^{2+} stores. (1,4,5)IP_3 has two metabolic fates; hydrolysis to (1,4)IP_2 as an initial step in the hydrolysis to inositol, and phosphorylation to (1,3,4,5)IP_4 by a Ca^{2+}-activated (1,4,5)IP_3 kinase. There is considerable debate as to the possible role of (1,3,4,5)IP_4 in mediating Ca^{2+} entry.

Lithium ions bind uncompetitively to the enzyme–substrate complexes at the phosphatase steps hydrolysing (1,4)IP_2 and IP. By preventing the recycling of inositol into phosphoinositides, Li^+ may moderate inositol phosphate-mediated signal transduction, which may account for its therapeutic action in manic depression.

5.4.4 Diacylglycerol

The second product of PI-PLC activity is the membrane-associated second messenger DAG, which plays a central role in the activation of PKC isoenzymes (*see* Section 5.5.3). The lifetime of DAG in the membrane is limited, since it is a substrate for diacylglycerol kinase which generates phosphatidic acid as a first stage in the recycling of the messenger into phospholipid. Two possible additional sources of DAG have been mentioned above: phosphatidylcholine hydrolysis by PC-selective PLC, and phospholipase D activity to generate phosphatidic acid, followed by hydrolysis to DAG (Fig. 5.7). The significance of these pathways in the CNS is unclear.

5.5 Second messenger-activated protein kinases

In this section (stage IV of Fig. 5.1) we shall consider two of the major second messenger-activated protein kinases, protein kinase A (PKA) and PKC. Two further classes play essential roles in neuronal modulation, Ca^{2+}/calmodulin-dependent protein kinases (which are activated by Ca^{2+} and will be considered after Ca^{2+} homeostasis has been introduced in Section 5.7.2)

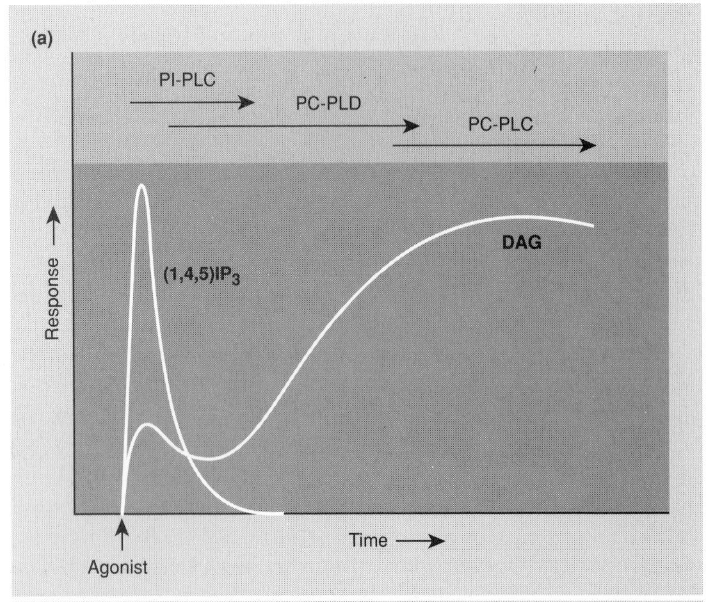

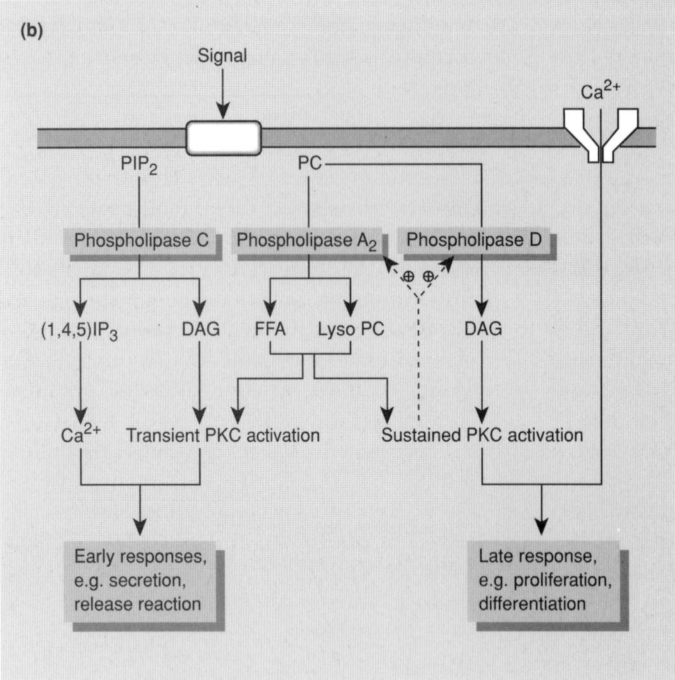

Fig. 5.7 Time courses of elevation in second messengers generated by signal-activated phospholipases. (a) Frequently, a discrepancy is seen in the time course of elevation of the two second messengers generated by PI-PLC, $(1,4,5)IP_3$ and diacylglycerol (DAG). The second phase of DAG elevation, which occurs once $(1,4,5)IP_3$ has returned to basal, has been suggested to originate from phospholipase D (PLD) action on PC, initially generating phosphatidic acid which could then be hydrolysed to DAG. Delayed induction or activation of a PC-PLC could further prolong the DAG elevation. (b) Possible interactions between Ca^{2+}, DAG, fatty acids (particularly arachidonate) and lysophospholipids (lysoPC) in the transient and sustained activation of PKC. Adapted from Nishizuka (1992a).

and tyrosine kinases which are involved in several plastic and developmental cascades and will be discussed in context in Section 13.2.2.

5.5.1 cAMP-dependent protein kinases

In the resting state, PKA is a tetramer comprised of two regulatory (R) and two catalytic (C) subunits (*see* Fig. 5.5). R–C interaction occurs via a pseudosubstrate site on each of the R-subunits which occlude the active sites of the C-subunits. Two molecules of cAMP bind cooperatively to each R-subunit, causing a conformational change which releases the constitutively active catalytic subunits. The cooperativity means that a small change in cAMP concentration

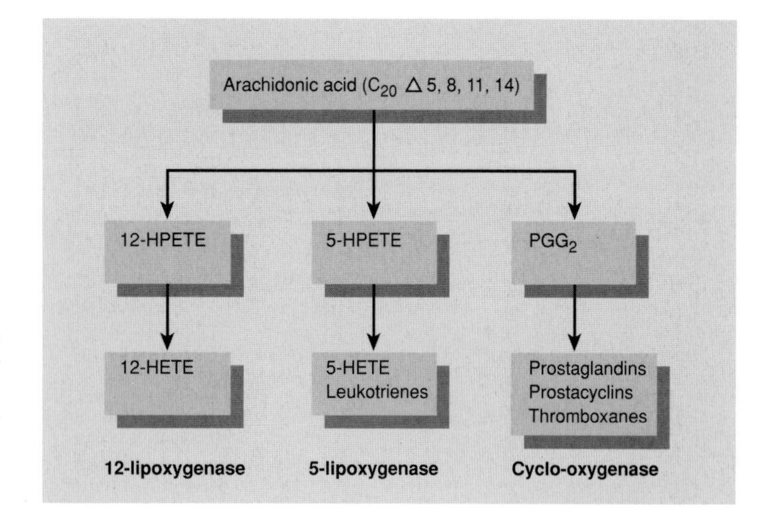

Fig. 5.8 Pathways of arachidonate metabolism. Arachidonate can be generated by phospholipase A_2 activity. Further metabolism can occur via the the 5- and 12-lipoxygenase and cyclo-oxygenase pathways, producing prostaglandins, prostacyclins and thromboxanes.

can cause a large enhancement in kinase activity.

Four R isoforms and three C isoforms are known. Type II PKA is present postsynaptically where it associates with the postsynaptic density via a specific 79 kDa A-kinase anchoring protein. PKA phosphorylates a wide range of target proteins. Among those of neuronal origin is tyrosine hydroxylase, the first enzyme in the pathway of catecholamine synthesis from tyrosine in catecholaminergic terminals (*see* Section 11.1.2).

5.5.2 Neuronal PKC isoenzymes

PKC was originally defined as a Ser/Thr protein kinase activity which was dependent on the addition of Ca^{2+}, DAG and acidic phospholipid such as phosphatidylserine (PS). This last is not required as a substrate, but rather to provide a amphiphilic environment to aid interaction with DAG during *in vitro* experiments. *In situ* this phospholipid requirement is met by the membrane phospholipids.

Subsequent molecular cloning and enzyme isolation has established a extended family of related kinases of 70–80 kDa (Table 5.3). Three groups of genes have been designated on the basis of sequence homology (Fig. 5.9): the cPKC ('conventional') group includes PKC-α, PKC-β_I and PKC-β_{II} (the latter two alternatively spliced forms of the same primary transcript) and PKC-γ. Of the nPKC ('new') family, PKC-δ, PKC-ε and PKC-ε′ (alternatively spliced) are found in brain, as is PKC-ζ from the aPKC ('atypical') family which will be discussed in Section 13.2.2

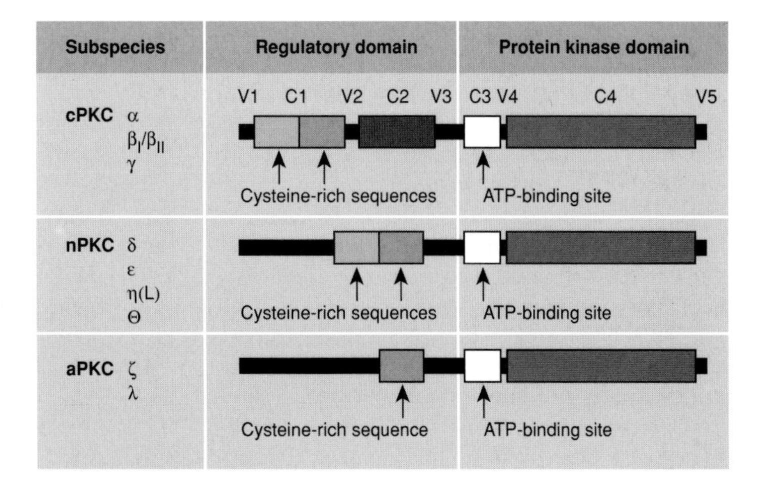

Fig. 5.9 Protein kinase C (PKC) isoenzymes. Major structural features of PKC subspecies. Conserved (C) and variable (V) regions of the cPKC group are shown. β_I and β_{II} are alternatively spliced from the same gene. The C2 region (only present in cPKCs) appears to be related to the Ca^{2+} sensitivity of the enzyme, and is lacking in the Ca^{2+}-independent nPKC and aPKC groups. aPKCs with only one cysteine-rich sequence are not activated by DAG or phorbol esters. Data from Nishizuka (1992b).

Table 5.3 Protein kinase C (PKC) subspecies in mammalian tissues.

Group	Subspecies	Molecular mass (kDa)	Activators	Expression in brain
cPKC	α	76 799	Ca^{2+}, DAG, PS, FFA, lysoPC	Yes
	β_I	76 790	Ca^{2+}, DAG, PS, FFA, lysoPC	Yes
	β_{II}	76 933	Ca^{2+}, DAG, PS, FFA, lysoPC	Yes
	γ	78 366	Ca^{2+}, DAG, PS, FFA, lysoPC	Yes (specific)
nPKC	δ	77 517	DAG, PS	Yes
	ϵ	83 474	DAG, PS, FFA	Yes
	ϵ'		DAG, PS, FFA	Yes
	η(L)	77 972	?	No
	θ	81 571	?	Largely non-neuronal
aPKC	ζ	67 740	PS, FFA, PIP_3?, not DAG	Yes
	λ	67 200	?	No

Data from Asaoka *et al.* (1992).
FFA, *cis*-unsaturated fatty acid; lysoPC, lysophosphatidylcholine; PS, phosphatidylserine.

in the context of receptor tyrosine kinases. nPKC isozymes are insensitive to Ca^{2+} but are activated by DAG and PS. In addition PKC-ε is activatable by *cis*-unsaturated fatty acids. The aPKCs are currently not well characterized: PKC-ζ does not respond to Ca^{2+}, DAG or phorbol esters but may respond to PIP_3 generated in response to receptor tyrosine kinase activation (*see* Section 13.2.2).

As with other proteins, the distribution of PKC isoforms in the brain can be investigated by a variety of techniques. *In situ* hybridization labels the soma of cells which express the isoform but does not identify the final location of the protein which may be in distal dendrites or terminals. Immunocytochemical analysis with type-specific antibodies can identify the regional distribution of the isoform (Fig. 5.10), while immuno-gold electron microscopy can additionally report subcellular distribution. Finally Western blotting or chromatographic separation can identify isoforms in synaptosomes or cultured cells.

Most CNS neurones contain α, β_{II} and γ cPKCs; however, a detailed analysis of cerebellar distribution reveals some specificity (Fig. 5.10). PKC-α is expressed in the CNS and peripheral tissues. In the cerebellum it is found both in Purkinje cells and granule cells. PKC-β_{II} is also expressed both in the CNS and peripheral tissues. In the cerebellum it is mainly found in granule cells, but is not expressed by Purkinje cells. PKC-γ is only expressed in the CNS. In the cerebellum, it is found in Purkinje cells but not in granule cells. As a generalization of PKC-ε PKC-α and β_I are found in terminals and/or growth cones, while PKC-B_{II} and PKC-γ are postsynaptic and can be associated with the postsynaptic density.

5.5.3 Activation of PKC

The 'conventional' cPKCs can be activated by Ca^{2+}, DAG, PS and either *cis*-unsaturated fatty acids or lysophospholipids (*see* Table 5.2). DAG *in vitro* greatly decreases the Ca^{2+} requirement for the enzyme. In addition arachidonate (the major fatty acid generated by PLA_2 can activate PKC, alone in the case of PKC-γ, or synergistically with DAG in the case of PKC-α or PKC-β_{II}. PKC-γ is more sensitive to stimulation by DAG than the other conventional kinases. cPKCs have a regulatory and a catalytic domain (*see* Fig. 5.9). The regulatory domain contains a cysteine-rich 'Zn^{2+}-finger' sequence which is the site of DAG and phorbol ester binding, activation of PKC resulting in the 'unmasking' of the regulatory domain from the catalytic site.

In vitro activation in intact cells or synaptosomes may be achieved by the addition of high micromolar concentrations of DAG or nanomolar concentrations of tumour-promoting

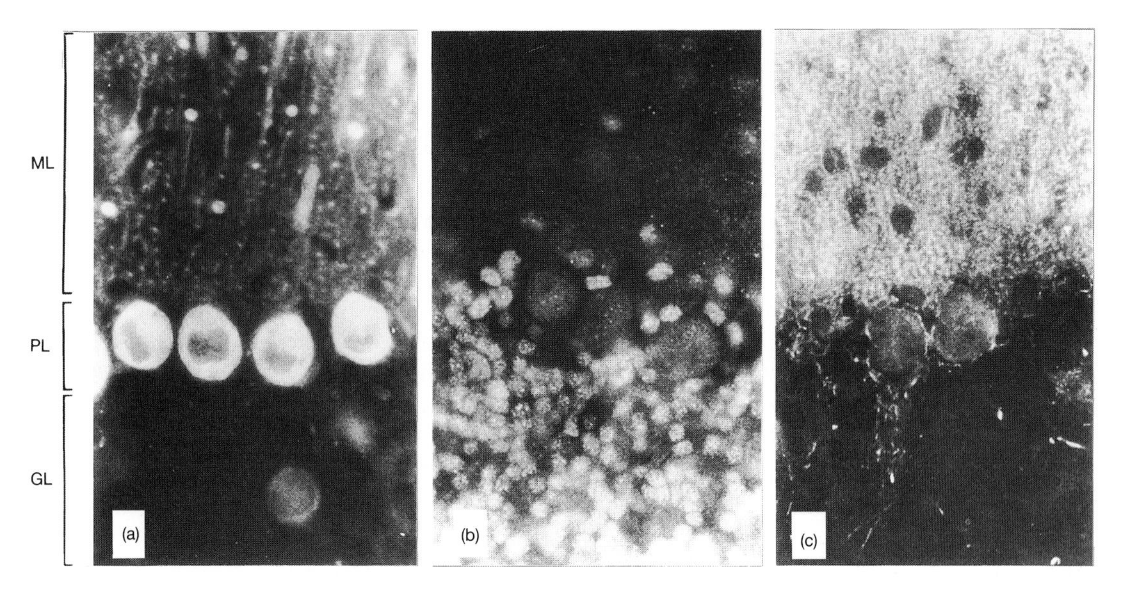

Fig. 5.10 Distribution of protein kinase C (PKC) isoforms in the cerebellum. Immunofluorescent staining of sections of cerebellar cortex with (a) monoclonal antibodies to PKC-γ showing localization in Purkinje cell bodies; (b and c) polyclonal antibodies to respectively (b) PKC-β_I (showing localization in granule cell bodies) or (c) PKC-β_{II} (probably localized in granule cell terminals). GL, granular layer (granule cell bodies); ML, molecular layer (Purkinje dendrites and granule cell terminals); PL, Purkinje (cell body) layer. Data from Ase *et al.* (1988).

phorbol esters which mimic the effects of DAG, but being non-metabolized give much more long lasting effects than DAG, which is rapidly removed by phosphorylation by DAG kinase to give phosphatidate. In the activated form, PKC is much more exposed to proteolytic degradation by Ca^{2+}-activated proteases such as *calpain* which attack the 'hinge' between the regulatory and catalytic regions. The products of proteolysis are a protein kinase which is independent of Ca^{2+} or phospholipid ('M-form') and a phospholipid-dependent phorbol ester binding protein. This proteolysis does not appear to occur during the normal physiological activation of the kinases by DAG.

When PKC in synaptosomes or cultured neurones is activated, subsequent hypotonic lysis of the preparation followed by separation of soluble and membrane fractions reveals an apparent 'translocation' of PKC to the membrane relative to inactivated controls. There may be a physical redistribution of the kinase, although it is also possible that the DAG merely stabilizes an existing weak association of the kinase with the membrane.

5.6 Modulation of voltage-activated ion channels

In this section we continue our progression through the sequence of events outlined in Fig. 5.1 by investigating stage V, the role of receptors in the indirect modulation of voltage-activated ion channels, both by the control of phosphorylation and by a more direct *membrane-delimited* interaction between G-proteins and ion channels.

It is necessary to distinguish at this stage between ion channel modulation at the cell body and at individual nerve terminals. The firing of an action potential down the axon from the axon hillock depends on the achievement of a critical level of depolarization at this site. This in turn is an integral in space and time of the effects of the ionotropic receptors and voltage-activated ion channels active on the cell soma and dendrites. G-protein-mediated metabotropic receptors can modulate the activity of these components and hence control the frequency or pattern of firing of a neurone. However once an action potential or train of action potentials is propagated down an axon it will usually invade all the terminals and would be expected to lead to an equal release of transmitter from each.

A fine control of depolarization at *individual* terminals is possible because of the presence of presynaptic metabotropic receptors, which may modulate Ca^{2+} entry (and hence exocytosis) in two ways. First the receptor may regulate the presynaptic Ca^{2+}-channel itself, by affecting the probability of opening, the voltage-dependency or some other kinetic parameter. Alternatively the receptor may modulate the activity of presynaptic Na^{+}- or K^{+}-channels and hence control the duration or intensity of the action potential. This would then indirectly regulate Ca^{2+} entry through the voltage-activated Ca^{2+}-channels and hence the release of transmitter (Fig. 5.11). It must be emphasized that metabotropic receptors do not override the requirement for an action potential to activate the ion channels but rather regulate the subsequent kinetics of channel opening. Modulation of ion channel activity is the major mechanism by which presynaptic receptors modulate the release of transmitters.

5.6.1 Mechanisms for presynaptic inhibitory receptor action

Most presynaptic autoreceptors (those responsive to the transmitter released by the terminal) and heteroreceptors decrease the release of neurotransmitter (*see* Table 5.3). There is evidence (*see* Section 5.6.5), at least at the cell body, for a direct 'membrane-delimited' interaction between the G-protein and the ion channel, without the involvement of cytoplasmic second messengers.

The two mechanisms discussed above — an increased activity of voltage-activated K^{+}-channels, leading to a hyperpolarization and decreased action potential intensity or duration or alternatively a decreased Ca^{2+}-channel con-

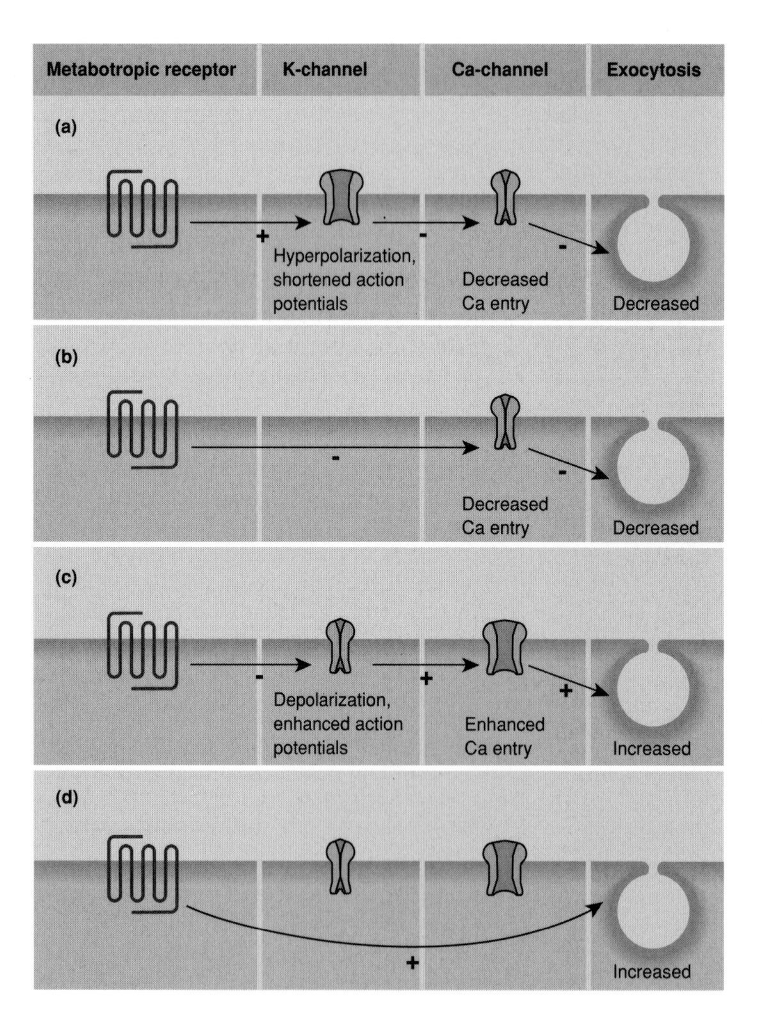

Fig. 5.11 Possible mechanisms for the modulation of exocytosis by presynaptic metabotropic receptors. G-proteins activated by presynaptic receptors may control ion channels by a direct 'membrane-delimited' mechanism or via generated second messengers and kinases. The loci at which regulation can occur include (a) activation of presynaptic K^{+}-channels leading to hyperpolarization, shortened action potentials and decreased release; (b) a more direct inhibition of Ca^{2+}-channels would have the same inhibitory effect; (c) K^{+}-channel inhibition would lead to enhanced action potentials and facilitated exocytosis; (d) direct modulation of some component in the Ca^{2+} secretion coupling mechanism.

ductance in response to an unchanged action potential — cannot directly be distinguished in CNS terminals which are too small for electrophysiology. However, with synaptosomes, one criterion which effectively eliminates a pure K^+-channel locus for an inhibitory presynaptic receptor is the ability to see inhibition when the synaptosomes are stimulated with elevated KCl, which provides a clamped depolarization in which activation of additional K^+-channels should not influence Ca^{2+}-channel opening (*see* Section 3.3.1). If inhibition is seen under these conditions a Ca^{2+}-channel locus, or even a direct modulation of Ca^{2+}-secretion coupling is indicated (Fig. 5.12).

K^+-channel modulation *can* be seen when synaptosomes are depolarized with 4-aminopyridine (4AP), which initiates spontaneous action potentials in the preparation (*see* Section 3.3.1). Thus PKC activation increases glutamate release evoked by 4AP but is totally ineffective when release is evoked by KCl, indicating an effect on K^+-channels controlling action potentials (Fig. 5.12).

Most presynaptic receptors are metabotropic and modulate the activity of voltage-activated ion channels. Like any generalization there are exceptions, and one which has been widely investigated is the neuronal nicotinic receptor (*see* Section 10.3.2) which has a partial presynaptic location and can evoke the release of ACh from synaptosomes in the absence of any other means of depolarization. Activation of this receptor would thus depolarize the terminal and evoke release in the absence of an invading action potential. The physiological consequences of this are still not fully investigated.

5.6.2 Postsynaptic regulation of voltage-activated ion channel activity

In contrast to the presynaptic terminal, the ionic events associated with metabotropic receptor activation on the cell body can be examined by direct electrophysiology. As at the terminal, a major effect of the metabotropic receptors is to modulate the kinetics of the voltage-activated channels. The mechanisms may be more diverse

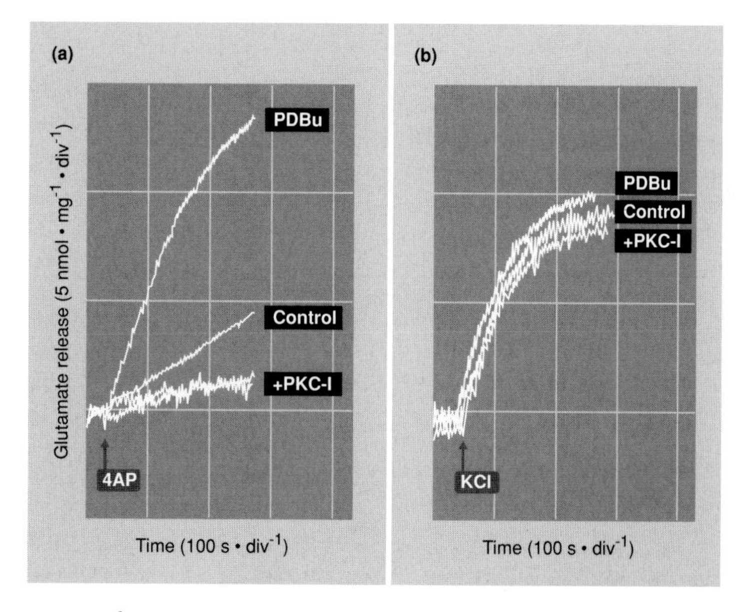

Fig. 5.12 Modulation of Ca^{2+}-dependent 4-aminopyridine (4AP)-evoked glutamate release by protein kinase C (PKC) in synaptosomes. Synaptosomes are induced to release glutamate by depolarization evoked by (a) 4AP, which initiates spontaneous 'action potentials' in synaptosomes, or (b) by KCl which causes a clamped depolarization. In the former case the release of glutamate can be strongly modulated by either activating PKC with phorbol dibutyrate (PDBu) or inhibiting PKC activity with a PKC inhibitor (PKC-I). However KCl evoked glutamate release is independent of PKC activity. This indicates that PKC does not modulate Ca^{2+} entry or Ca^{2+} secretion coupling but that it exerts a strong control over the 4AP-evoked 'action potentials' perhaps by inhibiting K^+-channels. Data from Coffey *et al.* (1993).

than at the terminal since a variety of direct and indirect mechanisms for coupling the activated Gα to the channel may exist. These include:

1 A classical cytoplasmic second messenger cascade, involving activation (via G_S) or inhibition (via G_i) of adenylyl cyclase, leading to altered cAMP levels, modulation of a cAMP-dependent protein kinase and finally altered phosphorylation of the channel itself.

2 Activation of phospholipase C by G_q or G_{11}, liberation of DAG, activation of PKC and channel phosphorylation as discussed for the nerve terminal.

3 Direct '*membrane-delimited*' interaction of G-proteins with ion channels.

4 Other second messenger systems.

The field is complex and in some cases contradictory, so only a few illustrative examples for each mechanism will be chosen.

5.6.3 cAMP-mediated regulation of voltage-activated channels

An example of cAMP-mediated K^+-channel regulation is found in *Aplysia* sensory neurones, where a 5-HT receptor, coupled via G_s to adenylyl cyclase, elevates cAMP which, via PKA, phosphorylates and inhibits voltage-dependent K^+ S-channels. The resulting increase in excitability is an element in the plastic (adaptive) changes which can be induced in *Aplysia* and which will be discussed in Chapter 13. The nature of the kinetic modification to the *Aplysia* K^+-channel has been investigated by addition of PKA to detached membrane patches in which single-channel conductances are being determined. It was found that the number of functional channels decreased, but the open probability of those channels still functional was unchanged. Mammalian large-conductance Ca^{2+}-activated K^+-channels (BK, *see* Section 4.5.1) reconstituted into lipid bilayers, can also be modulated by cAMP-dependent phosphorylation, which alters their sensitivity to Ca^{2+} and voltage.

5.6.4 PKC-mediated regulation of ion channels

Ca^{2+}-, Na^+- and K^+-channels often contain consensus sequences which may be phosphorylated *in vitro* by PKC. One channel for which phosphorylation may play a major role is the delayed rectifier K^+-channel (*see* Section 4.5.1). Whole cell patch-clamp analysis of cerebellar Purkinje and granule cells show that PKC activation causes an inhibition of the delayed rectifier K^+ conductance. A similar regulation via the delayed rectifier operates at the squid giant synapse and can be postulated for the mammalian presynaptic PKC-mediated effects discussed above.

There is no consistent effect of PKC activation on neuronal Ca^{2+}-channels. Depending on the cell, an inhibition, no effect, an activation or even the appearance of new Ca^{2+}-channels (in *Aplysia*) has been reported. Many inhibitors of PKC have limited specificity, although peptide inhibitors of PKC which can be infused by means of a patch electrode are more discriminating and will, for example, block the ability of NA, GABA and phorbol esters to inhibit somatic Ca^{2+}-channels in dorsal root ganglion cells. Since chronic phorbol ester exposure causes down-regulation of PKC, it is necessary to distinguish between a direct activation of PKC and the removal of a tonic inhibitory effect by down-regulation. It is therefore important that more physiological PKC-mediated agonists be shown to have comparable effects.

5.6.5 Membrane-delimited modulation of voltage-activated channels

Although cyclase inhibition via G_i can be monitored in neuronal cell bodies, a more significant mechanism may be a direct *membrane-delimited* interaction between the activated Gα and the channel itself, usually resulting in an increase in the activity of K^+-channels (causing hyperpolarization) and/or to decrease Ca^{2+}-channel activity (Fig. 5.13). Membrane-delimited interactions were proposed when it was found, with the advent of patch-clamp analysis of receptor/channel interactions, that receptor-mediated regulation of both Ca^{2+} and K^+ conductances could be observed in isolated patches in the complete absence of cytoplasm and hence of soluble second messengers and kinases. G-proteins were however required since receptor-mediated effects could often be inhibited by prolonged exposure to pertussis toxin (which ADP ribosylates and hence inhibits G_i and G_o). Additionally GTP-γ-S (which causes a long lasting activation of G-proteins) could mimic the

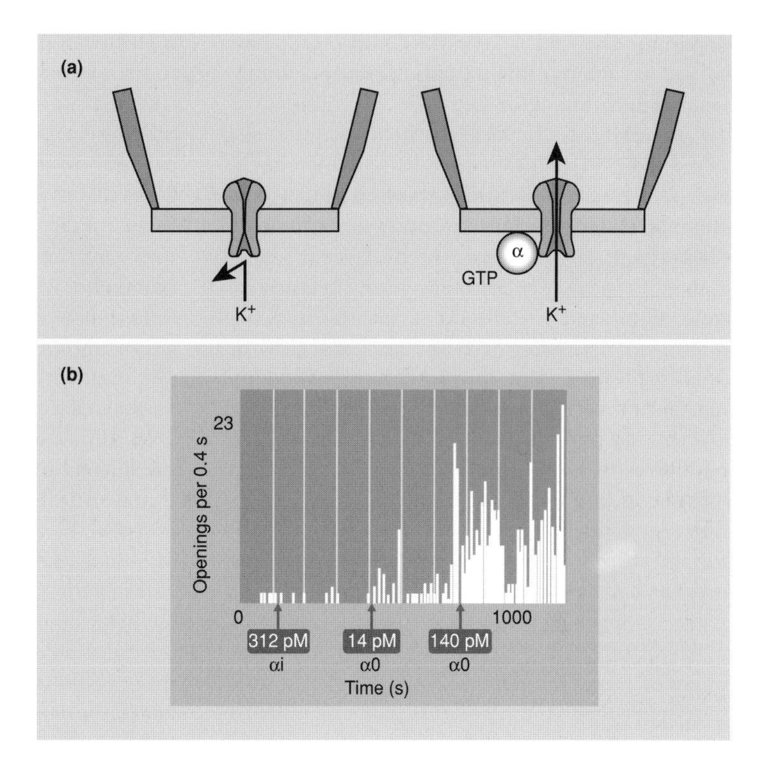

Fig. 5.13 Cell-attached patch and direct regulation of ion channels. (a) K^+-channels in isolated inside-out membrane patches from hippocampal pyramidal cells do not open at -80 mV unless GTP-γ-S activated recombinant $G\alpha_o$- subunits are added. (b) The frequency of opening is greatly enhanced by the application of activated $G\alpha_o$ to the cytoplasmic face of the isolated patch while $G\alpha_i$ is inactive. Note that there is no cytoplasm present and therefore diffusible second messenger systems can be eliminated. From Van Dongen *et al.* (1988), reproduced with permission from *Science*. © 1988 by the AAAS.

action of the transmitter on isolated patches, or after injection into neurones, in the absence of any detectable change in second messenger levels. Finally purified activated G-protein α-subunits could directly influence channels activity in isolated patches.

While there is not yet definitive proof that a G-protein subunit interacts *directly* with a channel, rather than by a membrane-phase second messenger system, two lines of evidence support such an interaction. Firstly in the cell-attached mode, channels can be modulated by agonists applied within the patch electrode but not by agonists added to the bath. This means that if a membrane-bound second messenger is being generated by receptors located outside the patched area its range is not sufficient to allow diffusion laterally within the membrane to affect channels within the patch.

Secondly regulatory interactions of G-proteins with N-channels can be detected as a change in the affinity of ω-conotoxin-GVIA binding, in a similar way that agonist affinity decreases when GTP/GDP exchange occurs on G-proteins bound to metabotropic receptors (*see* Section 5.3). Thus in membranes from a neuroblastoma × glioma cell line, addition of the non-hydrolysable GTP analogue GMP-PNP causes a decreased ω-conotoxin-GVIA binding.

Membrane-delimited interaction is not unique to N-channels, as N-, L- and P-channels in the CNS each appear to be inhibitable by interactions with G-proteins. In a few cases the nature of the inhibition has been investigated: presynaptic noradrenaline receptors inhibit the N-channels which mediate Ca^{2+} entry into sympathetic varicosities by restricting the channel opening to a medium-probability mode, similar to the partially-inhibited state which the channels assume after a few seconds of continuous depolarization.

The most detailed information regarding K^+-channel modulation by G-proteins is available for the cardiac inwardly-rectifying K^+ current $[I_{K(IR)}]$ which is activated by ACh and adenosine via a member of the G_i family. Activated α-subunits of G_{i1-3} can activate K^+-channels if applied to the cytoplasmic face of an isolated patch. The analogous current in central and peripheral neurones can be activated by a large number of metabotropic receptors including muscarinic, α_2-adrenergic, dopamine-D_2, 5-HT_{1A}, GABA-B, opiates (μ and δ), adenosine-A_1, neuropeptide Y and somatostatin utilizing G_i or G_o.

Ca^{2+} currents are inhibited by the same range of presynaptic receptors as activate K^+-channels, indeed there is frequently ambiguity as to which channel is modified to produce an inhibitory response. It is not clear which class of Ca^{2+}-channel is primarily affected, since the survival time of N- and L-channels in isolated patches is limited; however T-type channels can be inhibited by dopamine (applied to an outside-out patch) or by GTP-γ-S (inside-out patch) in the absence of any soluble cofactors. The typical inhibition of Ca^{2+} current with optimal transmitter concentrations ranges from 30 to 70% in different cells.

The receptors which activate K^+-channels and inhibit Ca^{2+}-channels are also those which inhibit adenylyl cyclase and reduce cAMP by pertussis-sensitive G-proteins. It is therefore important to eliminate a conventional second messenger-mediated regulation, for example, by showing that agents which directly modify cAMP levels do not affect channel reactivity. In addition it is necessary to eliminate an indirect coupling via PI-PLC, DAG and PKC, not least because this could be a membrane-confined messenger system. However most of the agonists that inhibit Ca^{2+}-channels do not stimulate PI hydrolysis, and activators and inhibitors of PKC do not influence this coupling.

5.7 Cytoplasmic free Ca^{2+} concentrations

We have now discussed at length the complex modes by which neuronal receptors may activate and regulate voltage-activated Ca^{2+}-channels. It is now time to consider how the free cytoplasmic $[Ca^{2+}]_c$ is controlled by the concerted activity of the plasma membrane receptors, transporters and pumps (stage VI in Fig. 5.1). We shall then be in a position to consider the contribution made by Ca^{2+} transport across internal membranes (*see* Section 5.8).

5.7.1 Measurement of $[Ca^{2+}]_c$

The standard method for determining intracellular free Ca^{2+} concentration uses fura-2 or a related fluorescent Ca^{2+} chelator. Ca^{2+} binding is a function of the free Ca^{2+} concentration, $[Ca^{2+}]_c$, and the dissociation constant for the chelator; in the case of fura-2, half-maximal binding occurs at about 200 nM $[Ca^{2+}]_c$. The excitation spectrum of fura-2 is shifted some 40 nm to the left when Ca^{2+} binds (Fig. 5.14), allowing the proportion of free fura and Ca^{2+}–fura (and hence the Ca^{2+} concentration) to be calculated from the ratio of emission intensity following excitation of 340 and 380 nm. An important feature of this ratiometric determination is that the signal is independent of dye loading or variations in thickness of the cell.

In some experiments fura-2 is introduced into single cells by microinjection. However a more common technique is to rely on the presence of non-specific esterases within the cytoplasm of most cells; fura-2 possesses five carboxyl groups, making the molecule hydrophilic and therefore membrane impermeant. If each of these carboxyls are esterified to form acetoxymethyl (AM) esters, the resultant fura-2-AM becomes sufficiently hydrophobic to cross the plasma membrane. Once in the cytoplasm non-specific esterases cleave off the AM groups sequentially, regenerating fura-2 which is too hydrophilic to escape from the cell. After typically a 30 min incubation much of the original fura-2-AM has accumulated in the cytoplasm as free fura-2.

Fura-2 can be loaded in this way into synaptosomes in suspension and into monolayers of cultured cells plated onto coverslips. The emission from the latter can be visualized by a fluorescence microscope with a photomultiplier allowing fast transients to be obtained from a single illuminated cell (Fig. 5.15).

In view of its extensive cytoplasmic chelation, Ca^{2+} diffuses only slowly within the cytoplasm. This, together with the highly elongated and asymmetric structure of neurones allows Ca^{2+} signals to be localized to specific regions within the neurone. The fluorescence microscopy technique mentioned above can be modified by using a sensitive video camera to capture images of the cell alternatively excited at 340 and 380 nm. This in turn allows an associated computer to calculate dynamic false-colour images of the cell giving spatial and temporal resolution of the free Ca^{2+} concentration. In Plate 2 (between pages 132 and 133) the response of a cultured neurone to electrical stimulation is shown, it is apparent that intracellular gradients exist, since

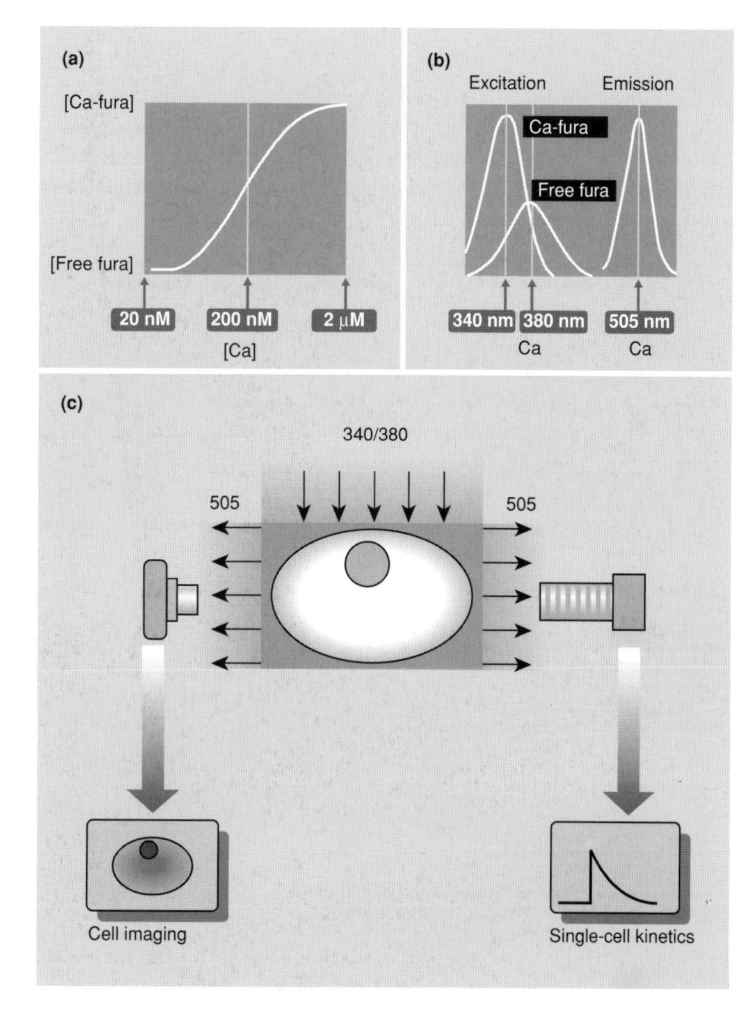

Fig. 5.14 Fura-2 and the determination of spatial and temporal gradients of Ca^{2+}. (a) Fura-2 is a fluorescent Ca^{2+} chelator which is 10% saturated with Ca^{2+} at 20 nM and 90% saturated at 2 μM. (b) Free fura emits maximally when excited at 380 nm, while Ca^{2+}-fura has maximal emission when excited at 340 nm. The ratio of emission when alternately excited at 340 and 380 nm is therefore a measure of the Ca^{2+} concentration in equilibrium with the chelator. (c) Fura-2 is a penta-carboxylic acid and is therefore impermeant. However its acetoxymethyl ester will cross the plasma membrane and be hydrolysed by non-specific cytoplasmic esterase, regenerating fura-2 which is now trapped and accumulates in the cytoplasm. Single cells or populations are illuminated alternately at 340 and 380 nm and the emitted 505 nm light is detected by either a photomultiplier (to allow calculation of a time course but not spatial resolution) or a video camera (allowing a dynamic false-colour image of $[Ca^{2+}]_c$ to be generated).

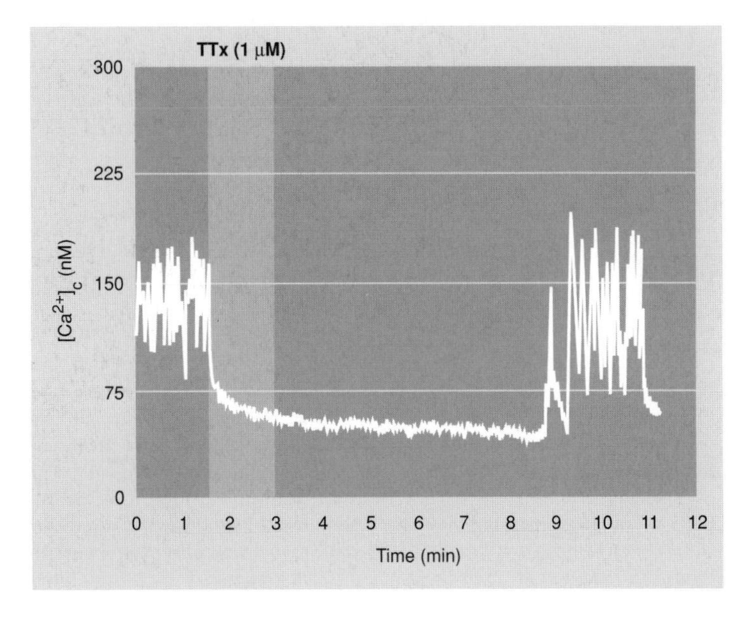

Fig. 5.15 Spontaneous fluctuations in $[Ca^{2+}]_c$ observed in a single hippocampal neurone in culture. $[Ca^{2+}]_c$ in a single cultured hippocampal neurone loaded with fura-2 was determined by limiting the excitation illumination to the soma of a single cell and monitoring the emission with a photomultiplier. The experiment shows that the cell undergoes spontaneous fluctuations in $[Ca^{2+}]_c$ and that these are prevented by the Na^+-channel inhibitor tetrodotoxin. Data from Bleakman *et al.* (1992), by permission of Macmillan Press.

$[Ca^{2+}]_c$ in growth cones rises before that in the cell soma.

For measuring $[Ca^{2+}]_c$ in individually injected neurones within brain slices, confocal microscopy (where a laser beam is focused and scanned in a defined focal plane) can be used to give resolution perpendicular to the plane of the sample. The indicator fluo-3 is often used in order to be compatible with available lasers (Fig. 5.16). An alternative technique which is very sensitive to changes in $[Ca^{2+}]_c$ but is less quantitative involves the microinjection of aequorin, a protein from the jellyfish *Aequorea forskalea* which becomes chemiluminescent when it chelates Ca^{2+} (*see* Fig. 7.8).

5.7.2 Basal $[Ca^{2+}]_c$ and the response to depolarization

Fura-2 experiments show that the concentration of free calcium in the cytoplasm of a resting neurone is typically about 0.1 μM or 10^4 times lower than in the surrounding medium. This is maintained by the dynamic balance between Ca^{2+} expulsion across the plasma membrane (predominantly by the Ca^{2+}-ATPase discussed in Section 4.3.2) and a constitutive inward leak. It should be evident that only the expulsion of the ion from the cell can regulate $[Ca^{2+}]$ in the long-term, although as we shall see (*see* Section 5.8.2) the temporary accumulation and release of Ca^{2+} from internal stores plays an important role in intracellular signalling, while the mito-

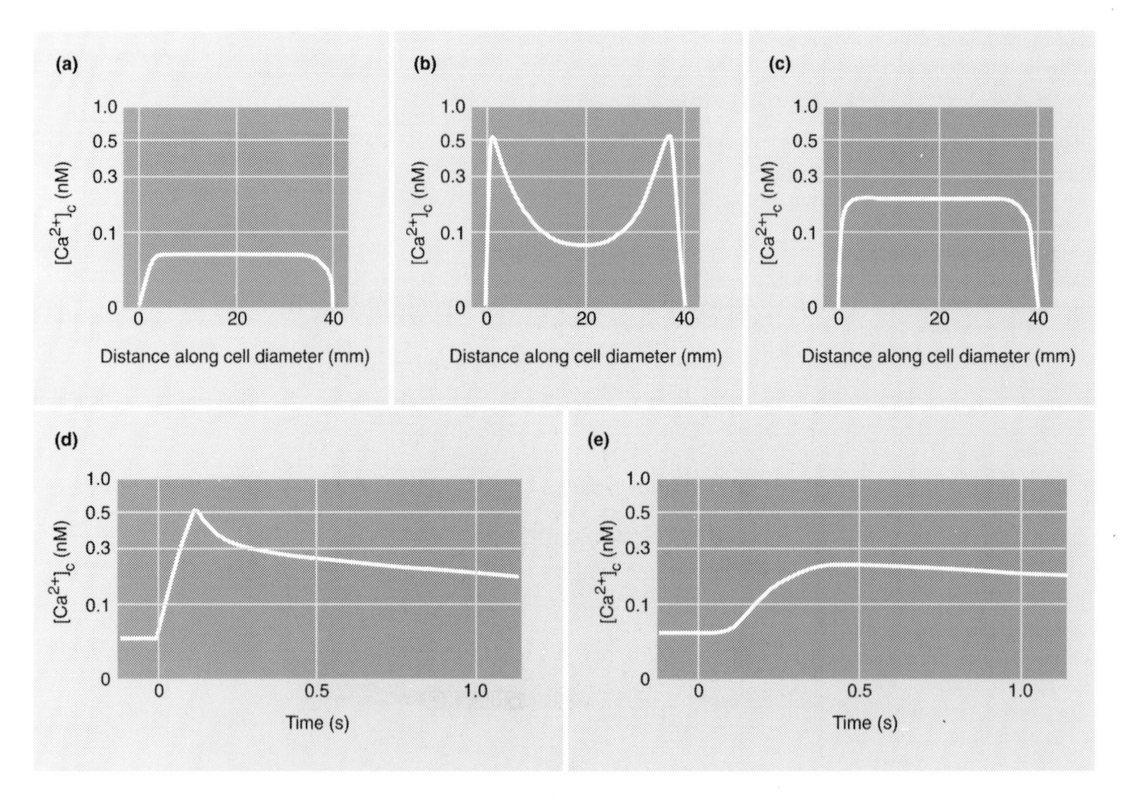

Fig. 5.16 Confocal microscopy and the kinetics of Ca^{2+} diffusion in a single neurone. A dissociated bullfrog sympathetic neurone was patch clamped in whole cell mode (resting potential −70 mV) and loaded with the long wavelength Ca^{2+} indicator fluo-3. (a) The cell was repetitively scanned across its diameter with the focused beam of an argon laser. (b) The cell was depolarized by a 100 ms step to +10 mV; the profile of $[Ca^{2+}]_c$ shown is that obtained at the end of this period, note the gradient of Ca^{2+} from the plasma membrane towards the centre. (c) One second after restoring the resting potential the gradient has decayed and $[Ca^{2+}]_c$ is uniformly elevated. (d, e) Time courses of $[Ca^{2+}]_c$ elevation under the plasma membrane and at the centre of the cell respectively. Data from Hernandez-Cruz *et al.* (1990).

chondria may be important in protecting the cell against temporary pathogenic Ca^{2+} overloads.

If the activity of the pump is sensitive to $[Ca^{2+}]_c$, then a steady-state cycling allows the cell to stabilize the resting $[Ca^{2+}]_c$ at a *set-point* at which the activity of the efflux pathway balances that of the inward leak. The resting $[Ca^{2+}]_c$ now is self-regulating since any displacement from the set-point by a transient influx of Ca^{2+} increases the activity of the pump above that of the leak until the set-point is reattained. Little is known about this constitutive inward leak of Ca^{2+} across the neuronal membrane. However, its presence can be deduced from the steady increase in $[Ca^{2+}]_c$ which is seen when cytoplasmic ATP is depleted.

The fura-2 signal in response to a clamped depolarization depends on the nature of the Ca^{2+}-channels which are activated. A transiently activated Ca^{2+}-channel shows a spike followed by a decline to a plateau (Fig. 5.17) as Ca^{2+} is pumped out of the cell or into internal stores. A non-inactivating Ca^{2+}-channel (such as the L-channel) in contrast shows a maintained plateau.

5.8 Calcium- and inositol phosphate-induced Ca^{2+} release

Internal non-mitochondrial Ca^{2+} stores contribute to Ca^{2+} signalling within the neurone as in many other cells (stage VII of Fig. 5.1). The neuronal cell body and dendritic tree, but not most terminals, contain cisternae formed from smooth endoplasmic reticulum, which accumulate Ca^{2+} by a Ca^{2+}-ATPase distinct from that at the plasma membrane since it is insensitive to calmodulin and probably transports $2Ca^{2+}$ per ATP hydrolysed (Fig. 5.18). This Ca^{2+}-ATPase can be selectively inhibited by *thapsigargin* resulting in depletion of the internal stores. Ca^{2+} storage by the cisternae is enhanced by the presence in their lumen of a high-capacity, low affinity Ca^{2+}-chelating protein related to the calsequestrin found in muscle sarcoplasmic reticulum.

Two types of cisternae can be distinguished containing either an *(1,4,5)IP$_3$ receptor* or a *ryanodine receptor* (Fig. 5.18). A family of (1,4,5)IP$_3$ receptors has now been defined, derived both

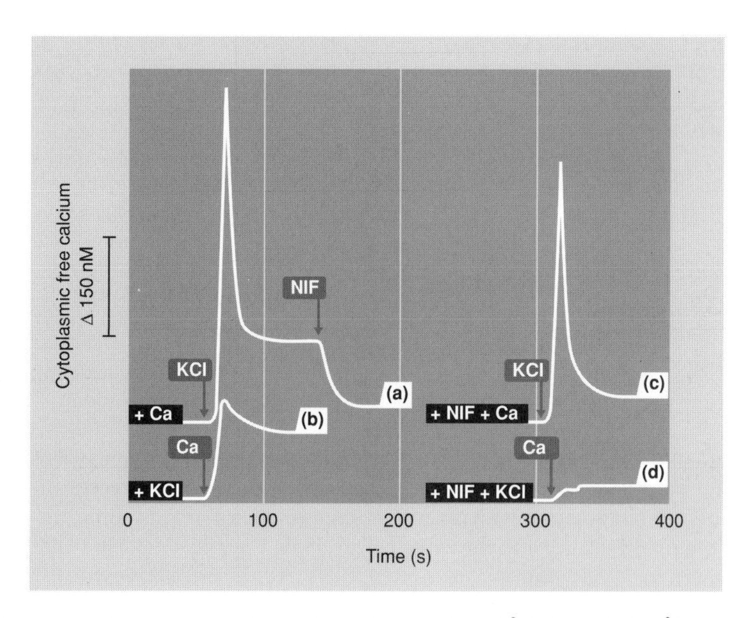

Fig. 5.17 Fura-2 responses due to inactivating and non-inactivating Ca^{2+}-channels. $[Ca^{2+}]_c$ was determined for a population of cerebellar granule cells cultured on a coverslip. (a) Response to 30 mM KCl showing a transient spike generated by an inactivating Ca^{2+} conductance and a steady plateau largely sensitive to the L-channel inhibitor nifedipine (NIF). (b) Depolarization prior to the addition of Ca^{2+} abolishes the transient component demonstrating that it is sensitive to voltage inactivation. (c) The transient component is insensitive to NIF. (d) The combination of predepolarization and NIF leaves only a slight residual $[Ca^{2+}]_c$ response. Data from Courtney *et al.* (1990).

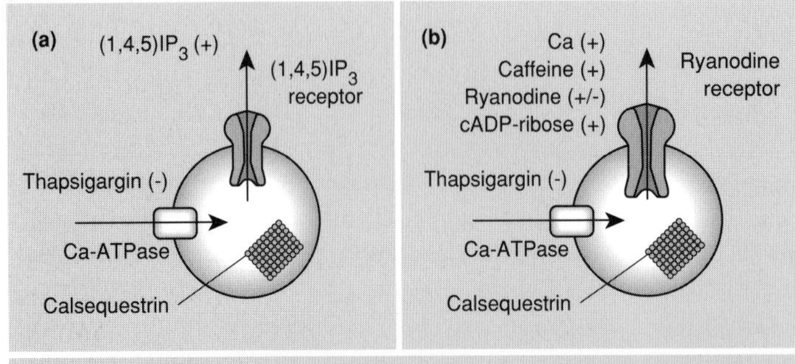

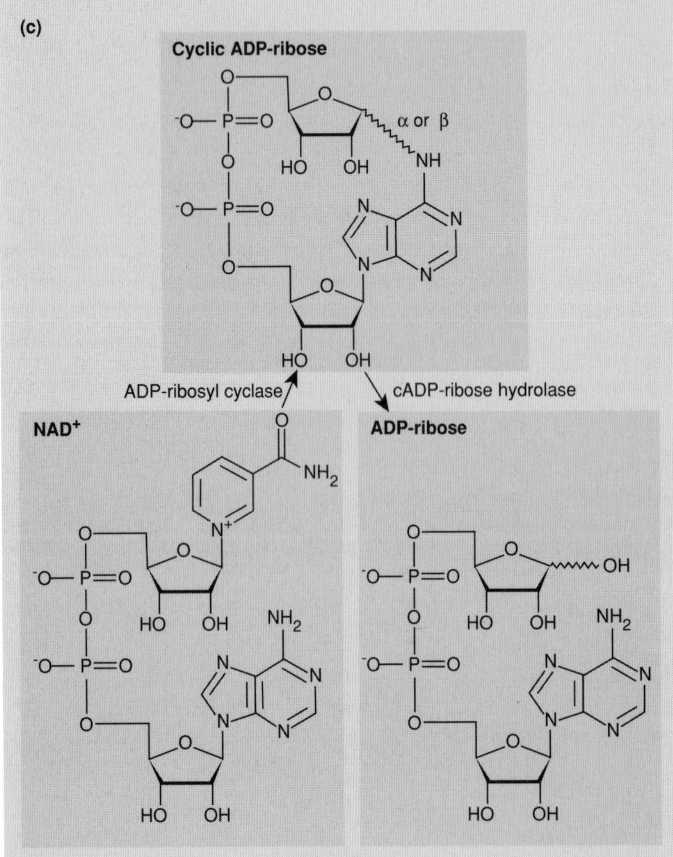

Fig. 5.18 Intracellular Ca^{2+} stores. (a) $(1,4,5)IP_3$-sensitive Ca^{2+} stores. Thapsigargin inhibits the Ca^{2+}-ATPase and prevents Ca^{2+} accumulation; calsequestrin-related proteins provide high-capacity, low-affinity binding sites. (b) Ca^{2+}-releasable Ca^{2+} stores possess the larger ryanodine receptor which can be activated synergistically by Ca^{2+} and cADP-ribose as well as *in vitro* by caffeine and ryanodine (which also inhibits at low concentration). (c) cADP-ribose metabolism.

from separate genes and by alternative splicing, which show distinct patterns of expression. The proteins have some 2700 amino acids with a very large N-terminal domain projecting into the cytoplasm and with seven or eight putative transmembrane domains close to the C-terminus. The receptor associates as a tetramer in the membrane and the resulting channel is activated by the binding of $(1,4,5)IP_3$ to a cytoplasmic site close to the N-terminus. Reconstituted channels in lipid bilayers show a transient (10 ms) opening of a *c.* 20 pS channel on binding $(1,4,5)IP_3$. Phosphorylation of two serine residues on the cerebellar $(1,4,5)IP_3$ receptor by PKA can uncouple $(1,4,5)IP_3$ binding from Ca^{2+}-channel opening, while heparin and elevated (1 μM) Ca^{2+} both inhibit $(1,4,5)IP_3$ binding to the receptor.

In the second class of cisternae ryanodine receptor Ca^{2+}-channels (Fig. 5.18) replace the $(1,4,5)IP_3$ receptor. RYR1 and RYR2 have been characterized in muscle, where they function as sarcoplasmic reticulum Ca^{2+} release channels. Neuronal ryanodine receptors resemble the cardiac RYR2. The receptors are enormous proteins of 500 kDa (some 5000 amino acid residues), showing considerable homology with the (1,4,5)IP_3 receptor in their C-terminal domains. They assemble as large homotetramers with a characteristic square appearance in the electron microscope. Nanomolar concentrations of ryanodine (a plant alkaloid) open the channels, while higher concentrations close them. The channel can also be activated by *caffeine* at high millimolar concentrations.

The physiological activation mechanism for the ryanodine receptor is currently debated. By analogy to skeletal muscle, the brain ryanodine receptor might be closely coupled to the plasma membrane, allowing changes in plasma membrane potential to directly trigger internal release. Alternatively there is evidence that increased cytoplasmic Ca^{2+} triggers release from the cisternae, and this *Ca^{2+}-induced Ca^{2+} release* may account for the propagation and maintenance of waves of $[Ca^{2+}]_c$ which are observed particularly in astrocytes and which may involve an interchange of Ca^{2+} between (1,4,5)IP_3 releasable stores and caffeine-sensitive stores. Recently a novel second messenger *cyclic ADP-ribose* (cADPR) (Fig. 5.19) has been described in a number of cells, including dorsal root ganglion neurones. cADPR is synthesized from NAD^+ by the enzyme ADP-ribosyl cyclase. Increases in concentration of the messenger have been detected in pancreatic β-cells during secretion, and addition of cADPR to permeabilized cells releases Ca^{2+} from ryanodine-sensitive, (1,4,5)IP_3-insensitive stores. cADPR and Ca^{2+} may act synergistically in releasing caffeine-sensitive Ca^{2+} stores.

Activation of the (1,4,5)IP_3 or ryanodine receptors in intact cells leads to a characteristic transient elevation of $[Ca^{2+}]_c$ which is independent of external Ca^{2+} followed by a plateau elevation which is maintained for the duration of the stimulus (Fig. 5.19). Although the spike is seen in the absence of external Ca^{2+} this is needed for the pools to refill after emptying. The release of Ca^{2+} is not all-or-none but seems to be *quantized*; the proportion released depending on the precise level of (1,4,5)IP_3. This could either be because the receptor sensitivity depends on the Ca^{2+} concentration within the intracellular store, or because a spectrum of receptor subtypes might exist differing in their responsiveness to (1,4,5)IP_3.

5.8.1 Inositol phosphates and Ca^{2+} entry

The sustained plateau elevation in response to inositol phosphate generation is abolished by performing the experiment in the absence of external Ca^{2+} (Fig. 5.19) and is associated with entry of Ca^{2+} across the plasma membrane. There is controversy both as to the signal which initiates this Ca^{2+} entry and the nature of the *second messenger-activated channel (SMAC)* which mediates it. (1,4,5)IP_3 (or the (1,3,4,5)IP_4 described below) could activate a plasma membrane SMAC; there could be a parallel coupling between a receptor-activated G-protein and a SMAC of the type described above for K^+- and voltage-activated Ca^{2+}-channels. Finally there could be some signalling between the internal Ca^{2+} stores and the SMAC allowing the plasma membrane to sense the emptying internal store and activate Ca^{2+} entry via the SMAC (Fig. 5.19).

There is some evidence in favour of this last hypothesis, since the emptying of internal stores by the Ca^{2+}-ATPase inhibitor thapsigargin promotes entry across the plasma membrane. How does this signalling operate? One possibility is that the internal stores are closely linked to the plasma membrane, such that the secondary phase of SMAC-mediated entry occurs directly into these stores rather than the bulk cytoplasm. This direct coupling is not now favoured, since it appears possible for the internal stores to be refilled from the cytoplasm rather than via a plasma membrane SMAC. Instead one is left with the need to identify the message by which the state of filling of the internal stores is signalled to the plasma membrane.

(1,4,5)IP_3 may be phosphorylated to (1,3,4,5)-IP_4 by a kinase which can be activated by Ca^{2+}/calmodulin under conditions where $[Ca^{2+}]_c$ is elevated (*see* Fig. 5.6). This suggests that Ca^{2+} released by (1,4,5)IP_3 might promote formation of (1,3,4,5)IP_4; this can be observed in rat cerebral cortical slices where muscarinic stimulation increases (1,3,4,5)IP_4 accumulation. The role of (1,3,4,5)IP_4 is contentious; there have been suggestions that it may promote refilling of (1,4,5)IP_3-sensitive Ca^{2+} stores (Fig. 5.19) or may directly activate plasma membrane SMACs.

The accumulation of Ca^{2+} by internal stores must be accompanied by the cotransport of an anion in order to maintain electroneutrality

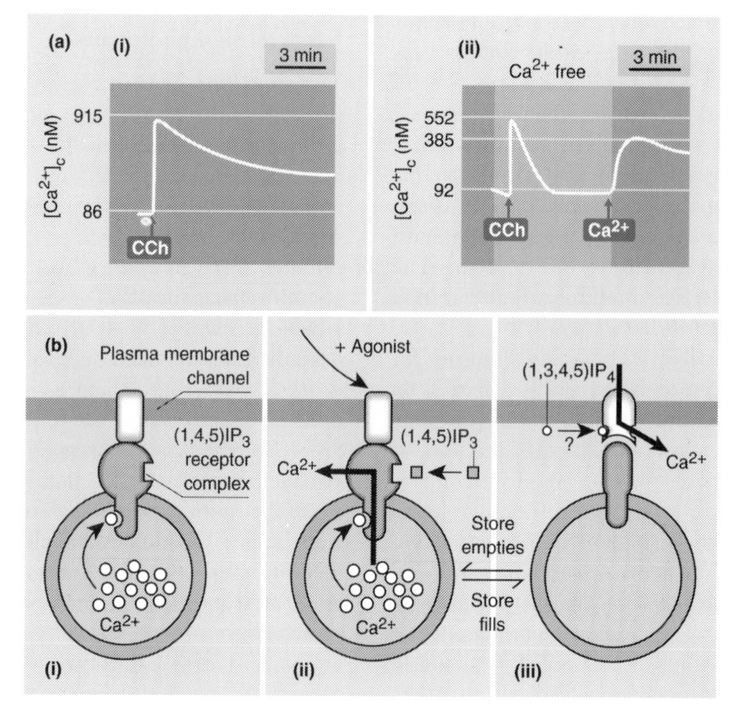

Fig. 5.19 Two phases of $(1,4,5)IP_3$ generated Ca^{2+} elevation. (a) Response of a neuroblastoma cell to carbachol (CCh), a muscarinic PLC-coupled agonist (i) in the presence of external Ca^{2+} addition of agonist gives a spike followed by a long-lasting plateau; (ii) in the initial absence of external Ca^{2+} a spike is still seen (due to release from internal stores) but this decays to baseline. Readdition of external Ca^{2+} in the continued presence of agonist gives a separate plateau phase. Adapted from Fatatis *et al.* (1992). (b) Conformational/capacitative model for $(1,4,5)IP_3$-induced Ca^{2+} entry in which conformational coupling is envisaged between $(1,4,5)IP_3$ receptor and a plasma membrane Ca^{2+}-channel. (i) In the resting state the store is full and internal Ca^{2+} maintains the conformation of the receptor such that it does not activate the plasma membrane channel. (ii) Agonist-generated $(1,4,5)IP_3$ releases the Ca^{2+} store generating a spike of internal Ca^{2+}. (iii) In the empty store the $(1,4,5)IP_3$ receptor no longer has Ca^{2+} bound to its lumenal domain so that it undergoes a conformational change which is transmitted to the plasma membrane channel allowing entry of external Ca^{2+}. Adapted from Berridge (1990) and Putney (1992).

across the membrane; conversely the efflux of Ca^{2+} would be accompanied by the parallel efflux of the anion. While the nature of this anion is unclear, it could clearly act as the elusive 'second messenger' signalling the state of filling of the internal stores. Evidence in support of this has come from the observation that a low molecular weight anionic factor liberated into the cytoplasm of a lymphocyte cell line when internal stores are depleted is capable of inducing Ca^{2+} uptake when added to a variety of cells.

5.8.2 Calcium oscillations and local gradients

The advent of cell imaging techniques which allow the continuous monitoring of $[Ca^{2+}]_c$ in individual cells has revealed the enormous complexity of intracellular Ca^{2+} signalling. Highly localized gradients can be formed in discrete areas of the cell and waves or transient oscillations can be propagated from one part of the cell to another. One particular aspect of this spatial heterogeneity — the highly localized Ca^{2+} elevation which triggers small synaptic vesicle exocytosis — will be reserved for Section 7.3; here we shall discuss the gradients which can be detected in neuronal soma and dendrites as well as in astrocytes.

Three factors allow the establishment and maintenance of large gradients of free Ca^{2+} in neurones and glia. Firstly at any one instant the vast majority of cytoplasmic Ca^{2+} is chelated by relatively immobile anionic groups on phospholipids and proteins. Thus a total Ca^{2+} concentration in the cytoplasm of the order of 1 mM results in a free concentration of less than 1 μM. Any gradient of free Ca^{2+} concentration will therefore be propagated very slowly — just as Ca^{2+} would travel very slowly down a cation-exchange column, spending most of its time bound to the stationary matrix. The second factor which aids the establishment of local gradients in neurones is the highly elongated geometry of the cell. For example, the postsynaptic membrane of most glutamatergic synapses is arranged as a dendritic spine (*see* Fig. 5.21) with a neck which will limit diffusion of Ca^{2+} and facilitate the establishment of a high local $[Ca^{2+}]_c$ within the spine.

The third factor in the generation and maintenance of $[Ca^{2+}]_c$ gradients is the presence of intracellular organelles capable of the reversible uptake and release of Ca^{2+}, as discussed above. Oscillations in individual cells and waves of elevated $[Ca^{2+}]_c$ moving through a syncitium of cells linked by gap junctions are particularly prominent in astrocytes, where they can be evoked by mechanical pressure or metabotropic glutamate receptor activation, although extensive coupling via gap junctions can also be detected in neurones during the period of circuit formation (prior to day 16 in the rat).

What then initiates and maintains these oscillations? There are currently two theories: *receptor-controlled models* propose that $(1,4,5)IP_3$ levels oscillate due to some periodicity prior to $(1,4,5)IP_3$ synthesis, such as a reversible PKC-mediated inhibition of the receptor or G-protein, while *second messenger-controlled models* propose that $[Ca^{2+}]_c$ can oscillate in response to a constant level of $(1,4,5)IP_3$, perhaps as an interplay between recycling $(1,4,5)IP_3$-releasable Ca^{2+} stores and Ca^{2+}-releasable Ca^{2+} stores (Fig. 5.20). It should be noted that both stores respond to their lumenal content of Ca^{2+}, becoming more sensitive to their respective releasing agents as they fill; this would evidently help to maintain an oscillatory cytoplasmic Ca^{2+} concentration.

5.8.3 Ca^{2+} responses in dendritic spines

Both ryanodine receptors and $(1,4,5)IP_3$ receptors are extraordinarily enriched in the Purkinje cells of the cerebellum, particularly associated with stacks of cisternae present in the dendritic spines (Fig. 5.21). The PLC-coupled metabotropic glutamate receptor (*see* Section 9.7) is also present in the adjacent plasma membrane, and glutamate released from the parallel fibres of synapsing granule cells produces a large localized increase in $[Ca^{2+}]_c$ which can be visualized by fura-2 image analysis. The strong indication is therefore that the $[Ca^{2+}]_c$ increase is $(1,4,5)IP_3$-mediated. Other components of the PI cycle, including PKC and a specific phospholipase C isozyme PI-phospholipase C are also present at high concentration.

A similar highly localized Ca^{2+} elevation is seen within dendritic spines of hippocampal neurones during synaptic transmission. Surprisingly $[Ca^{2+}]_c$ can remain elevated for several minutes after the termination of transmission and after any $[Ca^{2+}]_c$ response within the dendritic shaft has decayed. Ca^{2+}-induced Ca^{2+} release involving the ryanodine receptor (see above) could provide a regenerative means of prolonging the Ca^{2+} elevation, although it is not clear whether the appropriate spines possess this class of cisternae. The response could be initiated either by the glutamate metabotropic receptor generating $(1,4,5)IP_3$ as discussed above, or by a direct elevation of $[Ca^{2+}]_c$, for example, via the NMDA receptor (*see* Section 9.6.3).

A possible buffer of cytoplasmic Ca^{2+} under these conditions is the mitochondrion. Brain mitochondria possess two Ca^{2+} transport pathways; a Ca^{2+} uniporter accumulating Ca^{2+} in response to the 180 mV of interior negative membrane potential and a $Ca^{2+}/2Na^+$ exchanger which allows Ca^{2+} efflux, completing a Ca^{2+} circuit across the membrane (Fig. 5.22). Since the activity of the uniporter increases strongly with $[Ca^{2+}]_c$, uptake and efflux balance at a *set-point* between 0.5 and 1 μM. Any elevation of $[Ca^{2+}]_c$ above this level would be rapidly sequestered by the mitochondria and Ca^{2+} would be released when the $[Ca^{2+}]_c$ fell. This mechanism may play a vital role in protecting neurones and other cells against damage caused by excessive $[Ca^{2+}]_c$, but would also function in the present context to smooth and prolong a sudden Ca^{2+} transient in the dendritic spine.

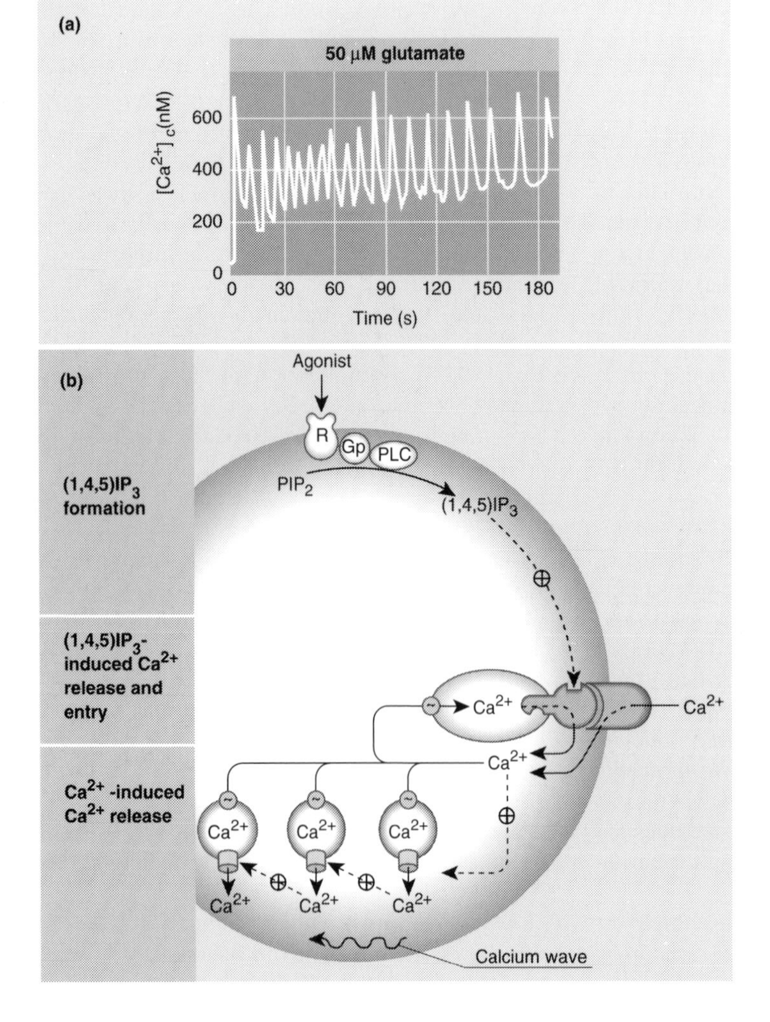

Fig. 5.20 Models for the initiation and maintenance of Ca^{2+} oscillations. (a) $[Ca^{2+}]_c$ oscillations are frequently observed in glia where they can be induced by a variety of stimuli. In this case oscillations are induced in a primary culture of rat glial cells by the addition of 50 μM glutamate, probably activating a PLC-coupled metabotropic receptor. Data from Charles *et al.* (1991). (b) Two-pool model for Ca^{2+} oscillations. $(1,4,5)IP_3$-evoked release of Ca^{2+} from $(1,4,5)IP_3$-sensitive stores and consequent uptake into the cell initiates a wave of Ca^{2+}-induced Ca^{2+} release from Ca^{2+}-sensitive stores. Adapted from Berridge (1990).

5.9 Ca^{2+}-modulated enzymes

The central role played by Ca^{2+} in the neurones is to trigger exocytosis. This will be discussed in detail in Chapter 7. However in addition distinct temporal and spatial patterns of Ca^{2+} concentration throughout the neurone can activate PKC isoenzymes and Ca^{2+}-calmodulin-dependent ion pumps, protein kinases and phosphatases, modulate ion channel activity, synaptic vesicle recruitment, neurite extension, growth cone motility and the induction of the persistent changes in synaptic transmission which occur in learning and can be recalled as memory (Fig. 5.23). The effects of these changes in $[Ca^{2+}]_c$ will be the central topic of much of the remainder of this book.

5.9.1 Calmodulin and Ca^{2+}/calmodulin-dependent effectors

The ubiquitous 15 kDa Ca^{2+}-binding regulatory protein calmodulin (CaM) is present in brain cytoplasm at 30–50 μM. Calmodulin binds to the dephosphorylated form of two major PKC substrates: the 87 kDa MARCKS protein (*see* Section 7.5.2) and GAP-43, also called B50 (*see* Section 7.5.1). When these proteins are phosphorylated by PKC their affinity for calmodulin decreases. While this control of free calmodulin availability can be demonstrated *in vitro* it has not yet been shown whether this is a significant regulatory mechanism *in situ*.

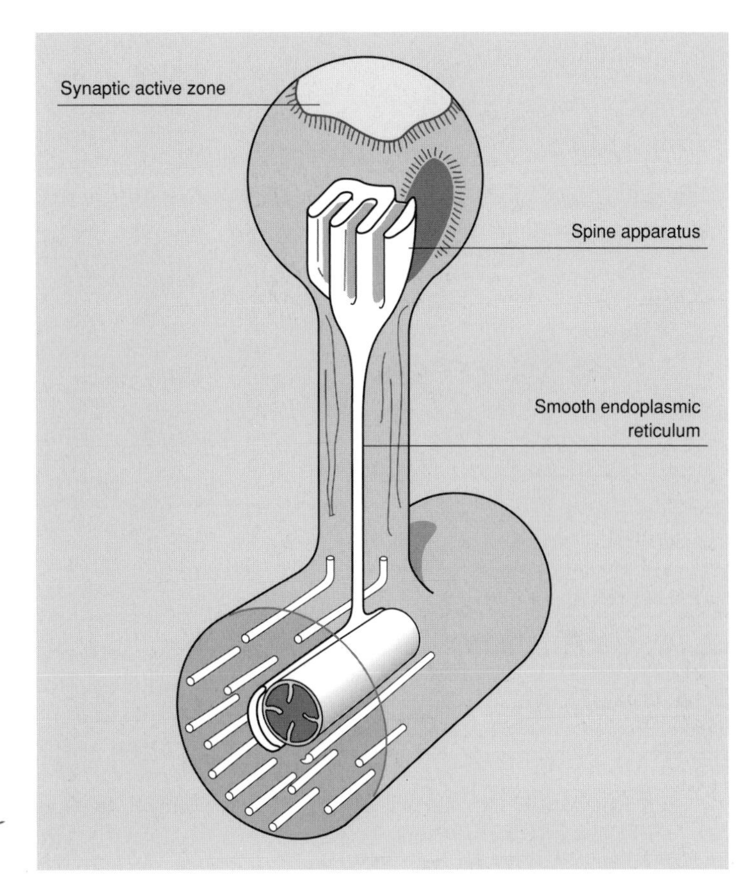

Fig. 5.21 The spine apparatus. Complex internal membrane specializations are found in dendritic spines. The spine apparatus consists of multiple flattened cisternae, separated by plates of dense material and is found in most large mushroom-shaped spines with large postsynaptic densities. Those in Purkinje cell spines contain high concentrations of (1,4,5)IP_3 receptors. The spine apparatus connects through a tubule with the smooth endoplasmic reticulum in the dendritic trunk. Adapted from Couteaux & Spacek (1985).

5.9.2 Ca^{2+}/calmodulin-dependent protein kinases

Type II Ca^{2+}*/calmodulin-dependent protein kinase (CaMKII)* is the predominant Ca^{2+}-dependent serine/threonine protein kinase in neurones and accounts for about 1% of total brain protein (Fig. 5.24). The kinase is also known as *multifunctional* Ca^{2+}*/calmodulin-dependent protein kinase*, reflecting its broad substrate specificity. CaMKII is a 500–600 kDa oligomer with approximately ten subunits made up predominantly of 54 kD α-subunits, with additional 60 kDa β-, β′- (alternatively spliced), γ- or δ-subunits. In addition to a catalytic region, each subunit possesses a regulatory sequence consisting of autoinhibitory and calmodulin-binding domains. Calmodulin binding neutralizes the effect of the autoinhibitory domain. The catalytic properties of individually expressed α- and β-subunits are similar to each other and to the enzymes as purified from brain. Thus a heteromeric oligomer is not required for activity.

An early event, which can precede phosphorylation of exogenous substrates, is the autophosphorylation of CaMKII on Thr^{286} (rat enzyme) of the α-subunit. This process is initially Ca^{2+}-dependent, but once a threshold level of phosphorylation is reached, further phosphorylation continues even in the presence of EGTA. Eventually up to 30 mol of phosphate can be incorporated by this means into a CaMKII holoenzyme (three per subunit). The phosphatases responsible for *in vivo* dephosphorylation of the autophosphorylated enzyme are currently uncharacterized. It is thought that such a stable switch to a Ca^{2+}-independent form of CaMKII might underlie some of the long-term changes observed during synaptic plasticity (*see* Section 13.5).

The mRNAs for the α-, β- and β′-subunits are most concentrated in the nervous system. The

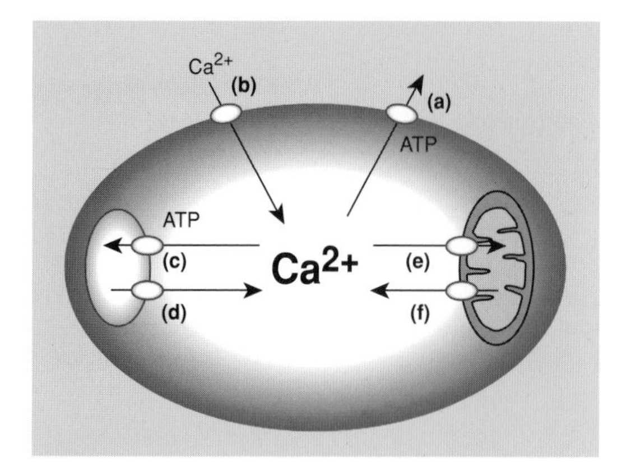

Fig. 5.22 Interactions between plasma membrane, mitochondria and non-mitochondrial stores in Ca^{2+} regulation. The plasma membrane achieves a steady-state at a $[Ca^{2+}]_c$ 'set-point' when Ca^{2+} is pumped out of the cytoplasm by the Ca^{2+}-ATPase (a) (which is highly sensitive to $[Ca^{2+}]_c$) at a rate which balances a constitutive inward 'leak' (b). This occurs typically at about 0.1 μM $[Ca^{2+}]_c$. Non-mitochondrial stores have a high affinity Ca^{2+}-ATPase (c) but a limited capacity to store the cation. Refilling of these stores after depletion (d) is facilitated by an elevated $[Ca^{2+}]_c$. Mitochondria have a uniport for Ca^{2+} uptake (e) (driven by the interior negative membrane potential and again highly sensitive to $[Ca^{2+}]_c$) and a constitutive outward leak (f). The set-point for mitochondrial Ca^{2+} transport is about 1 μM. Mitochondria have an enormous capacity to accumulate Ca^{2+} above their set-point as a complex with phosphate. This may be important in protecting neurones against pathological Ca^{2+} overload and in assisting a prolonged $[Ca^{2+}]_c$ elevation which can be observed, e.g. in dendritic spines.

α- and β-subunits are differentially distributed; for example in the cerebellum the α-subunit is expressed in Purkinje cell but not in granule cells. The most spectacular subcellular concentration of the enzyme is in postsynaptic densities (PSD) where α-subunits can account for 20–40% of the total protein, although its physiological substrates are not known. Specific binding proteins in the postsynaptic density may exist to anchor this soluble enzyme.

CaMKII copurifies with synaptosomes during subcellular fractionation. PSDs are frequently still attached to synaptosomes (*see* Fig. 2.5) and at least half of this activity is accessible to added ATP, suggesting its location on the broken postsynaptic membrane. Within the presynaptic cytoplasm some CaMKII activity is associated with synaptic vesicles. A major presynaptic target for type II CaM-kinase is *synapsin I* (*see* Section 7.4.6) as is tyrosine hydroxylase in catecholaminergic terminals. Other neuronal substrates include tubulin, and the microtubule-associated proteins *tau* and *MAP2*. Tau is of particular interest in view of the presence of abnormal patterns of tau phosphorylation in the neurofibrillary tangles isolated from the brains of Alzheimer's patients.

One approach to establish the role of an enzyme such as CaMKII is to examine the phenotype resulting from inhibition or deletion of the enzyme. Despite the widespread distribution of CaMKII, transgenic mice lacking the gene for the α-subunit of the enzyme not only

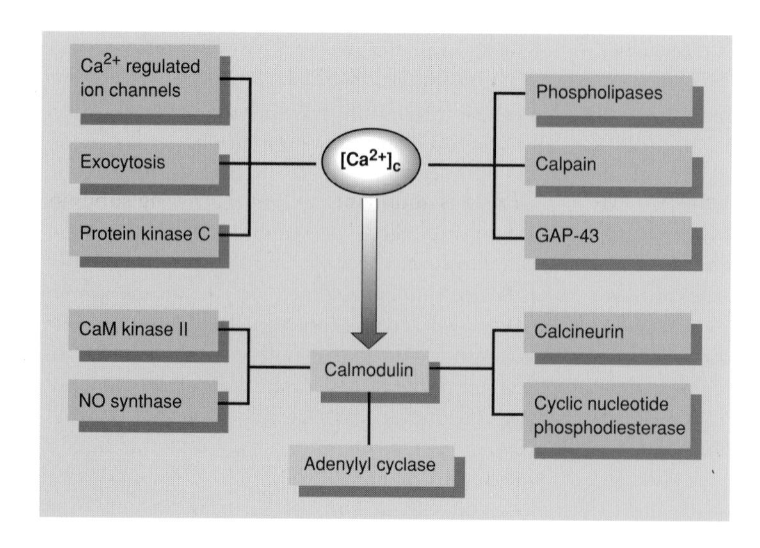

Fig. 5.23 Some of the major effectors regulated by $[Ca^{2+}]_c$ in neurones. An elevation in $[Ca^{2+}]_c$ due to activation of plasma membrane ion channels or release of internal Ca^{2+} stores can affect a large number of targets in addition to exocytosis itself: these include kinases, phosphatases, phospholipases, proteases, channels, etc.

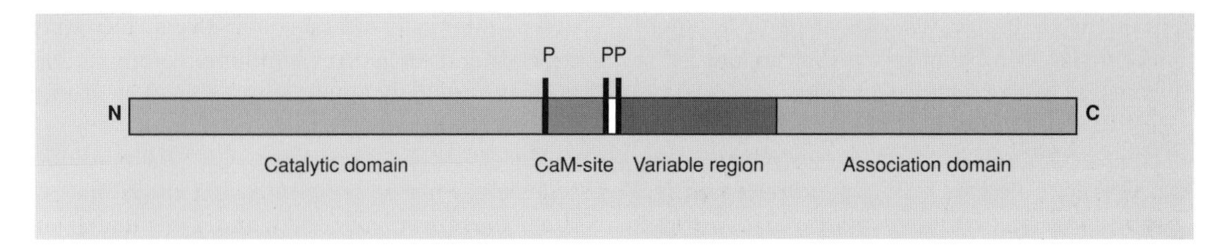

Fig. 5.24 Ca^{2+}/calmodulin (CaM)-stimulated protein kinase II. Major features common to the primary sequence of CaMKII subunits. The variable region accounts for most of the differences between the subunits. The association domain is involved in the binding of the enzyme to itself to form multimers and to target proteins.

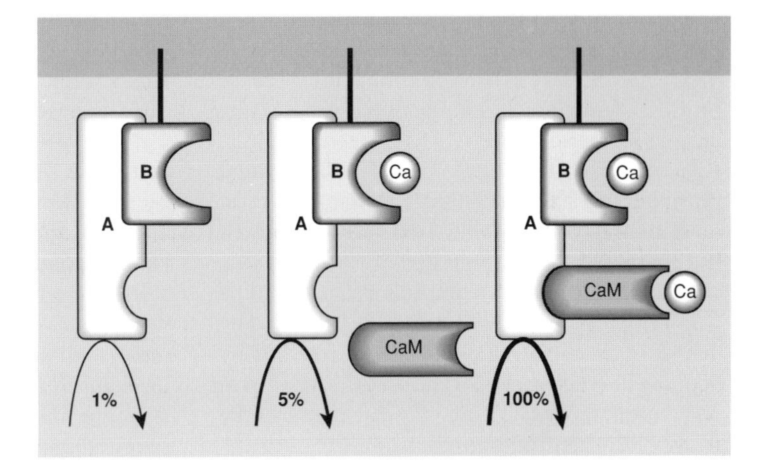

Fig. 5.25 Calcineurin, a Ca^{2+}-dependent phosphatase. The dimeric (A, B) enzyme is anchored to the membrane by a myristoyl moiety on the B-subunit. In the absence of Ca^{2+} the enzyme shows only 1% of maximal activity hydrolysing a protein phosphate (Pr-P). The binding of 4 mol of Ca^{2+} to the B-subunit only increases the activity of the catalytic A subunit to 5%. The further binding of Ca^{2+}-calmodulin (CaM) to the A-subunit gives the fully active form. Adapted from Armstrong (1989).

survive but develop normally, although they display difficulties in learning novel tasks. Protein kinase inhibitors can be used to investigate the functions of their substrates in intact terminals and cells. However, while an inhibitor may be specific when tested against a range of purified kinases, in the complex milieu of the cell other, uncharacterized, kinases may be inhibited, or bioenergetic functions disturbed, which if unrecognized can lead to false conclusions. A classic example is the *in vitro* calmodulin inhibitor trifluoroperazine, which in intact terminals proves to be a potent mitochondrial uncoupler, inhibiting transmission by collapsing ATP levels. The role of calmodulin-dependent events may also be investigated by exploiting Ba^{2+}, which does not activate calmodulin but can substitute for Ca^{2+} in a variety of processes including many forms of exocytosis (*see* Chapters 7 and 8).

5.9.3 Brain protein phosphatases

The level of phosphorylation of a protein is a dynamic balance between the activity of protein kinases and phosphatases. The major neutral serine/threonine phosphatases are classified as PrP-1, PrP-2A and PrP-2B (or calcineurin). Only the last of these is calmodulin-dependent, but it is appropriate to consider each phosphatase at this point.

PrP-1 is abundant in brain. Targeting subunits which couple PrP-1 to specific substrates may be responsible for the association of the phosphatase with brain particulate fractions. Two endogenous inhibitors I-1 and I-2 control PrP-1 activity. Certain brain regions contain an additional PrP-1 inhibitor, *DARPP-32* (dopamine and cAMP regulated phosphoprotein) which is abundant in basal ganglia, neurones containing dopamine D_1 receptors and cerebellar Purkinje cells. *PrP-2A* is cystosolic, stimulated by polycations and is abundant in brain. *Okadaic acid* is a potent inhibitor of PrP-2A and to a lesser extent PrP-1 but ineffective against calcineurin.

PrP-2B, calcineurin, is abundant in brain but has a narrow specificity. The phosphatase (Fig. 5.25) is activated by Ca^{2+}/calmodulin and is a heterodimer of two polypeptides; calcineurin-A (61 kDa), which exists in a number of isoforms, contains the catalytic and calmodulin binding sites while calcineurin-B (19 kDa) has Ca^{2+} binding sites homologous to those on calmodulin itself. The phosphatase may therefore be activated by binding both the Ca^{2+}/calmodulin complex and Ca^{2+} to the permanently attached B-subunit. This dual activation allows for a highly cooperative activation of the phosphatase by Ca^{2+} (compare the similar synergism of the plasma membrane Ca^{2+}-ATPase, Section 4.3.2). Calcineurin is the only known Ca^{2+}-activated phosphatase. The B-subunit is myristoylated on its N-terminal glycine which may serve to attach the complex to the membrane.

Calcineurin can dephosphorylate L-type Ca^{2+}-channels phosphorylated by PKA. The action of calcineurin may be either direct, or by dephosphorylating endogenous inhibitor proteins such as DARPP-32. Calcineurin can be irreversibly activated in the absence of Ca^{2+} or calmodulin by limited proteolysis of the calmodulin binding sites. *Calpains*, which are Ca^{2+}-activated proteases can accomplish this proteolysis, suggesting that this may have physiological relevance.

5.10 Ca^{2+}/calmodulin and nitric oxide

Nitric oxide (NO) is a free radical gas which rapidly combines with oxygen to give NO_2. In tissues its half-life is typically 5 s. NO is synthesized biochemically by the enzyme NO-synthase (NOS) which converts arginine into NO and citrulline (Fig. 5.26). NO was first investigated in a biochemical context as an 'endothelial relaxing factor' produced in endothelial cells and causing the relaxation of adjacent smooth muscle cells.

NOS has a widespread distribution in brain. Generally the enzyme is present in populations of interneurones rather than principle neurones. In the cerebellum, which has the highest concentration of NOS, the enzyme is expressed in basket and granule cells but not in Purkinje cells. In other brain regions the enzyme is very selectively expressed; in the cortex and hippocampus it has been reported that only some 2% of neurones express the enzyme.

The 150 kDa monomeric Ca^{2+}/calmodulin-requiring enzyme responds to elevations in $[Ca^{2+}]$ by producing NO and citrulline from arginine in an unusual five-electron oxidation (Fig. 5.27). The enzyme contains flavin, tetrahydrobiopterin and Fe, and can be inhibited by arginine derivatives including L-N^{G}-nitroarginine or L-N^{G}-methylarginine. NOS catalyses a partial reaction known as NADPH-diaphorase which produces a blue stain in the presence of nitroblue tetrazolium thus assisting the histochemical localization of the enzyme. In neurones NOS appears to be particularly sensitive to the Ca^{2+} elevations resulting from NMDA-receptor activation. A further regulation of NOS occurs by phosphorylation. PKC can inactivate NOS and this can be reversed by calcineurin, which being calmodulin-dependent will amplify the activation of NOS itself by $[Ca^{2+}]_c$.

NO is a curious 'neurotransmitter'. Being a membrane-permeant gas it requires no exocytotic mechanism nor any plasma membrane receptor. It is thus able to act as a local modulator whose range is limited by its half-life before conversion to NO_2.

The major target in the brain for NO is soluble *guanylyl cyclase* which also exists in membrane-bound forms. NO binds to the haem in this enzyme resulting in activation. cGMP regulates protein kinases, cyclic nucleotide phosphodiesterase and ion channels. In the cerebellum most NO is synthesized in the cell bodies of the granule cells in response to NMDA-receptor activation. Surrounding astrocytes respond with an increase in cGMP, and the mossy fibre terminals synapsing onto the granule cells also appear to respond with an increased cGMP. Thus, at least in the cerebellum, NO has some of the properties of a retrograde messenger feeding back from the postsynaptic to presynaptic membrane (*see* Section 13.5); however in the peripheral nervous system NO may be a more conventional forward-acting transmitter in certain non-adrenergic non-cholinergic synapses. The effects of NO can be blocked by haem which binds the gas with high affinity.

Soluble guanylyl cyclases are heterodimers and are *in vitro* targets not only for NO, but also

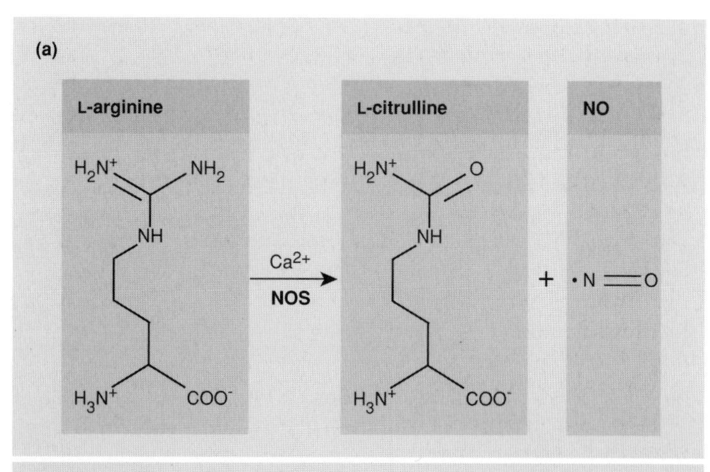

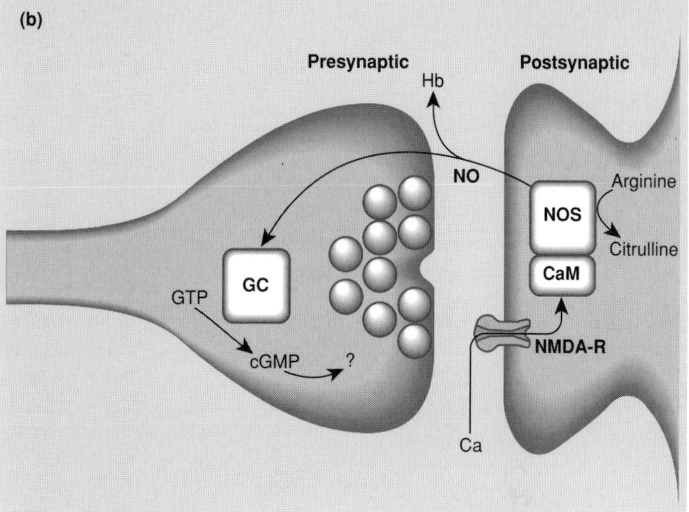

Fig. 5.26 Nitric oxide synthase (NOS) and cGMP. (a) NOS is a calmodulin-dependent enzyme activated by an increase in $[Ca^{2+}]_c$, e.g. via NMDA receptor (NMDA-R) activation. NOS generates NO from arginine. (b) The hypothesis that NO can act as a retrograde messenger proposes that NO diffusing to the presynaptic terminal activates guanylyl cyclase (GC) generating cGMP. Transcellular actions of NO can be inhibited *in vitro* by haemoglobin (Hb) which binds NO tightly. Specific actions of cGMP in the presynaptic terminal have yet to be demonstrated but general actions include ion channel activation (in retinal rod outer segments, *see* Section 6.2), activation of cGMP-dependent protein kinases, and both inhibition and activation of cAMP phosphodiesterases. Any GC in the NO-generating cell will not respond to the gas as GC is potently inhibited by an increase in $[Ca^{2+}]_c$. NO may also act on surrounding astrocytes.

for CO and the hydroxyl ion. In the cerebellum cGMP formation is greatest in Purkinje cells and glia. The cyclase is also present in granule cells, but since the enzyme is *inhibited* by elevated $[Ca^{2+}]$ this prevents cyclase in the same cell as the activated NOS from responding.

An additional target for NO is ADP-ribosyltransferase, which is activated by the gas, while excessive rates of NO synthesis *in vitro* lead to cytotoxicity, perhaps as a result of free radical OH formation. Whether a similar toxicity occurs *in vivo* in pathological conditions is unclear.

Further reading

Receptors and G-proteins

Bockaert, J. (1991) G proteins and G-protein-coupled receptors: structure, function and interactions. *Curr. Op. Neurobiol.* **1**, 32–42.

Hepler, J.R. & Gilman, A.G. (1992) G proteins. *Trends Biochem. Sci.* **17**, 383–387.

Hibert, M.F., Trumppkallmeyer, S., Bruinvels, A. & Hoflack, J. (1991) Three-dimensional models of neurotransmitter G-binding protein-coupled receptors. *Mol. Pharmacol.* **40**, 8–15.

Hille, B. (1992) G protein-coupled mechanisms and nervous signalling. *Neuron* **9**, 187–195.

Lefkowitz, R.J. (1992) G proteins: the subunit story thickens. *Nature* **358**, 372–372.

Lledo, P.M., Homburger, V., Bockaert, J. & Vincent, J.D. (1992) Differential G protein-mediated coupling of D_2 dopamine receptors to K^+ and Ca^{2+} currents in rat anterior pituitary cells. *Neuron* **8**, 455–463.

Milligan, G. (1983) Regional distribution and quantitative measurement of the phosphoinositidase C-linked guanine nucleotide binding proteins $G_{11\alpha}$ and $G_{q\alpha}$ in rat brain. *J. Neurochem.* **61**, 845–851.

Tota, M.R., Candelore, M.R., Dixon, R.A.F. & Strader, C.D. (1991) Biophysical and genetic analysis of the ligand-binding site of the β-receptor. *Trends Pharmacol. Sci.* **12**, 4–6.

PKA and cAMP

Carr, D.W., Stofko-Hahn, R.E., Fraser, I.D.C., Cone, R.D. & Scott, J.D. (1992) Localization of the cAMP-dependent protein kinase to the postsynaptic densities by *A-Kinase Anchoring Proteins*. Characterization of AKAP 79. *J. Biol. Chem.* **267**, 16 816–16 823.

Phospholipases A_2, C and D

Burgoyne, R.D., Cheek, T.R. & O'Sullivan, A.J. (1987) Receptor-activation of phospholipase A_2 in cellular signalling. *Trends Biochem. Sci.* **12**, 332–333.

Katz, A., Wu, D. & Simon, M.I. (1992) Subunits β-gamma of heterotrimeric G protein activate β2 isoform of phospholipase C. *Nature* **360**, 686–689.

Liscovitch, M. (1992) Crosstalk among multiple signal-activated phospholipases. *Trends Biochem. Sci.* **17**, 393–399.

Needleman, P., Turk, J., Jakshik, B.A., Morrison, A.R. & Lefkowith, J.B. (1986) Arachidonic acid metabolism. *Annu. Rev. Biochem.* **55**, 69–102.

Nishizuka, Y. (1992a) Hypercellular signalling by hydrolysis of phospholipids and activation of protein kinase C. *Science* **258**, 607–614.

Protein kinase C

Aderem, A. (1992) The MARCKS brothers: A family of protein kinase C substrates. *Cell* **71**, 713–716.

Asaoka, Y., Nakamura, S., Yoshida, K. & Nishizuka, Y. (1992) Protein kinase C, calcium and phospholipid degradation, *Trends Biochem. Sci.* **17**, 414–417.

Ase, K., Saito, N., Shearman, M.S., Kikkawa, U., Ono, Y., Igarashi, K., Tanaka, C. & Nishizuka, Y. (1988) Distinct cellular expression of βI and βII-subspecies of protein kinase C in rat cerebellum. *J. Neurosci.* **8**, 3850–3856.

Farago, A. & Nishizuka, Y. (1990) Protein kinase C in transmembrane signalling. *FEBS Lett.* **268**, 350–354.

Kanoh, H., Yamada, K. & Sakane, F. (1990) Diacylglycerol kinase: a key modulator of signal transduction? *Trends Biochem. Sci.* **15**, 47–50.

Liyanage, M., Frith, D., Livneh, E. & Stabel, S. (1992) Protein kinase C group B members PKC-δ, -ε, -ζ and PKC-L(η). Comparison of properties of recombinant proteins *in vitro* and *in vivo*. *Biochem. J.* **283**, 781–787.

Nishizuka, Y. (1992b) Signal transduction: Crosstalk. *Trends Biochem. Sci.* **17**, 367–367.

Saito, N., Itouji, A., Totani, Y., Osawa, I., Koide, H., Fujisawa, N., Ogita, K. & Tanaka, C. (1993) Cellular and intracellular localization of ε-subspecies of protein kinase C in the rat brain; presynaptic localization of the ε-subspecies. *Brain Res.* **607**, 241–248.

Suzuki, T., Okumura-Noji, K., Tanaka, R., Ogura, A., Nakamura, K., Kudo, Y. & Tada, T. (1993) Characterization of protein kinase C activities in postsynaptic density fractions prepared from cerebral cortex hippocampus and cerebellum. *Brain Res.* **619**, 69–75.

Wetsel, W.C., Khan, W.A., Merchenthaler, I., Rivera, H., Halpern, A.E., Phung, H.M., Negro-Vilar, A. & Hannun, Y.A. (1992) Tissue and cellular distribution of the extended family of protein kinase C isoenzymes. *J. Cell Biol.* **117**, 121–133.

Modulation of ion channels

Anwyl, R. (1991) Modulation of vertebrate neuronal calcium channels by transmitters. *Brain Res. Rev.* **16**, 265–281.

Bergamaschi, S., Govoni, S., Battaini, F., Trabucchi, M., Del Monaco, S. & Parenti, M. (1992) G protein modulation of omega-conotoxin binding sites in neuroblastoma X glioma NG 108–15 hybrid cells. *J. Neurochem.* **59**, 536–543.

Brown, A.M. & Birnbaumer, L. (1990) Ionic channels and their regulation by G protein subunits. *Annu. Rev. Physiol.* **52**, 197–213.

Brown, D.A. (1990) G-proteins and potassium currents in neurons. *Annu. Rev. Physiol.* **52**, 215–242.

Coffey, E.T., Sihra, T.S. & Nicholls, D.G. (1993) Protein kinase C and the regulation of glutamate exocytosis from cerebrocortical synaptosomes. *J. Biol. Chem.* **268**, 21 060–21 065.

Dolphin, A.C. (1990) G protein modulation of calcium currents in neurons. *Annu. Rev. Physiol.* **52**, 243–255.

Linden, D.J., Smeyne, M., Sun, S.C. & Connor, J.A. (1992) An electrophysiological correlate of protein kinase C isozyme distribution in cultured cerebellar neurons. *J. Neurosci.* **12**, 3601–3608.

Schultz, G., Rosenthal, W., Hescheler, J. & Trautwein, W. (1990) Role of G proteins in calcium channel modulation. *Annu. Rev. Physiol.* **52**, 275–292.

Shearman, M.S., Sekiguchi, K. & Nishizuka, Y. (1989) Modulation of ion channel activity: a key function of the protein kinase C family. *Pharmacol. Rev.* **41**, 211–237.

Sternweis, P.C. & Pang, I.H. (1990) The G-protein-channel connection. *Trends Neurosci.* **13**, 122–126.

Surprenant, A., Horstman, D.A., Akbarali, H. & Limbird, L.E. (1992) A point mutation of the α_2-

adrenoceptor that blocks coupling to potassium but not calcium currents. *Science* **257**, 977–980.

Sweeney, M.I. & Dolphin, A.C. (1992) 1,4-Dihydropyridines modulate GTP hydrolysis by G_o in neuronal membranes. *FEBS Lett.* **310**, 66–70.

Van Dongen, M.J., Codina, J., Olate, J., Mattera, R., Joho, R., Birnbaumer, L. & Brown, A.M. (1988) Newly identified brain potassium channels gated by the guanine nucleotide binding protein Go. *Science* **242**, 1433–1436.

$[Ca^{2+}]_c$ determination

Burgoyne, R.D. & Cheek, T.R. (1991) Locating intracellular Ca stores. *Trends. Biochem. Sci.* **16**, 319–320.

Tsien, R.Y. (1988) Fluorescence measurement and photochemical manipulation of cytosolic free calcium. *Trends Neurosci.* **11**, 419–424.

Tsien, R.Y. (1989) Fluorescent probes of cell signalling. *Annu. Rev. Neurosci.* **12**, 227–253.

Calcium homeostasis

Baimbridge, K.G., Celio, M.R. & Rogers, J.H. (1992) Calcium-binding proteins in the nervous system. *Trends Neurosci.* **15**, 303–308.

Grover, A.K. & Khan, I. (1992) Calcium pump isoforms: diversity, selectivity and plasticity. *Cell Calcium* **13**, 9–17.

Kennedy, M.B. (1989) Regulation of neuronal function by calcium. *Trends Neurosci.* **12**, 417–420.

Mayer, M.L. & Miller, R.J. (1990) EAA pharmacology: excitatory amino acid receptors, second messengers and regulation of intracellular Ca^{2+} in mammalian neurons. *Trends Pharmacol. Sci.* **11**, 254–260.

Pietrobon, D., Di Virgilio, F. & Pozzan, T. (1990) Structural and functional aspects of calcium homeostasis in eukaryotic cells. *Eur. J. Biochem.* **193**, 599–622.

Inositol phosphates

Berridge, M.J. (1993) Inositol trisphosphate and calcium signalling. *Nature* **361**, 315–325.

Downes, C.P. (1988) Inositol phosphates: a family of signal molecules. *Trends Neurosci.* **11**, 336–338.

Downes, C.P. & Macphee, C.H. (1990) Myo-inositol metabolites as cellular signals. *Eur. J. Biochem.* **193**, 1–18.

Fisher, S.K., Heacock, A.M. & Agranoff, B.W. (1992) Inositol lipids and signal transduction in the nervous system: an update. *J. Neurochem.* **58**, 18–38.

Internal Ca^{2+} stores and $[Ca^{2+}]_c$

Berridge, M.J. (1990) Calcium oscillations. *J. Biol. Chem.* **265**, 9583–9586.

Bleakman, D., Harrison, N.L., Colmers, W.F. & Miller, R.J. (1992) Investigations into neuropeptide Y-mediated presynaptic inhibition in cultured hippocampal neurones of the rat. *Br. J. Pharmacol.* **107**, 334–340.

Charles, A.C., Merrill, J.E., Dirksen, E.R. & Sanderson, M.J. (1991) Intercellular signaling in glial cells — calcium waves and oscillations in response to mechanical stimulation and glutamate. *Neuron* **6**, 983–992.

Clapham, D.E. (1993) Cellular calcium: a mysterious new influx factor. *Nature* **364**, 763–764.

Courtney, M.J., Lambert, J.J. & Nicholls, D.G. (1990) The interactions between plasma membrane depolarization and glutamate receptor activation in the regulation of cytoplasmic free calcium in cultured cerebellar granule cells. *J. Neurosci.* **10**, 3873–3879.

Couteaux, R. & Spacek, J. (1988) Specializations of subsynaptic cytoplasms. In *Cellular and Molecular Basis of Synaptic Transmission*, H. Zimmermann, ed. Berlin: Springer, pp. 25–50.

Enkvist, M.O.K. & McCarthy, K.D. (1992) Activation of protein kinase C blocks astroglial gap junction communication and inhibits the spread of calcium waves. *J. Neurochem.* **59**, 519–526.

Fatatis, A., Bassi, A., Monsurrò, M.R., Sorrentino, G., Mita, G.D., Di Renzo, G.F. & Annunziato, L. (1992) Lan-1: A human neuroblastoma cell line with M_1 and M_3 muscarinic receptor subtypes coupled to intracellular Ca^{2+} elevation and lacking Ca^{2+} channels activated by membrane depolarization. *J. Neurochem.* **59**, 1–9.

Ferris, C.D. & Snyder, S.H. (1992) Inositol phosphate receptors and calcium disposition in the brain. *J. Neurosci.* **12**, 1567–1574.

Galione, A. (1993) Cyclic ADP-ribose: a new way to control calcium. *Science* **259**, 325–326.

Garcia-Sancho, J., Alvarez, J., Montero, M. & Villalobos, C. (1992) Ca^{2+} influx following receptor activation. *Trends Pharmacol. Sci.* **13**, 12–13.

Garyantes, T.K. & Regehr, W.G. (1992) Electrical activity increases growth cone calcium but fails to inhibit neurite outgrowth from rat sympathetic neurons. *J. Neurosci.* **12**, 96–103.

Hallam, T.J. & Rink, T.J. (1989) Receptor-mediated Ca^{2+} entry: diversity of function and mechanism. *Trends Pharmacol. Sci.* **10**, 8–10.

Hernandez-Cruz, A., Sala, F. & Adams, P.R. (1990) Subcellular Ca transients visualized by confocal microscopy in a voltage-clamped vertebrate neuron. *Science* **247**, 858–862.

Irvine, R.F. (1989) How do inositol 1,4,5-trisphosphate and inositol 1,3,4,5-tetrakisposphate regulate intracellular Ca? *Biochem. Soc. Trans.* **17**, 6–9.

Meldolesi, J., Clementi, E., Fasolato, C., Zacchetti, D. & Pozzan, T. (1991b) Ca influx following receptor activation. *Trends Pharmacol. Sci.* **12**, 289–292.

Miller, R.J. (1992) Neuronal Ca: getting it up and keeping it up. *Trends Neurosci.* **15**, 317–319.

Putney, J.W. (1990) The integration of receptor-regulated intracellular calcium release and calcium entry across the plasma membrane. *Curr. Top. Cell Regn.* **31**, 111–127.

Putney, J.W. (1992) Inositol phosphates and calcium entry. In *Advances in Second Messenger and Phosphoprotein Research*, Vol. 26, J.W. Putney ed. New York: Raven Press, pp. 143–159.

Rossier, M.F. & Putney, J.W. (1991) The identity of the calcium-storing, IP_3-sensitive organelle in non-muscle cells: calciosome, endoplasmic reticulum or both? *Trends Neurosci.* **14**, 310–314.

Taylor, C.W. & Marshall, I.C.B. (1992) Calcium and inositol 1,4,5-trisphosphate receptors: a complex relationship. *Trends Biochem. Sci.* **17**, 403–407.

Wictome, M., Henderson, I., Lee, A.G. & East, J.M. (1992) Mechanism of inhibition of the calcium pump of sarcoplasmic reticulum by thapsigargin. *Biochem. J.* **283**, 525–529.

CaMKII and calmodulin

Colbran, R.J. (1992) Regulation and role of brain calcium/calmodulin-dependent protein kinase II. *Neurochem. Int.* **21**, 469–497.

Liu, Y. & Storm, D.R. (1990) Regulation of free calmodulin levels by neuromodulin: neuron growth and regeneration. *Trends Pharmacol. Sci.* **11**, 107–111.

MacNicol, M. & Schulman, H. (1992) Multiple Ca^{2+} signalling pathways converge on CaM kinase in PC12 cells. *FEBS Lett.* **304**, 237–240.

Rostas, J.A.P. & Dunkley, P.R. (1992) Multiple forms and distribution of calcium/calmodulin-stimulated protein kinase II in brain. *J. Neurochem.* **59**, 1191–1202.

Schulman, H. & Hanson, P.I. (1993) Multifunctional Ca^{2+}/calmodulin-dependent protein kinase. *Neurochem. Res.* **18**, 65–77.

Protein phosphatases

Armstrong, D.L. (1989) Calcium channel regulation by calcineurin, a Ca-activated phosphatase in mammalian brain. *Trends Neurosci.* **12**, 117–122.

Cohen, P. (1992) Signal integration at the level of protein kinases, protein phosphatases and their substrates. *Trends Biochem. Sci.* **17**, 408–413.

Cohen, P., Holmes, C.F. & Tsukitani, Y. (1990) Okadaic acid: a new probe for the study of cellular regulation. *Trends Biochem. Sci.* **15**, 98–102.

Nitric oxide and cGMP

Bredt, D.S. & Snyder, S.H. (1992) Nitric oxide, a novel neuronal messenger. *Neuron* **8**, 3–11.

Garthwaite, J. (1991) Glutamate, nitric oxide and cell-cell signalling in the nervous system. *Trends Neurosci.* **14**, 60–67.

Goy, M.F. (1991) cGMP: the wayward child of the cyclic nucleotide family. *Trends Neurosci.* **14**, 292–299.

McCall, T. & Vallance, P. (1992) Nitric oxide takes centre-stage with newly defined roles. *Trends Pharmacol. Sci.* **13**, 1–6.

Snyder, S.H. (1992) Nitric oxide: first in a new class of neurotransmitters. *Science* **257**, 494–496.

Vincent, S.R. & Hope, B.T. (1992) Neurons that say NO. *Trends Neurosci.* **15**, 108–113.

6

Sensory receptors

6.1 Introduction

Sensory receptors are responsible for the input into the nervous system from the classic senses of sight, hearing, touch, taste and smell as well as from sensors monitoring the function of the body's internal organs. Sensory receptors may respond to photons (sight), to mechanical stress (hearing, touch) or to chemical stimuli (taste, smell). The appropriate stimulus to the sensory receptors generates a hyperpolarization or depolarization; and the intensity of the stimulus is coded in the frequency of firing. Vertebrate photoreceptors do not have axons, but directly modulate their spontaneous release of transmitter in response to light-evoked changes in membrane potential.

6.2 Photoreceptors

The human retina contains some 120 million rod photoreceptors responsible for high-sensitivity, black and white vision together with 6.5 million cone cells containing pigments with maximal absorption in the blue, green or red. Rods and cones are divided into outer and inner segments (Fig. 6.1). The outer segments absorb photons, hyperpolarize the membrane and thus reduce the rate of spontaneous transmitter release from the inner segment which synapses onto the bipolar and horizontal cells.

The vertebrate rod cell has been most closely investigated. Most of the volume of the rod outer segment is filled with an ordered stack of disks separated from each other by narrow (15 nm) layers of cytoplasm. The major protein of the disks is the integral 39 kDa *rhodopsin* which contains the chromophore *11-cis-retinal* and shows strong structural homology to agonist-coupled 7-transmembrane helix receptors. The cofactor retinal, with a maximal absorbance at about 500 nm, is linked to the ε-amino group of lys_{296} of the apoprotein *opsin* via a protonated Schiff's base. This lysine is located in the middle of the seventh transmembrane region indicated by hydropathy plots (Fig. 6.2). In the dark the *cis* conformation of the chromophore is stabilized by its local environment. However absorption of a photon causes isomerization to the all *trans* form, a process which requires $125\,kJ \cdot mol^{-1}$ (a photon in the visible range might typically have $200\,kJ \cdot mol^{-1}$) probably to move the positive Schiff's base away from a fixed anionic residue in the protein (a 0.5 nm movement of the retinal ring occurs relative to the Schiff's base).

Rhodopsin possesses structural homology not only to G-protein-coupled metabotropic receptors but also to bacteriorhodopsin, a light-driven proton pump found in a halophilic bacterium. It must however be stressed that bacteriorhodopsin shows little sequence homology with visual rhodopsin, although they both contain a retinal bound as a Schiff's base which undergoes a *cis-trans* isomerization on absorbing a photon. The ability of bacteriorhodopsin to form regular two-dimensional pseudo-crystalline arrays has allowed a three-dimensional structure to be determined. This model is the basis for the proposed structures of visual rhodopsin and 7-transmembrane receptors in general (Fig. 6.2). The topology of visual rhodopsin is such that the N-terminus is located within the lumen of the disk while the 'intracellular' loops associated with G-protein interactions are exposed to the cytoplasm.

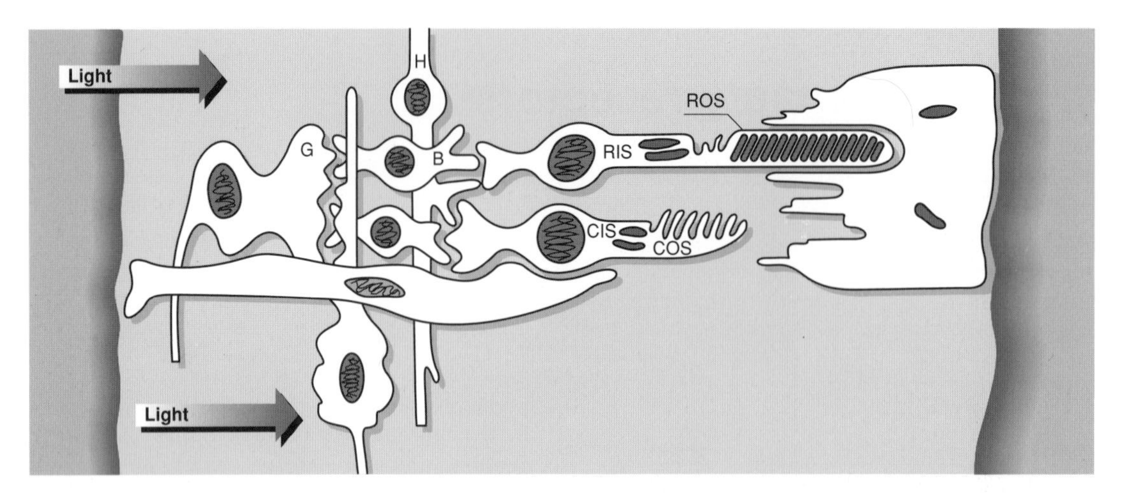

Fig. 6.1 A simplified view of the vertebrate retina. Note that light must traverse several cells before striking the outer segments. Not all cell types are shown. From Shichi (1989). B, bipolar cell; CIS, cone inner segment; COS, cone outer segment; G, ganglion cell; H, horizontal cell; RIS, rod inner segment; ROS, rod outer segment.

The energy in the initial light-induced displacement of the retinal relaxes by inducing conformational changes in the neighbouring residues which are propagated within 0.1 ms through the receptor to the cytoplasmic domains. *In vitro* this may be followed by the dissociation of the retinal prosthetic group into the membrane lipids, although the dissociation is much too slow to be significant for normal visual transduction.

The light-induced conformational change creates a cytoplasmic binding site for *transducin* (G_t) a trimeric G-protein homologous to G_s. In the dark G_t is weakly attached to the cytoplasmic face of the disk (Fig. 6.3). A single rhodopsin molecule activated by a photon can activate over 1000 molecules of G_t per second. The released $G_t\alpha$ in turn activates a disk membrane associated *cGMP phosphodiesterase* (PDE). PDE has two polypeptides of 88 and 84 kDa and in the dark is inhibited by the binding of two further inhibitory 13 kDa peptides ('I'). To activate PDE, two molecules of $G_t\alpha$ bind to the 'I' subunits and remove them from the complex. Activation of the phosphodiesterase rapidly lowers the concentration of cGMP in the outer segment cytoplasm; the amplification inherent in this cascade allows photoexcitation of one rhodopsin molecule to induce the hydrolysis of at least 10^5 cGMP molecules.

In the dark cGMP maintains a population of plasma membrane cation-selective channels in an active state; as a result Na^+ entry partially depolarizes the membrane potential to about −40 mV. The cloned cation channel has a molecular weight of 63 000 with multiple transmembrane domains and a cytoplasmic segment with a putative cGMP binding domain homologous to that of cGMP-dependent protein kinase. The steady inward flux of Na^+ through the channels in the dark (about 50 pA per cell) must of course be charge-balanced by other ion fluxes. The plasma membrane of the inner segment contains constitutively active K^+-channels, and efflux of K^+ through these channels maintains overall electroneutrality. A high density of Na^+/K^+-ATPases in the inner segment is also present to pump Na^+ back out of the cell and K^+ in. The effect of light is thus to cause a graded decrease in the Na^+ conductance of the outer segment plasma membrane. This allows the cell to hyperpolarize towards the K^+ equilibrium potential (about −70 mV), decreasing the spontaneous exocytosis of transmitter released onto synapses with horizontal and bipolar cells (*see* Fig. 6.1). Bipolar cells have large presynaptic terminals (up to 7 μm in diameter) permitting patch-clamp analysis of Ca^{2+}-channels. Sensitivity to DHPs was found indicating a dominant role for L-channels, contrasting with the non-L specificity found in most other CNS terminals (*see* Section 7.3).

The transduction steps for the light→dark transition are of course as important as those for the dark→light response. Termination is much

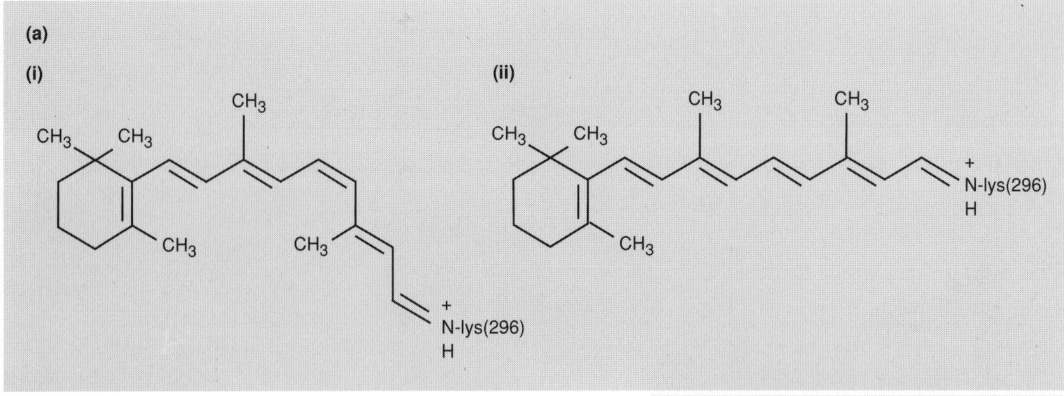

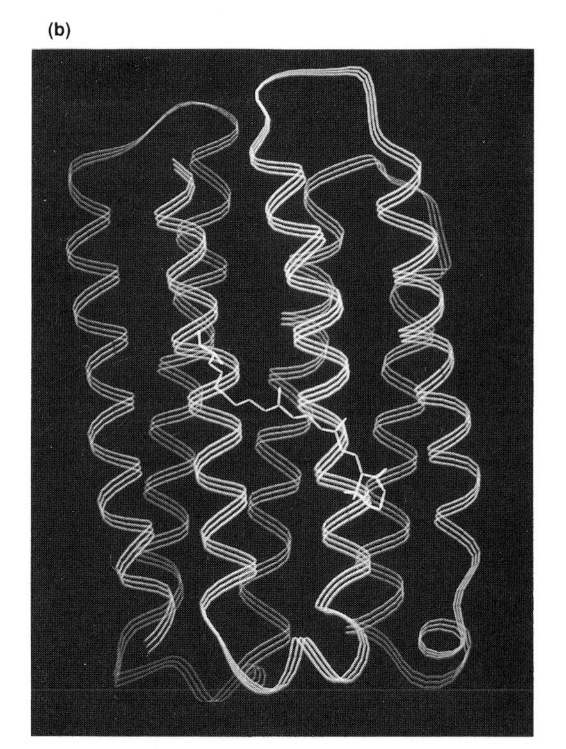

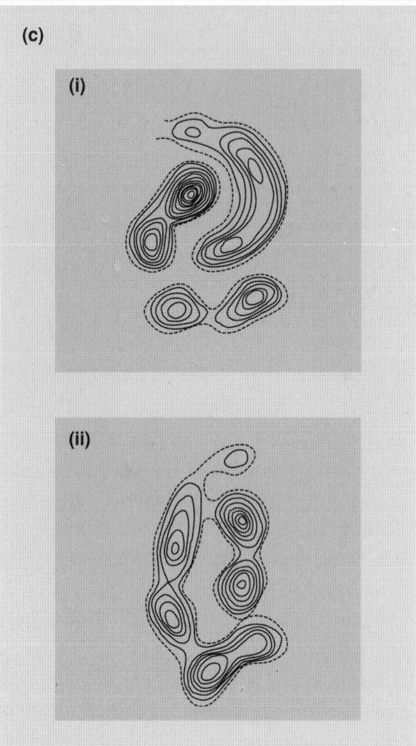

Fig. 6.2 Visual rhodopsin and its homology to bacteriorhodopsin. (a) Absorption of a photon by the chromophore 11-*cis* retinal (i) causes rotation of the 11-*cis* double bond to the *trans* conformation (ii). Retinal is linked via a Schiff's base to Lys_{296} of the apoprotein opsin. (b) The structure of bacteriorhodopsin, the ribbon diagram shows the backbone of the polypeptide with its seven α-helices. From Henderson *et al.* (1990), by permission of Academic Press. (c) Projection structure of visual rhodopsin (i) and bacteriorhodopsin (ii) at 9 Å resolution, note that while the structures are related, rhodopsin is wider and less elongated. From Schertler *et al.* (1993), reproduced with permission from *Nature*. © 1993 Macmillan Magazines Ltd.

more than simply a switching off of the above transduction pathway, which would not account for the fast decay of the response or the slower adaptation to prolonged illumination. These processes are controlled by phosphorylation of rhodopsin and by $[Ca^{2+}]_c$.

The cGMP-dependent ion channels can conduct Ca^{2+} as well as Na^+, the divalent cation contributing about 15% of the total ion current. The resulting steady-state $[Ca^{2+}]_c$ achieved between this influx and Ca^{2+} extrusion pathways, notably an atypical Ca^{2+}, Na^+, K^+

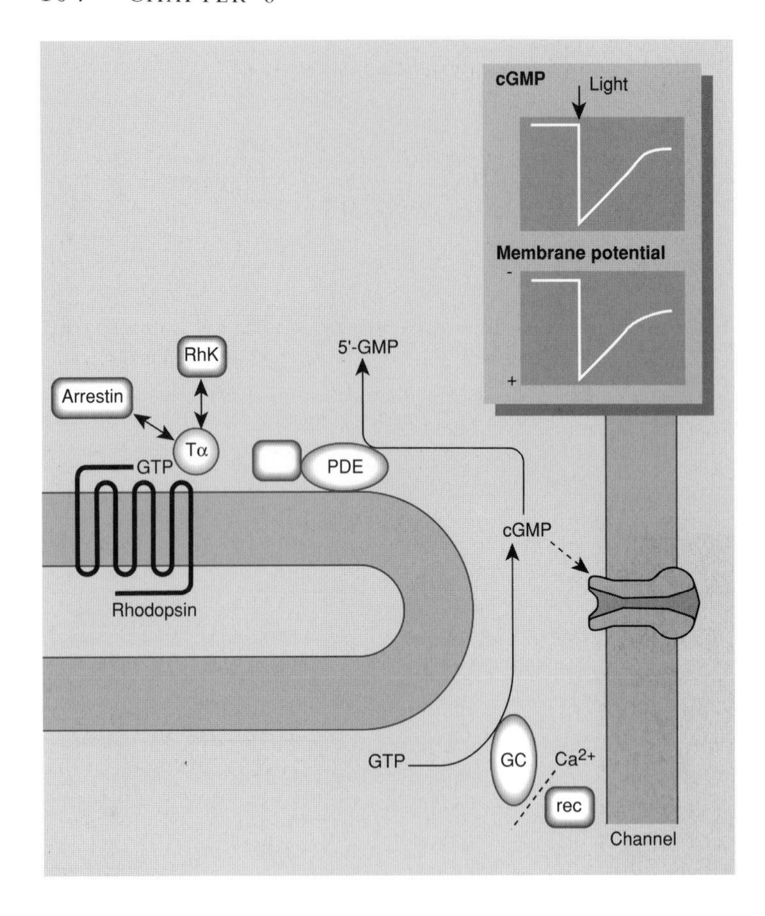

Fig. 6.3 Phototransduction. Absorption of a photon by rhodopsin causes a conformational change which allows transducin (T) to bind; GTP-activated Tα dissociates and activates cGMP phosphodiesterase (PDE). Hydrolysis of cGMP causes the closure of cGMP-dependent cation channels in the plasma membrane. Decreased Na^+ entry causes hyperpolarization which decreases spontaneous transmitter release at the inner segment synapse. Ca^{2+} entry through the cation channels is also inhibited. The lowered $[Ca^{2+}]_c$ activates guanylyl cyclase (GC) by allowing recoverin (rec) to bind to the enzyme, restoring cGMP levels. Arrestin and rhodopsin kinase (RhK) desensitize rhodopsin during high light exposure by dissociating transducin from rhodopsin.

exchanger (which shows no homology to the cloned cardiac Na^+/Ca^{2+} exchange, *see* Section 4.2.1) helps to control the gain of the visual transduction cascade. In the dark $[Ca^{2+}]_c$ is typically 300 nM which produces a five-fold inhibition of the *guanylyl cyclase* located on the cytoskeletal *axoneme* filaments, and which is responsible for regenerating cGMP. During prolonged light the closure of the plasma membrane ion channel allows $[Ca^{2+}]_c$ to be lowered to about 10 nM, with a time-constant of about 0.5 s, by the transport pathways removing Ca^{2+} from the cytoplasm. The 23 kDa Ca^{2+}-binding protein *recoverin* binds to guanylyl cyclase under these low $[Ca^{2+}]_c$ conditions, resulting in a five-fold activation of the enzyme and increased production of cGMP, which opposes the light-evoked phosphodiesterase activity and decreases the sensitivity of the photoreceptor cell.

The eye has an enormous capacity to adapt to differing light levels, for example, moonlight is only about a millionth as intense as full sunlight. Apart from changes in the diameter of the iris and in signal processing by the ganglion cells, which will not be considered here, there are changes in the responsiveness of rods and cones themselves. A high level of activated transducin in response to bright light can activate a 65 kDa *rhodopsin kinase*. The kinase (which does not require activated G-protein) can phosphorylate activated (but not inactive) rhodopsin preventing it from interacting with transducin. Rhodopsin kinase associates with the membrane via a 15C isoprenoid side chain. Phosphorylation of rhodopsin occurs at C-terminal serine and threonine residues. A 48 kDa inhibitor protein *arrestin*, which is abundant in the rod cytoplasm, assists this desensitization by blocking access of transducin to the activated rhodopsin. These processes are closely analogous to that of the β_2-adrenoceptor discussed in more detail in Section 11.4.1. The process of recovery requires the dephosphorylation of rhodopsin by a type 2A phosphatase as well as rapid hydrolysis of GTP associated with $G_{t\alpha}$.

Colour-sensitive cones contain variants of

rhodopsin in which the retinal absorption spectrum is altered by modification of the protein environment, giving absorption maxima at 445 nm (blue), 535 nm (green) or 570 nm (red). The structure of the outer segment of the cone differs from that of the rod, as the flattened disks are invaginations of the plasma membrane rather than being separate. The pathway of visual transduction is believed to be similar to that in rods, although distinct G-proteins may be involved. Cones provide less amplification than rods, which can be activated by a single photon. Thus although each rod cell contains some 10^8 rhodopsin molecules, the cell can register the absorption of a single photon by just one of the molecules, while coincident signals from five rods can produce a conscious signal. The transmitter released by rods and cones is believed to be glutamate.

6.3 Chemoreceptors, smell and taste

The mammalian olfactory system can distinguish between thousands of odorants and can detect particularly pungent examples at concentrations of a few parts per trillion. The rat possesses about 10^6 olfactory neurones (Fig. 6.4) each of which extends a single apical dendrite to the epithelial surface where it terminates in some ten cilia extending into the mucus of the nasal cavity. The neurone's unbranched axon projects to the olfactory bulb at the front of brain.

A gene family of between 100 and 1000 olfactory has been identified using the polymerase chain reaction and primer oligonucleotides based on the sequence of other G-protein-coupled receptors. Expression is limited to sensory olfactory neurones and the receptors are localized to

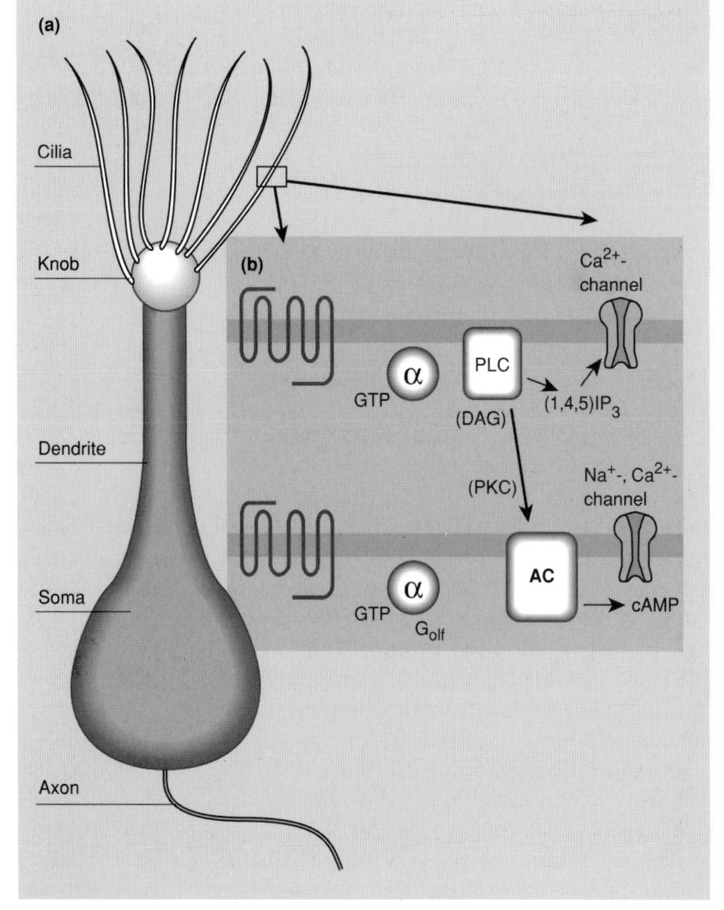

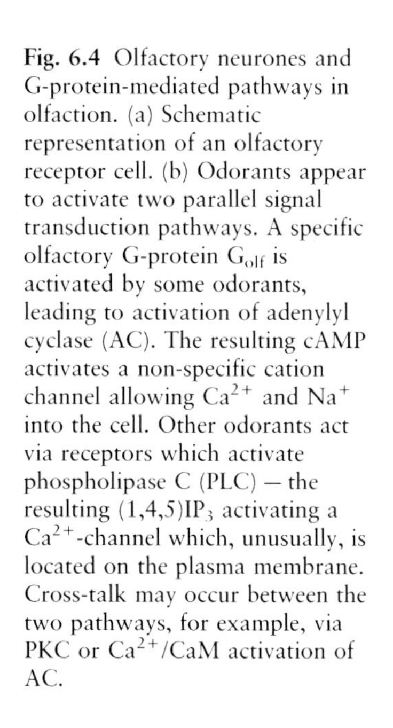

Fig. 6.4 Olfactory neurones and G-protein-mediated pathways in olfaction. (a) Schematic representation of an olfactory receptor cell. (b) Odorants appear to activate two parallel signal transduction pathways. A specific olfactory G-protein G_{olf} is activated by some odorants, leading to activation of adenylyl cyclase (AC). The resulting cAMP activates a non-specific cation channel allowing Ca^{2+} and Na^+ into the cell. Other odorants act via receptors which activate phospholipase C (PLC) — the resulting $(1,4,5)IP_3$ activating a Ca^{2+}-channel which, unusually, is located on the plasma membrane. Cross-talk may occur between the two pathways, for example, via PKC or Ca^{2+}/CaM activation of AC.

the cilia of the neurones. It is possible that odorants do not interact directly with the receptors, but rather as a complex with *odorant-binding proteins* which are secreted by glands at the front of the nose.

The cilia contain high concentrations of an olfactory-neurone specific type III adenylyl cyclase together with a novel G-protein subunit $G_{olf}\alpha$ which can mediate cyclase activation *in vitro* (Fig. 6.5). The cilia also express a cyclic nucleotide-activated non-specific cation channel, highly homologous to the photoreceptor cell channel (*see* Section 6.2). The C-terminal region of the channel has a cyclic nucleotide binding consensus sequence which interestingly has a higher affinity for cGMP than cAMP. In parallel with these pathways, an $G_o\alpha$-mediated $(1,4,5)IP_3$ signalling pathway has been reported. Both nucleotide and $(1,4,5)IP_3$ pathways appear to act by activating non-specific cation channels in the plasma membrane allowing Ca^{2+} to enter the cilia.

Interactions occur between the two pathways; activation of PKC by phorbol esters increase the electrical response of olfactory receptor cells to cAMP-generating odorants, while increased $[Ca^{2+}]_c$ also potentiates the cAMP-linked response, probably via calmodulin activation of adenylyl cyclase.

Desensitization of odorant receptors appears to be mediated by a similar mechanism to that observed in visual transduction and β-adrenergic mediated regulation, namely phosphorylation by a βARK (β-adrenoceptor kinase) isoform (*see* Section 11.4.1) and arrestin-mediated decoupling of the G-protein interaction.

An involvement of cGMP raises the possibility of a role for nitric oxide (NO) and NO-synthase (NOS). NOS is highly expressed in the olfactory bulb, and cGMP increases in response to high concentrations of odorants. These responses are abolished by NOS inhibitors. Since NOS is activated by an increase in $[Ca^{2+}]_c$ (*see* Section 5.10), while NO has the properties of a short-lived intercellular messenger, it is possible that it functions to recruit adjacent neurones during intense stimuli.

Receptors for taste are much more restricted than for smell. Despite this, there is still considerable debate as to the precise mechanisms by which taste is transduced. Combinations of four basic responses (to sweetness, bitterness, sourness and saltiness) are believed to underlie our sense of taste. Highly diverse mechanisms have been proposed to underlie the sense of taste (Fig. 6.5). A *salty* taste may be evoked by the inward current through amiloride-sensitive voltage-insensitive cation channels which may be activated by direct interaction with the ions, rather than via a conventional receptor. A *sour* taste is evoked by acids, and taste cells responding to protons may do so by eliciting an inward H^+ current through the amiloride-sensitive Na^+-channels which may also be responsible for detecting saltiness, although pH-dependent blockade of K^+-channels has also been reported. *Bitter* stimuli causes an increase in Ca^{2+} release from internal stores in single rat tongue taste cells, suggestive of a $(1,4,5)IP_3$-mediated pathway. Finally *sweetness* may be transduced via a G-protein and cAMP linked pathway. It must however be emphasized that to date none of these mechanisms are definitively established.

6.4 Mechanoreceptors, hearing and touch

Mechanoreceptors on sensory nerve fibres respond to movement created by touch, by distortion of hairs or, in the inner ear, by sound vibrations. The molecular mechanism of mechanoreceptors is as yet unclear, although stretch-activated ion channels have been described which open in response to distortion of the plasma membrane.

The senses of hearing and balance are both transduced by *hair cells* in the vestibular apparatus and cochlea respectively. Hair cells (Fig. 6.6) possess an apical surface with a hexagonally arranged bundle of *stereocilia* of graded length composed of dense paracrystalline actin, linked at their tips and attached to an underlying rigid actin matrix known as the *cuticular plate*. In addition the cell has a single distinct microtubule-containing *kinocilium* extending directly from the plasma membrane. In mammalian cochlea the kinocilium disappears soon after birth.

Hair cells release transmitter continuously, but mechanical deflection of the bundle towards the tallest stereocilium opens ion channels in the plasma membrane producing a depolarization and excitation of afferent fibres synapsing with the hair cells, while the opposite deflection results in hyperpolarization and decreased release of the

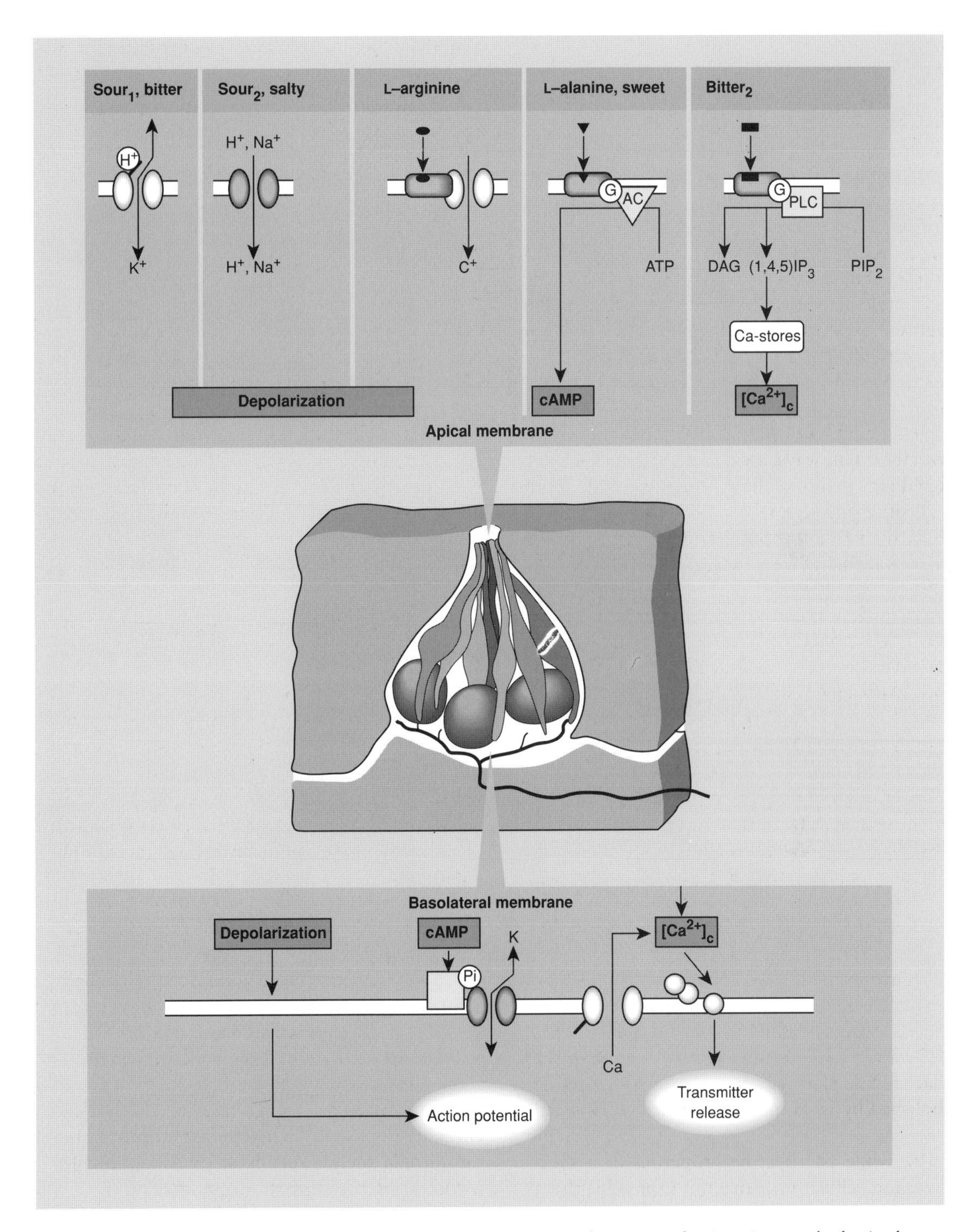

Fig. 6.5 Taste transduction. A summary of the chemosensory transduction mechanisms in taste buds. At the apical membrane sour (acidic) and some bitter compounds block K^+-channels causing depolarization; protons and Na^+ ions from sour and salty stimuli can directly depolarize by entry through an amiloride-sensitive cation channel; arginine activates a cation channel; alanine and sweet compounds activate a cAMP coupled receptor, while some bitter compounds activate PLC. At the basolateral membranes of the taste receptor cells each of these transduction mechanisms can depolarize the membrane allowing transmitter release to occur onto the nerve bundles. Adapted from Roper (1992), by permission of the Society for Neuroscience.

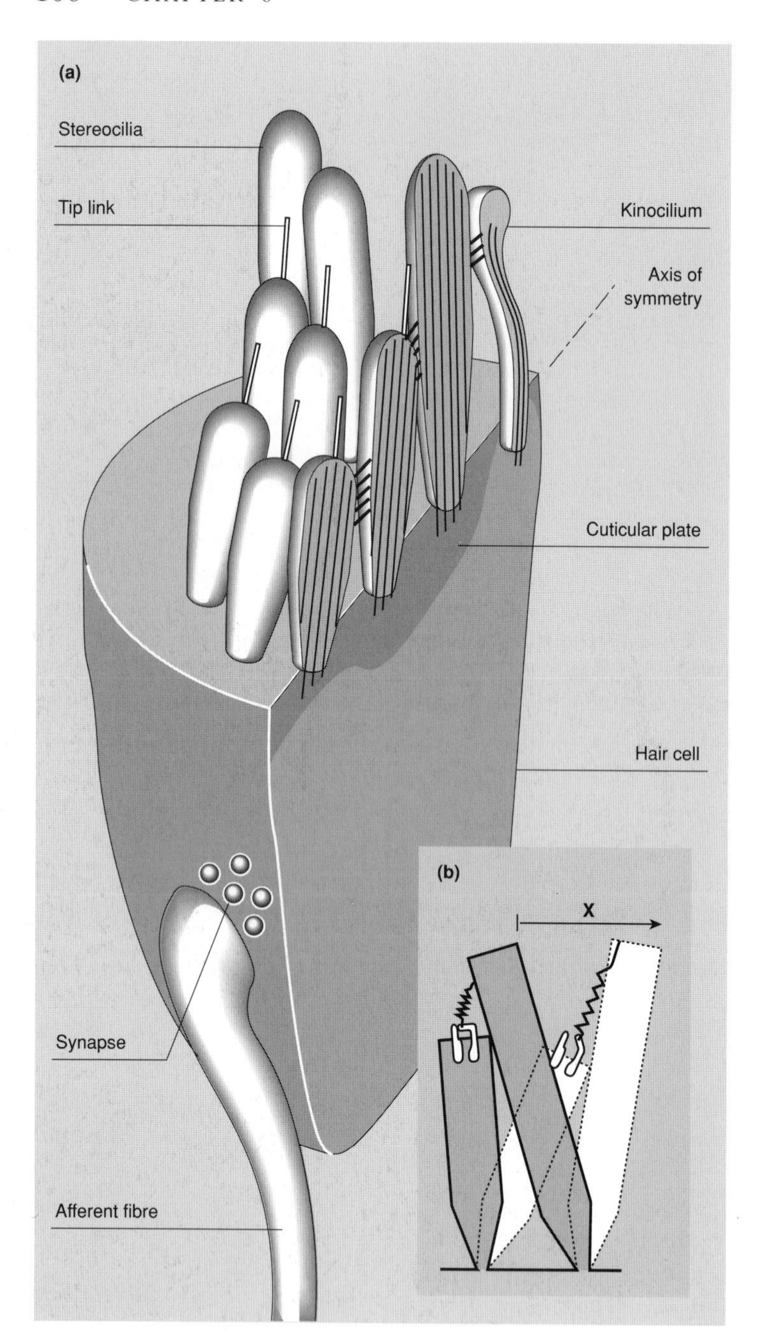

Fig. 6.6 Sound transduction. (a) A hair cell is shown in cross-section. Stereocilia are graded in length, embedded in a rigid cuticular plate and attached to each other via tip links. (b) In the tip-link model of transduction, mechanical deflection of the stereocilia would directly open cation channels located at their tips. The resulting depolarization would lead to the exocytosis of transmitter onto afferent fibres shown in (a). From Pickles & Corey (1992), by permission of Elsevier Trends Journals.

transmitter (generally glutamate). Deflection of the kinocilium has no effect.

In the vestibular apparatus this deflection is caused by the inertia of fluid in the ducts when the head is subjected to angular acceleration, while in the cochlea the stereocilia of an individual hair cell resonates at a precise frequency related to its position within the cochlea.

Further reading

Visual transduction

Chabre, M. & Deterre, P. (1989) Molecular mechanism of visual transduction. *Eur. J. Biochem.* **179**, 255–266.

Heidelberger, R. & Matthews, G. (1992) Calcium influx and calcium current in single synaptic terminals of goldfish retinal bipolar neurones. *J. Physiol. (Lond.)* **447**, 235–256.

Henderson, R., Baldwin, J.M., Ceska, T.A., Zemlin, T.A., Beckman, E. & Downing, K.H. (1990) Model for the structure of bacteriorhodopsin based on high-resolution electron cryo-microscopy. *J. Mol. Biol.* **213**, 899–929.

Kaupp, U.B., Niidome, T., Tanabe, T., Terada, S., Bonigk, S., Stuhmer, W., Cook, N.J., Kangawa, K., Matsuo, H., Hirose, T., Miyata, T. & Numa, S. (1989) Primary structure and functional expression from cDNA of the rod photoreceptor cyclic GMP gated channel. *Nature* **342**, 762–766.

Lagnado, L. & Baylor, D. (1992) Signal flow in visual transduction. *Neuron* **8**, 995–1002.

Schertler, G.F.X., Villa, C. & Henderson, R. (1993) Projection structure of rhodopsin. *Nature* **362**, 770–772.

Shichi, H. (1989) Molecular biology of vision. In *Basic Neurochemistry*. G. Siegel, B.W. Agranoff, R.W. Albers & P. Molinoff (eds). New York: Raven Press, pp. 137–148.

Uhl, R., Wagner, R. & Ryba, N. (1990) Watching G-proteins at work. *Trends Neurosci.* **13**, 64–70.

Olfaction

Breer, H. & Shepherd, G.M. (1993) Implications of the NO/cGMP system for olfaction. *Trends Neurosci.* **16**, 5–9.

Dawson, T.M., Arriza, J.L., Jaworsky, D.E., Borisy, F.F., Attramadal, H., Lefkowitz, R.J. & Ronnett, G.V. (1993) β-Adrenergic receptor kinase-2 and β-arrestin-2 as mediators of odorant-induced desensitization. *Science* **259**, 825–829.

Frings, S. (1993) Protein kinase C sensitizes olfactory adenylyl cyclase. *J. Gen. Physiol.* **101**, 183–205.

Reed, R.R. (1992) Signalling pathways in odorant detection. *Neuron* **8**, 205–209.

Ronnett, G.V. & Snyder, S.H. (1992) Molecular messengers of olfaction. *Trends Neurosci.* **15**, 508–513.

Taste transduction

Kinnamon, S.C. (1988) Taste transduction: a variety of mechanisms. *Trends Neurosci.* **11**, 491–496.

Roper, S.D. (1992) The microphysiology of peripheral taste organs. *J Neurosci.* **12**, 1127–1134.

Sound transduction

Pickles, J.O. & Corey, D.P. (1992) Mechanoelectrical transduction by hair cells. *Trends Neurosci.* **15**, 254–259.

7

Mechanisms of exocytosis

7.1 Introduction

Regulated exocytosis is the process whereby vesicles fuse with the plasma membrane in a process which requires a specific signal, usually Ca^{2+}, to trigger release of their contents into the extracellular space. It must be distinguished from constitutive exocytosis where continuous fusion occurs to deliver membrane and membrane proteins to the plasma membrane of a variety of cells. However as will become apparent, regulated and constitutive secretion may use some common elements to control the intracellular membrane trafficking associated with vesicle production, transport and retrieval. This is particularly useful since a detailed genetic analysis is possible of constitutive secretion occurring in yeast.

Even within the constraints of regulated secretion there is no universal mechanism. Indeed it is reasonable that the submillisecond exocytosis of fast-acting neurotransmitters should differ in detail from relatively slow hormonal secretion (Fig. 7.1). Furthermore there are significant differences between the exocytosis of neurotransmitters such as GABA, glutamate and acetylcholine from *small electron-lucid synaptic vesicles (SSVs)* which undergo autonomous recycling at the nerve terminal and the release of neuropeptides from *large dense-core vesicles (LDCVs)* which are produced by the *trans-Golgi* network in the cell body, to merit a separate discussion, which will be reserved for Chapter 8. A single cycle of exocytosis, vesicle retrieval and refilling involves a number of stages, none of which is understood in full detail:

1 The triggering of exocytosis by Ca^{2+} entry.
2 The exocytotic fusion of the synaptic vesicle with the presynaptic membrane; a process for which less than 1 ms can generally be allowed.
3 The freeing and re-occupation of the release site within less than 100 ms to allow further release.
4 The retrieval of the synaptic vesicle from the membrane without allowing vesicle-specific and plasma membrane-specific proteins to randomize.
5 The processing and refilling of the synaptic vesicles.
6 The transport of the refilled vesicles through the cytoplasm and their interaction with the presynaptic cytoskeleton.
7 The mechanism for the ordered storage of large numbers of synaptic vesicles ready for exocytosis.

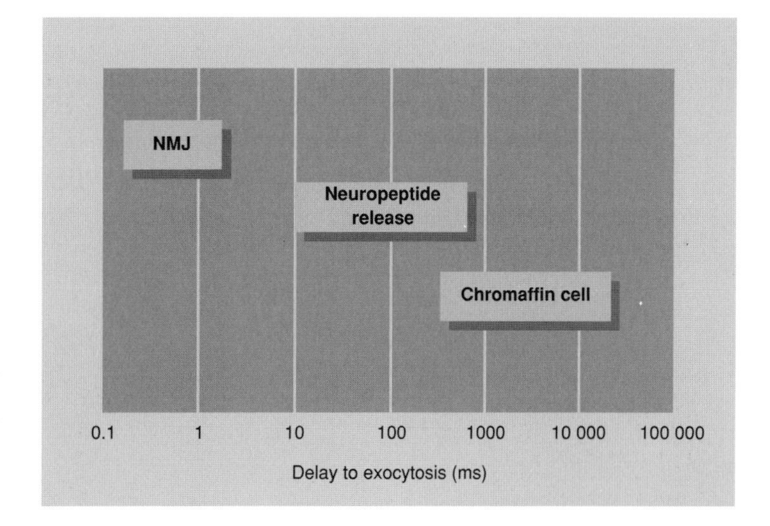

Fig. 7.1 The time scale of exocytosis in different cells. The approximate time delay between stimulus and the initiation of secretion can vary greatly between different cells, underlying the observation that there is no single universal mechanism of exocytosis. NMJ, neuromuscular junction.

8 The delivery of synaptic vesicles to the release site in readiness for the next cycle of release.

7.2 Morphology of the nerve terminal

More is known about the morphology of the vertebrate neuromuscular junction (NMJ) than any other terminal. However it is important to realize that this peripheral terminal differs in a number of respects from a typical CNS terminal. One of the main distinctions is that the NMJ is designed to release several hundred synaptic vesicles in response to a single action potential, whereas most CNS terminals may require several action potentials in order to release a single vesicle.

7.2.1 The neuromuscular junction

Motor neurones innervate striated muscle and trigger contraction by depolarizing the plasma membrane of the muscle cell triggering Ca^{2+} entry and release from internal stores. The vertebrate NMJ utilizes acetylcholine whereas in crustacea the corresponding transmitter is glutamate. The suitability of the vertebrate preparation for electrophysiology and electron microscopy has allowed a detailed study of the morphology of neurotransmission. Fast muscle fibres are singly innervated and possess long, branched axon terminals. In cross-section the terminal contains multiple, transverse *active zones* located opposite cylindrical infoldings of the postsynaptic membrane (Fig. 7.2). In addition to the dense packing of synaptic vesicles within the cytoplasm (a single terminal may contain 10^6 vesicles), a double row of synaptic vesicles is attached to the presynaptic membrane along the cytoplasmic face of each active zone. Freeze-fractured replicas of NMJ from frog fast fibres show the presence of large intramembranous particles as pairs of double rows on the presynaptic membrane at the active zones. These particles have been tentatively identified as the Ca^{2+}-channels which trigger exocytosis. The active zones are in register with the infoldings of the postsynaptic membrane while the nicotinic acetylcholine receptor and postsynaptic densities line the outer regions of the folds, but do not extend right to the bottom of the folds.

The NMJ was used in order to establish the *quantal* nature of neurotransmitter release; an action potential arriving at the terminal caused a release of acetylcholine sufficient to cause a large postsynaptic depolarization, or end-plate potential. In the absence of depolarization small (0.5 mV) random depolarizations or *miniature end-plate potentials* (*mepps*) are detected, which Katz and coworkers proposed were due to the random release of packets of ACh corresponding to the contents of individual synaptic vesicles. A normal action potential results in the synchronous release of several hundred quanta; if however Ca^{2+} entry is decreased by lowering external Ca^{2+} and increasing $[Mg^{2+}]$ then a statistical fluctuation in the magnitude of the end-plate potential can be detected as small numbers of discrete quanta are synchronously released.

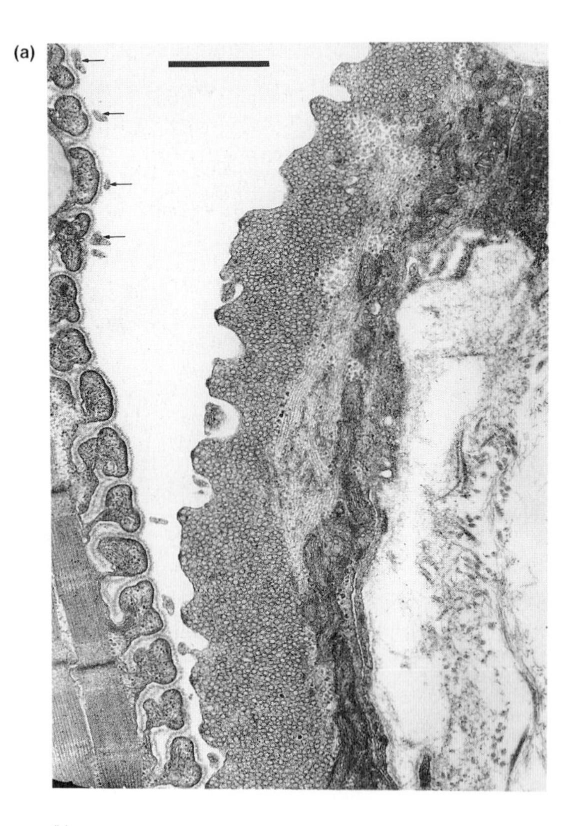

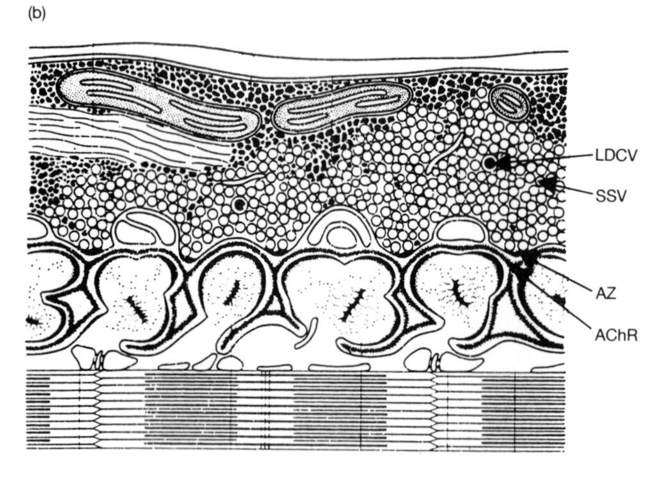

Fig. 7.2 The neuromuscular junction. (a) Electron micrograph of a frog neuromuscular junction in a longitudinal section after mechanical separation of pre- and postsynaptic portions, bar = 0.8 μm. (b) Diagram of the junction showing small synaptic vesicles (SSV), large dense-cored vesicles (LDCV), an active zone (AZ) and the location of nicotinic ACh receptors (AChR).

Even with the NMJ it is not easy to capture the act of neurotransmitter exocytosis under the electron microscope. Firstly although an action potential increases the rate of vesicle release from about $1 \cdot s^{-1}$ to as much as $100\,000 \cdot s^{-1}$, this is maintained for only a fraction of a millisecond resulting in the release of only about 200 synaptic vesicles ('quanta'). These are randomly scattered along the 600 μm length of the junction. Therefore the sample must be capable of being fixed within a few milliseconds of being stimulated while vesicle exocytosis is still continuing. Finally the plane of section must fortuitously pass through the mid-line of an exocytosing vesicle in order to give an unambiguous profile.

In order to show that vesicle fusion and quantal secretion are synchronous a quick-freezing technique is required. A neuromuscular

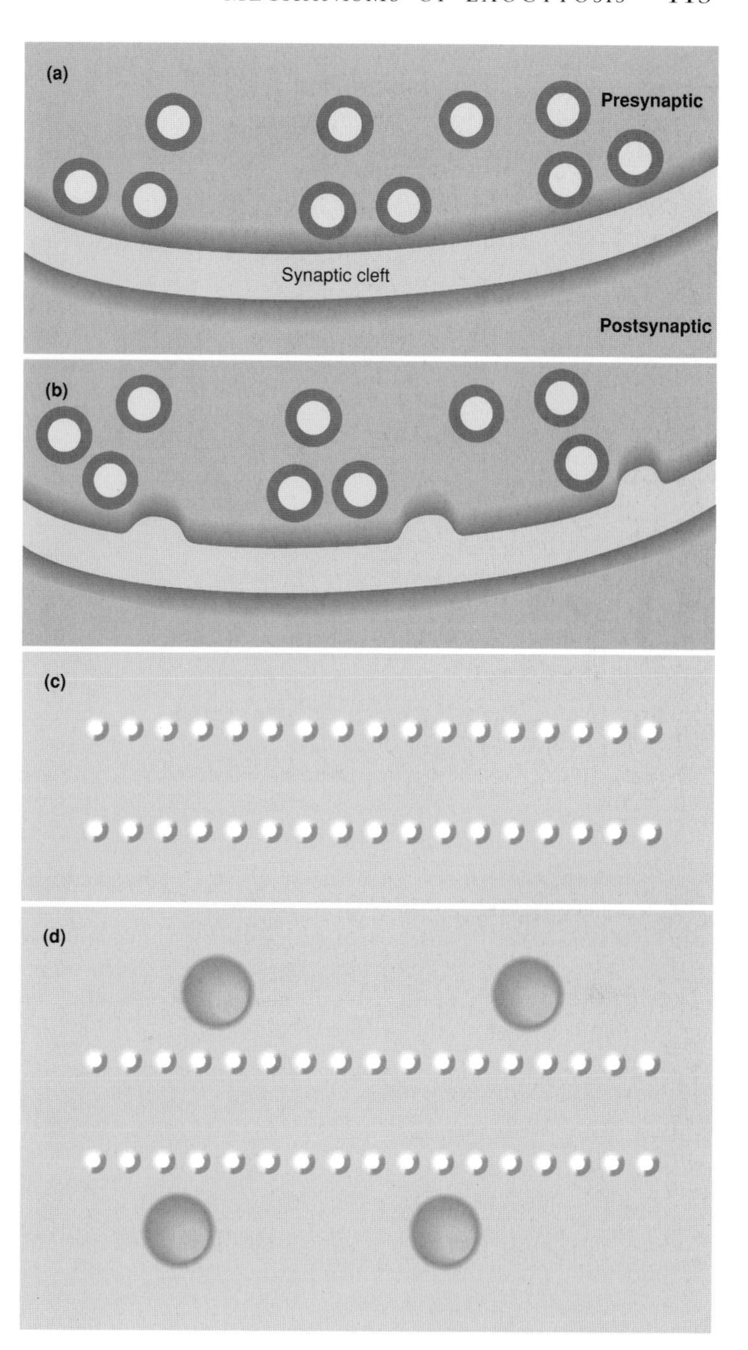

Fig. 7.3 Visualizing exocytosis at the neuromuscular junction. (a) Artist's impression of a freeze-substituted section through an unstimulated active zone of the frog neuromuscular junction, showing uniform 50 nm small synaptic vesicles in cross-section. (b) Impression of a similar section which has been electrically stimulated for 5 ms in the presence of 4-aminopyridine (4AP) to enhance action potentials and then immediately frozen between two liquid helium-cooled blocks. Small pockets are seen with the same diameter and curvature of synaptic vesicles which may represent collapsing vesicles. (c) Impression of a freeze-fracture view of a single unstimulated active zone showing two parallel rows of particles which may represent components of presynaptic Ca^{2+}-channels or the exocytotic apparatus. (d) Impression of a similar freeze-fracture stopped as in (b), after 5 ms of stimulation in the presence of 4AP. Note the appearance of pits consistent with exocytosing vesicles. From Heuser & Reese (1981), by permission of *Journal of Cell Biology*. © Rockefeller University Press.

preparation is given a single electric stimulation and within 1–2 ms is slammed against a copper block cooled with liquid helium. A 10 μm thick layer of tissue against the block freezes within 1 ms. The K^+-channel inhibitor 4-aminopyridine (*see* Section 3.3.1) greatly enhances quantal release by lengthening the duration of the presynaptic action potentials, and under these conditions many exocytotic profiles can be seen (Fig. 7.3). In the first studies the earliest fusions were seen at 5 ms; with subsequent technical improvements and the use of thin sections the time lag has been reduced to 2.5 ms, comparable to the time delay before quanta can be detected electrophysiologically.

A major problem inherent in the static images

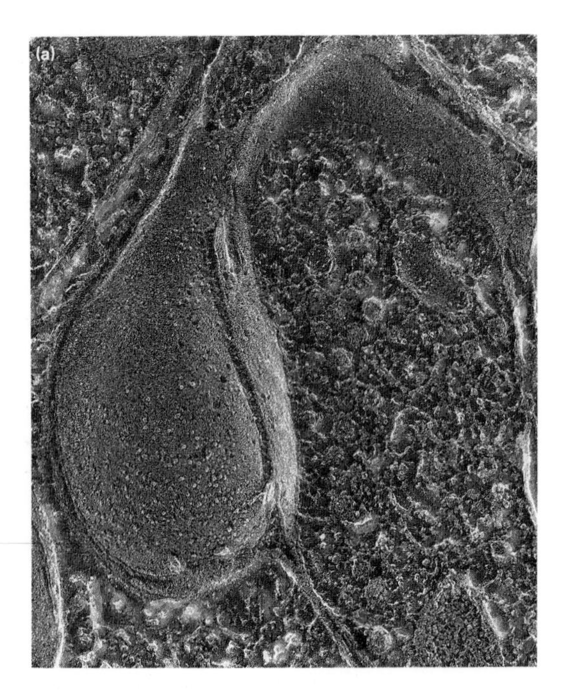

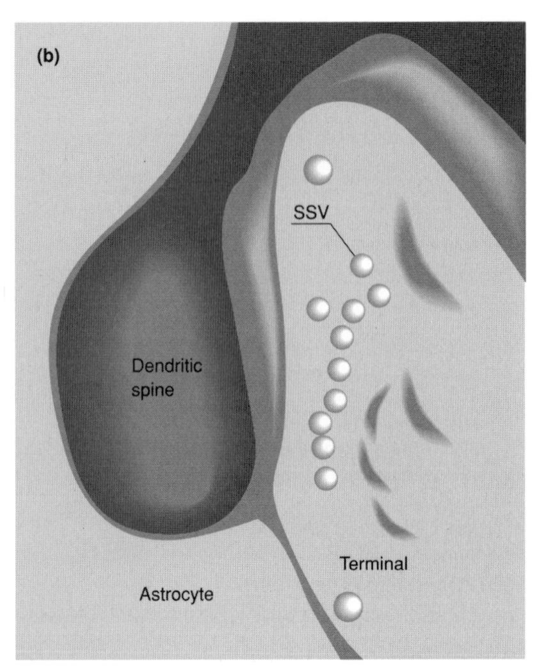

Fig. 7.4 A nerve terminal and dendritic spine from the mammalian cerebellum. Rapidly frozen, shallow-etched section of the cerebellar molecular layer, showing the synapse between a parallel fibre varicosity and a Purkinje dendritic spine. SSV, small synaptic vesicle. From Landis *et al.* (1988), reproduced with permission from *Neuron*. © Cell Press.

which are obtained under the electron microscope is how to distinguish between a vesicle caught in the act of exocytosis and an endocytotic profile. Thus prolonged KCl depolarization of the NMJ produces 'dimples' all over the terminal surface; does this represent a disorganization of specific active zone release under these conditions ('ectopic release') or just sites of endocytosis?

7.2.2 The CNS synapse

As shown in Fig. 5.22, the dendrites of neurones such as hippocampal pyramidal cells or cerebellar Purkinje cells possess *dendritic spines* protruding from the dendrite itself and forming the postsynaptic elements of the synapse. Figure 7.4 shows a glutamatergic synapse in the cerebellum between the varicosity of a granule cell parallel fibre and a Purkinje cell dendritic spine. The presynaptic terminal can be seen to be packed with small synaptic vesicles apparently enmeshed with cytoskeletal elements. The *active zones* within CNS terminals releasing fast-acting neurotransmitters differ in morphology from the linear arrays seen at the NMJ. In thin sections, active zones extend for about 0.5 μm in the plane of the membrane. Opposite the active zone is a postsynaptic specialization, the *postsynaptic density*, which contains receptors, while within the synaptic cleft itself a distinctive staining is seen.

7.2.3 Electrophysiological techniques for monitoring exocytosis

The most rapid, sensitive and physiologically relevant technique for monitoring the release of a neurotransmitter is to determine electrophysiologically the effect of the released transmitter on the postsynaptic cell in an intact synapse. Thus *excitatory (epsp)* or *inhibitory (ipsp) postsynaptic potential* changes can be determined when the transmitter binds to and activates the postsynaptic receptors. The postsynaptic electrode can be located either outside, but in close proximity to, the postsynaptic cell (in which case a *field epsp* is detected) or it may penetrate or seal onto the postsynaptic neurone. The presynaptic neurone can be fired by a field electrode which can be set to depolarize the presynaptic neurone at repeated intervals. This technique is of course limited to monitoring the release of transmitters which act on ionotropic rather than metabotropic receptors at the synapse in question.

One problem with this approach is how to quantify release. An empirical measure of synaptic transmission is to determine the slope of the recorded epsp (i.e. the initial rate of depolarization in $mV \cdot ms^{-1}$). However this does not distinguish between increased transmission as a result of enhanced transmitter release and that resulting from an enhanced postsynaptic efficiency, e.g. by an upregulation of postsynaptic receptors. In order to quantify the release of transmitter itself it is necessary to perform a quantal analysis as discussed above. However quantal analysis in the CNS is much more difficult than at the NMJ and there is considerable current debate concerning the validity of this technique.

7.2.4 The neurochemical approach

Since the temporal resolution of biochemical experiments with cultured neurones or synaptosomes is limited, it might be thought that this preparation could not contribute to our understanding of transmitter exocytosis occurring in a millisecond timescale. However as discussed in Chapter 2, both of these preparations contain all the machinery for the uptake, storage and exocytotic release of neurotransmitters and release neurotransmitter by mechanisms which respond to the same inhibitors and modulators as more complex preparations.

A number of fundamental questions can be addressed with the synaptosomal preparation. These include the nature of the presynaptic Ca^{2+}-channels; the protein chemistry of the exocytotic process itself; the role of protein phosphorylation in release; the loci for presynaptic neurotoxins; the energetic requirements of exocytosis, the nature of the machinery for membrane retrieval and recycling and whether different transmitters are released by different mechanisms.

7.2.5 Presynaptic Ca^{2+}-channels coupled to transmitter exocytosis

Due to the small size of most nerve terminals, electrophysiological studies have been impracticable, except in special cases (see below), and Ca^{2+}-channels coupled to transmitter release have therefore been investigated by monitoring $^{45}Ca^{2+}$ fluxes or fura-2 signals in response to depolarization, their modulation by pharmacological agents and neurotoxins and the correlation of Ca^{2+} entry to release. The electrophysiological channel classifications discussed in Chapter 5 are derived from studies on neuronal cell bodies. Thus the great majority of voltage-activated Ca^{2+}-channels in the CNS may be located on dendrites and nerve terminals and may include relatively few L-channels or conotoxin-sensitive N-channels. At the same time it is almost certainly premature to create a simple category of 'P-channel' based on sensitivity to incompletely characterized spider venom toxins (*see* Box 4.3).

The pharmacology of presynaptic Ca^{2+}-channels on the mammalian synaptosomal plasma membrane does not readily fit with the existing classifications. $^{45}Ca^{2+}$ entry in response to prolonged KCl depolarization is biphasic, with a transient rapid entry of ^{45}Ca followed by a slow uptake which does not inactivate (Fig. 7.5). This is reflected in the fura-2 response where an initial spike is followed by a plateau. In mammalian synaptosomes no consistent sensitivity to either dihydropyridines or ω-conotoxin-GVIA is seen for either phase, although chick brain synaptosomes show much more inhibition with the conotoxin. The transient Ca^{2+} entry into mammalian synaptosomes can be largely inactivated by the predepolarization of synaptosomes in the absence of external Ca^{2+}, with the result that when Ca^{2+} is re-added, entry only occurs through the non-inactivating pathway.

The exocytosis of different neurotransmitters may be coupled to distinct subclasses of release-coupled Ca^{2+}-channels. Furthermore, the Ca^{2+}-channel pharmacology for a given transmitter can differ between brain areas and species. For this reason the evidence for each transmitter class will be considered separately, and discussion will be largely confined to mammalian terminals.

Amino acids. Inhibition of the transient phase of ^{45}Ca entry by depolarization of synaptosomes in the absence of external Ca^{2+} allows the action of the subsequent non-inactivating 'plateau' phase of Ca^{2+} entry to be studied in isolation when the cation is re-added. The release of GABA and glutamate from synaptosomes is unaffected,

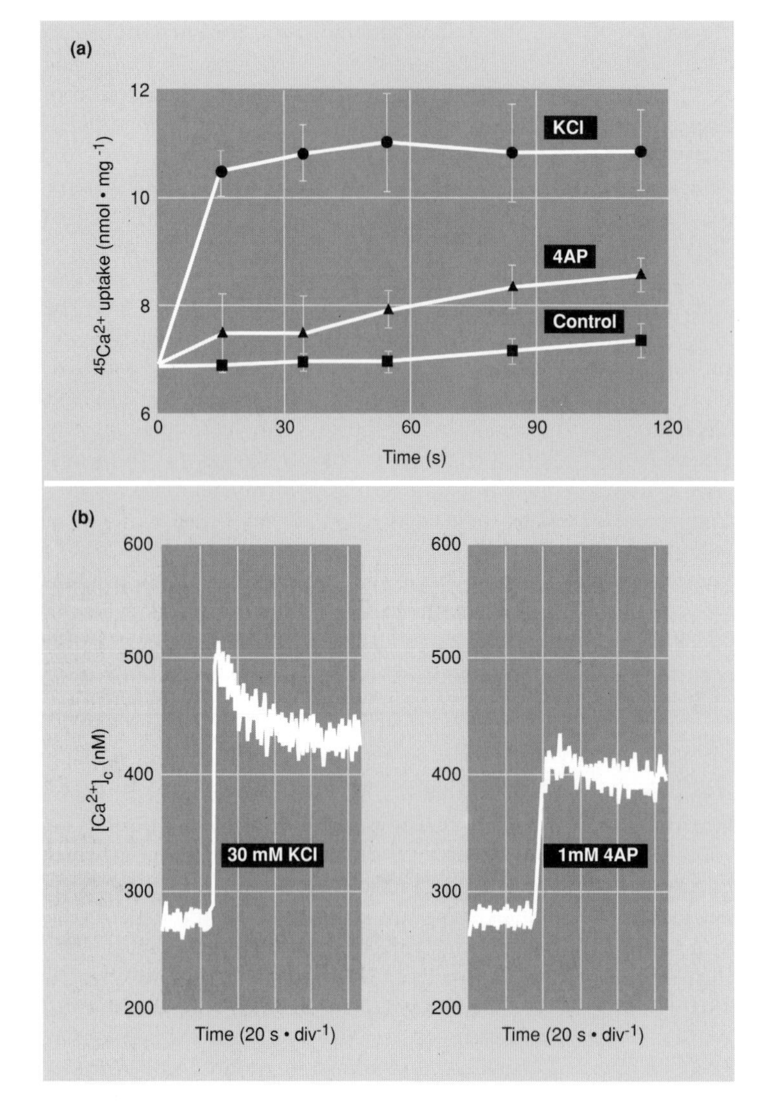

Fig. 7.5 ^{45}Ca and $[Ca^{2+}]_c$ responses of depolarized synaptosomes. Depolarization-evoked Ca^{2+} entry into synaptosomes can be monitored either with ^{45}Ca (a) or by the increase in $[Ca^{2+}]_c$ determined with fura-2 (b). Note that a large initial entry of ^{45}Ca is only seen with KCl and not with 4-aminopyridine (4AP) addition, although both agents evoked the release of transmitters. Data from Tibbs *et al.* (1989).

indicating that their release is coupled to the non-inactivating component. A non-inactivating Ca^{2+}-channel (or rather a channel whose probability of firing does not decrease with the time of depolarization) is consistent with presynaptic patch-clamp analysis of the large hippocampal mossy fibre terminals which release glutamate and prodynorphin-derived peptides and are just large enough for patch-clamp analysis of currents. The mammalian Ca^{2+}-channels associated with amino acid exocytosis are also resistant to dihydropyridines and ω-conotoxin-GVIA but sensitive to some agatoxins.

Type II transmitters. In contrast to the amino acids, catecholamine exocytosis from synaptosomes has been reported to correlate with the initial rapid phase of Ca^{2+} entry. Similarly, the more rapid phase of acetylcholine release from KCl-depolarized synaptosomes correlates with the initial phase of Ca^{2+} entry. Thus as a generalization (which is sure to have exceptions) amino acids and biogenic amines may be released by distinctive Ca^{2+}-channels. Consistent with this, ω-conotoxin-GVIA causes some inhibition of release of acetylcholine, noradrenaline, dopamine and 5-HT suggesting their linkage to presynaptic N-channels. This is not inconsistent with the failure to observe a convincing inhibition of Ca^{2+} entry into mammalian synaptosomes with ω-conotoxin-GVIA, since catecholaminergic synaptosomes represent only a minor fraction of terminal preparations from most brain areas.

Presynaptic Ca^{2+}-channels which have to date been examined electrophysiologically are those on unusually large terminals such as neurosecretosomes from rat pituitary, retinal bipolar cell terminals (both of which have properties characteristic of L-channels), those at the *squid giant synapse* (where sustained channels are found which are insensitive to DHPs or conotoxin) and the *chick ciliary ganglion calyx*, a large cholinergic terminal which can be patched in a 'whole-terminal' mode (Fig. 7.6), and which possess a high concentration of ω-conotoxin-GVIA-sensitive Ca^{2+}-channels on the presynaptic aspect of the terminal. It is interesting however that these last show very little voltage-dependent inactivation.

7.2.6 Ca^{2+} entry and its coupling to transmitter release

With synaptosomes, as indeed with any brain slice or cell preparation in which postsynaptic potentials are not measured electrophysiologically, it is not possible to determine the release from a single terminal in response to a single action potential (the chick ciliary ganglion may

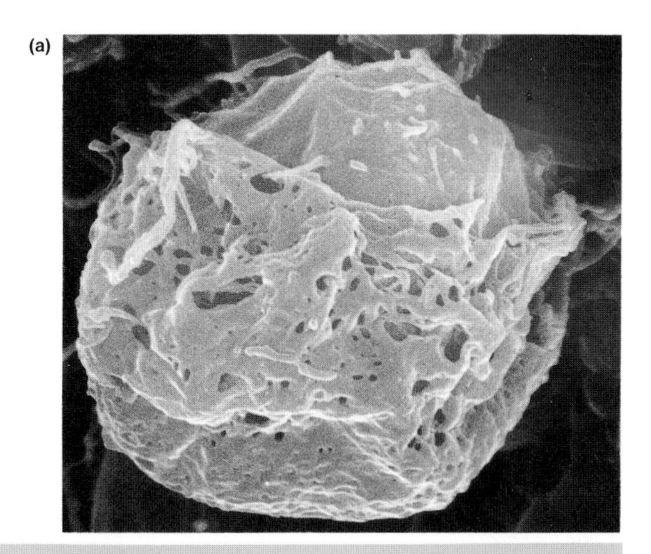

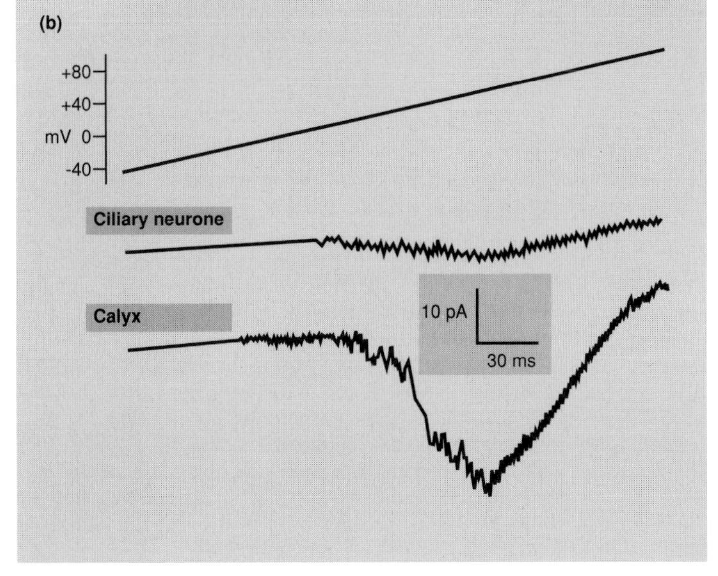

Fig. 7.6 The chick ciliary ganglion calyx. (a) Electron micrograph of a calyciform ending covering more than half of the surface of a ganglion cell. The nerve ending surface shows a lacework appearance. (Magnification ×7000.) From Fujiwara & Nagakuro (1989), by permission of Elsevier Science Ireland Ltd. (b) Whole cell/terminal Ba^{2+} currents recorded on the soma of a ciliary ganglion and at the presynaptic membrane of the calyx terminal during a steady voltage ramp from −40 to +80 mV. Note the much higher density of Ca^{2+}-channels on the terminal. Data from Stanley (1991).

be an exception to this). Instead, the synaptosomal population must be depolarized by one of the methods discussed easier (*see* Section 3.3.1) and the cumulative release of transmitter detected over a finite time interval, which can vary from ~50 ms if rapid mixing and sampling techniques are available to several minutes (Fig. 7.7). There are a number of factors which must be taken into consideration in order to relate these findings to physiological release:

1 The method of depolarization is important. Elevated KCl produces a 'clamped' depolarization for the duration of its application (*see* Section 3.3.1). Thus any Ca^{2+}-channels which

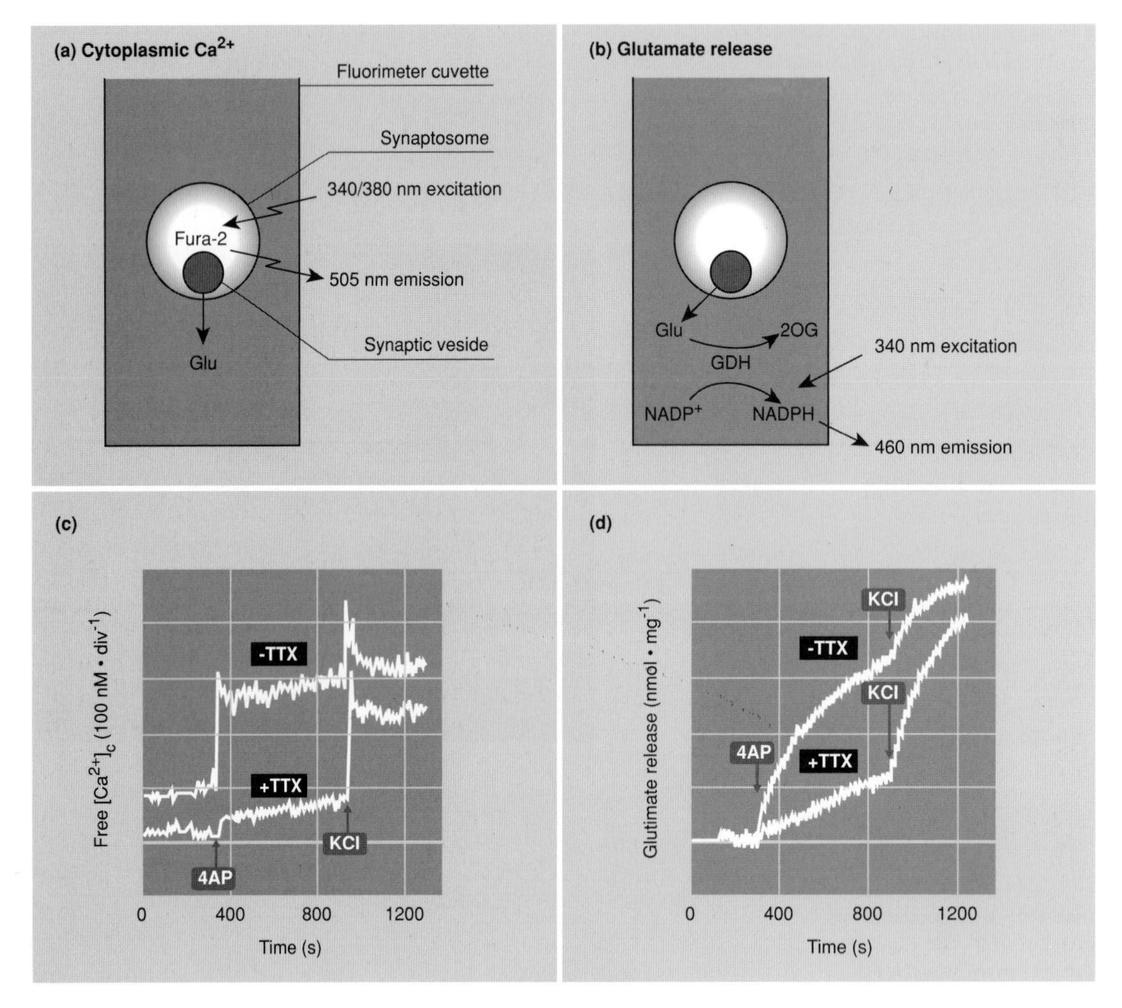

Fig. 7.7 Parallel fluorometric assays of synaptosomal cytoplasmic free Ca^{2+} and glutamate release. (a) Cytoplasmic free Ca^{2+}, $[Ca^{2+}]_c$, is determined in one aliquot of synaptosomes suspended in a fluorimeter cuvette after preloading with fura-2. (b) In parallel, a second synaptosomal resuspension is incubated in medium containing additionally L-glutamate dehydrogenase (GDH) and $NADP^+$. As glutamate is released it is oxidized to 2-oxoglutarate (2OG) by the enzyme and $NADP^+$ is reduced to NADPH whose fluorescence can be monitored. (c, d) A typical experiment: upper traces: 4-aminopyridine (4AP) elevates $[Ca^{2+}]_c$ by inhibiting presynaptic K_A^+-channels responsible for stabilizing the plasma membrane potential allowing spontaneous action potentials to develop. As a consequence glutamate is released. Lower traces: in the presence of the Na^+-channel inhibitor tetrodotoxin (TTX) to inhibit the action potentials, the 4AP-induced elevation in $[Ca^{2+}]_c$ and the release of glutamate are greatly reduced. However the $[Ca^{2+}]_c$ increase and glutamate release evoked by clamped depolarization by 30 mM KCl still occur in the presence of TTX. Glutamate release traces have been corrected for basal release. Adapted from Nicholls & Attwell (1990).

undergo voltage inactivation will only be operative in the first second of depolarization, as is seen with the highly biphasic ^{45}Ca entry into synaptosomes after KCl (*see* Fig. 7.5).

2 If depolarization is prolonged for more than a second, the likelihood is that any pool of synaptic vesicles which were docked at the presynaptic membrane ready for release would become depleted and require replenishment from a 'reserve pool' of vesicles, a process likely to be much slower than exocytosis itself. Therefore prolonged depolarization might be expected to result in a strongly biphasic rate of release. This is seen in the case of the KCl-evoked exocytosis of glutamate from synaptosomes, where 20% of the releasable glutamate is exocytosed within 2 s, while the residue is released with a half-life of about 60 s.

3 During an extended depolarization of synaptosomes, brain slices or cultured neurones there is a Ca^{2+}-independent component which is superimposed on the Ca^{2+}-dependent release, and is due to the reversal of the appropriate electrogenic Na^{+}-coupled neurotransmitter transporter in the plasma membrane. Some Ca^{2+}-independent release can be observed with most transmitters, but the effect is most prominent in the case of the amino acids (*see* Fig. 9.10) and correlates with the duration and extent of the collapse of the Na^{+} electrochemical potential across the plasma membrane.

7.3 Localized Ca^{2+} and the trigger for SSV exocytosis

Although the identity of the Ca^{2+} trigger which induces the fusion of synaptic vesicles with the plasma membrane is still debated (*see* Section 7.4), the submillisecond delay between depolarization and transmitter release which is determined by electrophysiology puts certain constraints on the properties that this entity must possess. Firstly the trigger must be closely associated with the Ca^{2+}-channel, since there is no time for Ca^{2+} to diffuse and equilibrate with the cytoplasm. Secondly it must be assumed that some vesicles are already docked at the release site, since there would be no time for vesicles to diffuse to the membrane. It is also unlikely that there is time for covalent modification such as phosphorylation to occur. Finally the frequency with which repetitive nerve stimulation can occur requires that Ca^{2+} must rapidly dissociate from the trigger.

At the NMJ, where some 200 vesicles are released per action potential, computer modelling has shown that a two-dimensional array of VACCs at the active zone opening synchronously in response to an action potential could maintain a highly localized submembrane Ca^{2+} concentration in the region of 100 μM for the duration of the channel opening. Experimental support for this localized Ca^{2+} has come from studies in which the giant presynaptic terminal of the squid is injected with fura-2 or with the Ca^{2+}-dependent photoprotein aequorin; on stimulation of the axon the Ca^{2+} signal was highest immediately underneath the presynaptic membrane (Plate 3, between pages 132 and 133). It should be emphasized that local gradients of free Ca^{2+} are possible because some 99.9% of total Ca^{2+} is at any one instant bound to relatively immobile cytoplasmic proteins, thus the diffusion of free Ca^{2+} is greatly slowed.

An elegant demonstration that the ability of a Ca^{2+} chelator to inhibit exocytosis depends more on the speed at which it will chelate Ca^{2+} rather than its absolute affinity for the cation has been performed on the squid giant synapse. Injection of the high-affinity chelator EGTA was ineffective, whereas less powerful but faster acting BAPTA derivatives reduced transmitter release. The K_d's of effective BAPTAs suggest that the peak $[Ca^{2+}]$ may be in excess of 100 μM and that the Ca^{2+}-binding trigger may act in less than 200 μs and be located very close to the Ca^{2+}-channels.

With synaptosomes it is possible to show that there is a close coupling between Ca^{2+} entry and glutamate release (Fig. 7.8). When the release of glutamate is correlated with the fura-2 signal during depolarization evoked by KCl a very steep relationship is obtained, with zero exocytosis at the resting $[Ca^{2+}]_c$ of about 200 nM, and maximal release at 400 nM. This either means that the triggering of release by Ca^{2+} requires only a slight change in $[Ca^{2+}]_c$ or that fura-2 is failing to detect the actual Ca^{2+} concentration at the release site. The latter appears to be the case: since fura-2 measures the averaged cytoplasmic Ca^{2+} in the whole synaptosomal preparation, and as it is saturated by 2–5 μM Ca^{2+}, it will not distinguish localized areas of high Ca^{2+} in the immediate vicinity of the VACCs. That these

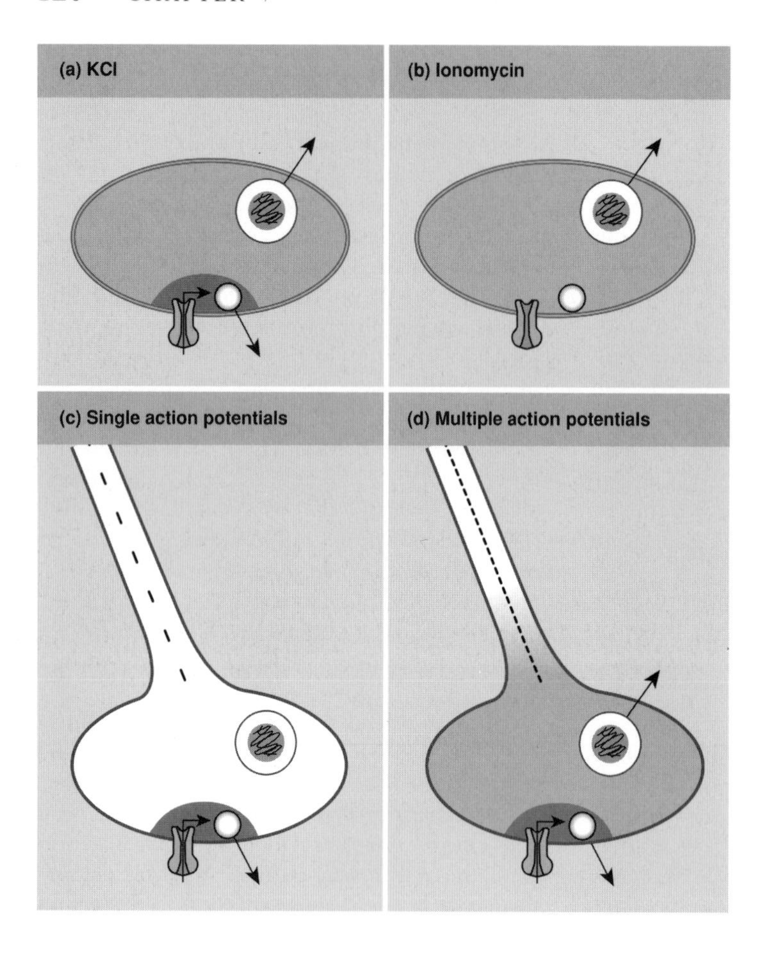

Fig. 7.8 Localized Ca^{2+} coupling occurs in CNS terminals for amino acid but not neuropeptide exocytosis. (a, b) KCl depolarization releases both amino acids (in small synaptic vesicles, SSVs) and neuropeptides (in large dense-core vesicles, LDCVs) from synaptosomes, whereas a uniform elevation in $[Ca^{2+}]_c$ by ionomycin releases only the peptide. This is interpreted as indicating the need for a 'hot-spot' of high $[Ca^{2+}]$ in the immediate vicinity of the Ca^{2+}-channel (which is not active in the ionomycin experiment) to release SSVs, while LDCVs are released by a modest elevation in bulk $[Ca^{2+}]_c$. See Verhage *et al.* (1991). (c, d) Single action potentials release only SSVs from intact neurones; repetitive stimulation is required to release neuropeptides. This is consistent with the synaptosome data if single action potentials generate the 'hot-spot' but are insufficient to elevate bulk $[Ca^{2+}]_c$.

exist, and are responsible for the triggering of amino acid release, can be inferred by the failure of the Ca^{2+}-ionophore ionomycin (which would not create 'hot-spots' of high local $[Ca^{2+}]$ at the active zone) to release glutamate even though the same bulk cytoplasmic concentration, is attained (Fig. 7.8). This suggests that the Ca^{2+} trigger for amino acid exocytosis has a low affinity for Ca^{2+} and requires a high Ca^{2+} concentration in the immediate vicinity of the mouth of the VACC.

A close proximity between Ca^{2+}-channels and the release trigger has great kinetic advantages both for the initiation and the termination of exocytosis. Onset of release can be very rapid because the Ca^{2+} has only a very short distance to travel after emerging from the mouth of the channel, and because it is not necessary to raise $[Ca^{2+}]_c$ in the entire terminal. Since $[Ca^{2+}]_c$ in the vicinity of the channel mouth can be very high, say 10–100 μM, the affinity of the trigger for Ca^{2+} can be relatively low; this in turn means that when the Ca^{2+}-channel closes and Ca^{2+} diffuses away from the channel, Ca^{2+} will rapidly dissociate from this low-affinity site, terminating exocytosis without the need to pump bulk Ca^{2+} out of the cytoplasm. It is of interest that the release of *neuropeptides* from CNS terminals differs and appears to utilize a high-affinity Ca^{2+} receptor which responds to bulk $[Ca^{2+}]_c$ (Fig. 7.8).

There are two forms of localized Ca^{2+} coupling which must be considered (Fig. 7.9). The first is that discussed above for the NMJ and the squid giant synapse, where it is necessary to activate synchronously an array of Ca^{2+}-channels in order to create a highly transient Ca^{2+} gradient under the membrane. However an alternative is that the Ca^{2+} gradient may be even more localized such that Ca^{2+} entering through a single Ca^{2+}-channel may impinge on the Ca^{2+} trigger at a single vesicle docking site and trigger

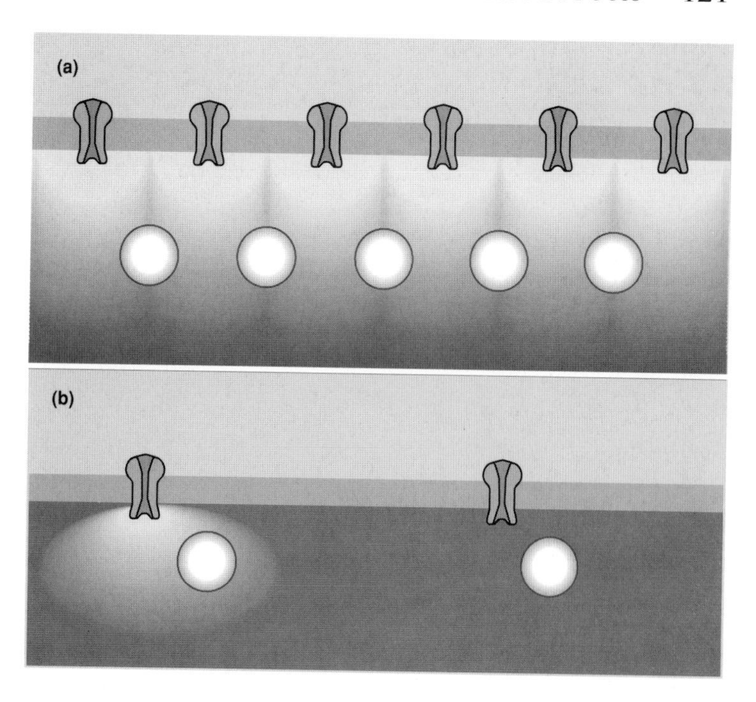

Fig. 7.9 Localized Ca^{2+} coupling schemes. Two variants of localized Ca^{2+} coupling to exocytosis. In (a) an array of Ca^{2+}-channels creates a two-dimensional microdomain of high Ca^{2+} immediately under the presynaptic membrane, triggering the synchronous release of many synaptic vesicles; this may occur at the neuromuscular junction and the squid giant synapse. (b) Individual Ca^{2+}-channels may create highly localized domains each controlling a single release site. In this way the release of precise numbers of quanta would be related to the probability of opening of individual channels. This mode may be applicable to low-quantal release CNS synapses where presynaptic control of the amount of released transmitter is essential.

exocytosis of one vesicle (Fig. 7.9). While the channel array is most appropriate at the NMJ to release synchronously large numbers of synaptic vesicles, it is a little difficult to see how this would allow Ca^{2+}-channel activity to exert the fine and precise control of the extent of exocytosis which appears to occur in CNS synapses, since these may demonstrate a low quantal release, perhaps only a single vesicle for several action potentials.

The unique morphology of the ciliary ganglion calyx (*see* Fig. 7.6) can help to answer this question. Although patch-clamp reveals that Ca^{2+}-channels are highly clustered on the presynaptic face of the calyx at a density of some 140 channels μm^{-2}, the stimulation of single Ca^{2+}-channels can trigger release of ACh, which can be detected by a sensitive luminescent assay. Thus the entry of only some 200 Ca^{2+} ions though one channel, which can be calculated to be sufficient to give a 20 nm diameter 'dome' within which Ca^{2+} is above 10 μM, can trigger exocytosis of a vesicle.

7.4 Protein chemistry of SSV exocytosis

The exocytotic–endocytotic cycle for SSVs in the nerve terminal (Fig. 7.10) requires the sequential filling of vesicles with transmitter, their specific translocation to the active zone of the presynaptic plasma membrane, docking of the vesicles with the plasma membrane ready for exocytosis, exocytosis itself, membrane retrieval and recycling. Although a complete picture of the protein chemistry of this cycle still eludes us, impressive advances have been made in recent years in the characterization of proteins specific to synaptic vesicles, the active zone, vesicle–cytoskeletal interactions and proteins involved in membrane traffic (Fig. 7.11). Once more it is necessary to emphasize that there is no universal exocytotic mechanism, and in particular that the release mechanism for small synaptic vesicles differs in several respects from that for chromaffin vesicles and LDCVs, which will be considered separately in Chapter 8.

The enormous speed of SSV exocytosis strongly suggests that a number of synaptic vesicles must be docked at the presynaptic membrane in advance of the arrival of the action potential. At the NMJ a rather precise number of vesicles are released in response to each action potential (about 200 out of a total complement of some 1000 000), which is much smaller than the total number of synaptic vesicles apparently docked at the presynaptic active zones, and indicates some precise form of interaction of this subpopulation with the membrane. The small pool can be

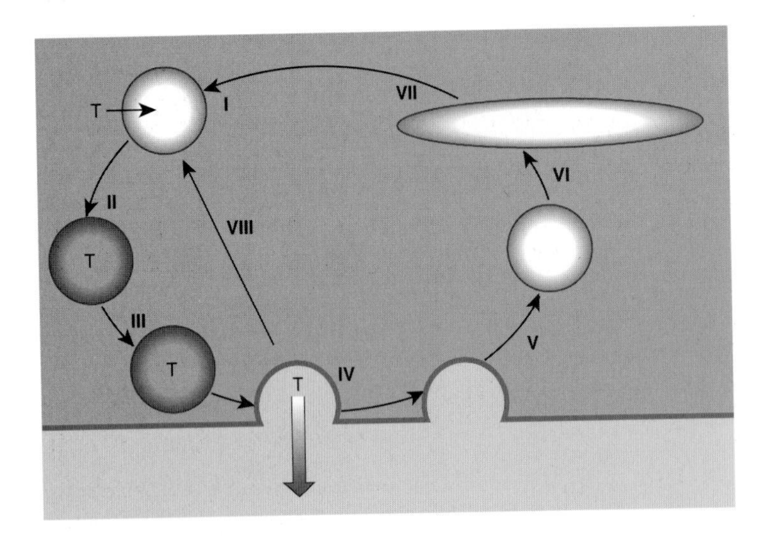

Fig. 7.10 Small synaptic vesicle movement in the nerve terminal. Vesicles accumulate transmitter (stage I), move to the plasma membrane (stage II), dock at the active zone (stage III) and release their contents in response to Ca^{2+} entry (stage IV). There is debate as to the relative importance of endocytosis via coated pits (stage V) and recycling via an endosomal intermediate (stages VI and VII) and direct retrieval of the vesicles after a brief 'kiss-and-run' contact with the presynaptic membrane (stage VIII).

replenished within 10 ms ready for the arrival of the next action potential. Simple lipid bilayer fusion is too slow by a factor of 1000 to account for the rate of exocytosis and is not significantly increased in model systems by potentially 'fusogenic' proteins. It is much more likely that an oligomeric fusion protein from the synaptic vesicle, or a complex formed from components of both the synaptic vesicle and presynaptic plasma membranes, forms a closed fusion pore across both membranes (Fig. 7.12). Fusion would be triggered by a conformational change induced by Ca^{2+} entering through closely coupled voltage-activated channels causing an opening of the fusion pore. If phospholipids diffused between the fusion pore monomers, the fusion 'neck' could dilate and allow an eventual collapse of the vesicle membrane into the plasma membrane. While it is important not to extrapolate automatically between the exocytosis of SSVs and larger secretory vesicles, the transient current flowing through such a fusion pore can be detected in whole-cell patched mast cells from beige mice, which contain enormous secretory vesicles (*see* Section 8.2.2).

7.4.1 Synaptic vesicle transporters

Although these will be considered in detail in the chapters devoted to specific neurotransmitter systems, a few points should be emphasized here. Firstly the catecholamines, histamine and serotonin probably all share a common vesicular transporter, with the result that the transmitter specificity of these terminals is determined by the biosynthetic enzymes present in that terminal. Similarly the inhibitory amino acid transmitters glycine and GABA share a common vesicular transporter, whereas the corresponding glutamate transporter is highly specific (in particular not transporting aspartate which coexists in the cytoplasm) so that the carrier now becomes the determinant of transmitter specificity. The driving force for the uptake of transmitters is the ΔpH or the membrane potential component of the proton electrochemical potential generated by the proton-translocating V-ATPase (*see* Section 4.3).

One synaptic vesicle transporter which may be generally present regardless of the transmitter is *SV2*. This is an 83 kDa glycoprotein (with glycosylation sites within the lumen of the vesicle) which has recently been cloned and shown to possess 12 putative transmembrane segments, cytoplasmic N- and C-termini and an overall homology to a family of bacterial sugar transporters. It has been suggested that SV2 may comprise the Cl^- transporter which is generally present in synaptic vesicles as a means of transmembrane charge neutralization.

7.4.2 SNAPs and SNAREs

The central questions concerning membrane traffic in general are the mechanisms of vesicle budding, or fission, membrane targetting and

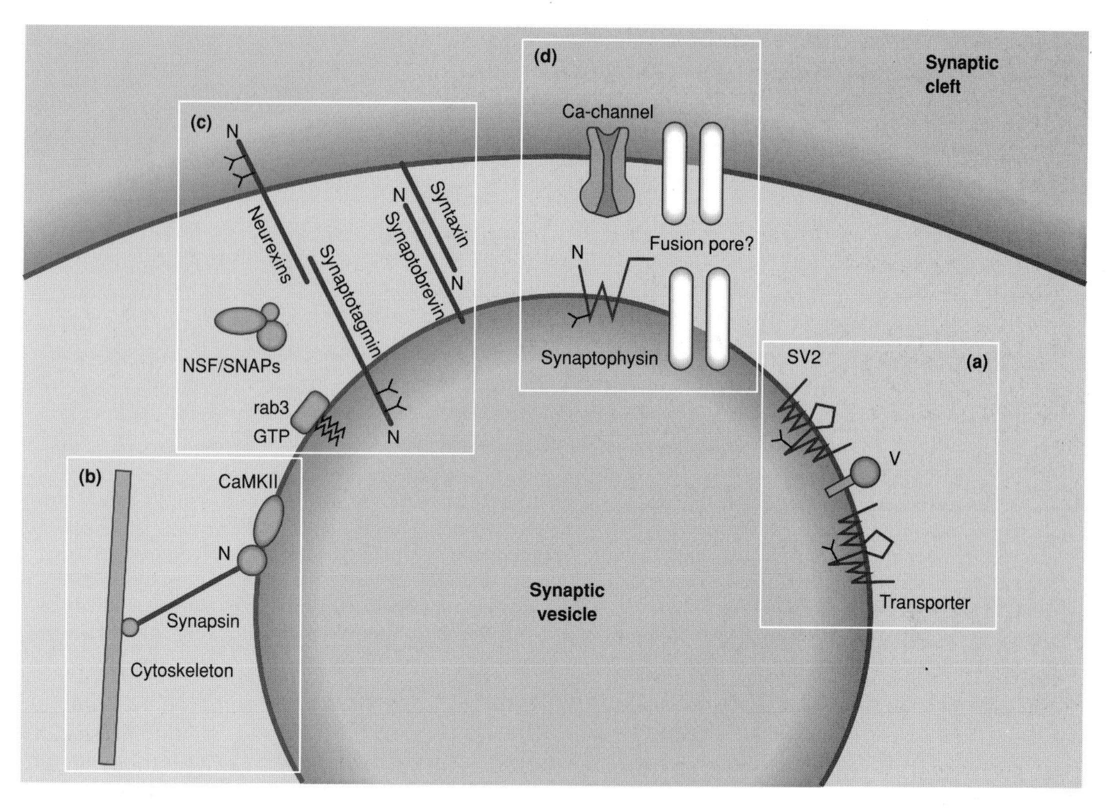

Fig. 7.11 An overview of proteins implicated in SSV targetting and exocytosis. (a) *Proteins implicated in vesicle refilling*: the V-ATPase (V) acidifies the lumen of the vesicle and generates a membrane potential positive inside; SV2 is a ubiquitous protein homologous to bacterial transporters and may possibly function as a Cl^- transporter neutralizing charge across the vesicle membrane; vesicular transmitter transporters typically have a 12-transmembrane structure. (b) *Proteins associated with vesicle–cytoskeletal interactions*: CaMKII is peripherally associated with vesicles and serves as an attachment point for synapsin I which can cross-link vesicles to F-actin. (c) *Proteins associated with targetting and docking*: rab3 is a small G-protein which may regulate membrane traffic (*see* Fig. 7.17); the *N*-ethylmaleimide-sensitive fusion protein (NSF) and soluble NSF attachment proteins (SNAPs) are ubiquitous cytoplasmic proteins which recognize specific SNAP-receptors (SNAREs) on pairs of membranes causing them to associate and fuse. Syntaxins A and B on the presynaptic membrane and synaptobrevin, also known as VAMP-2 function as SNAREs (*see* Fig. 7.13). Synaptotagmin is a ubiquitous vesicular protein which associates with presynaptic neurexins and syntaxin and may regulate release. (d) *Proteins associated with the fusion event*: the fusion pore has not been unequivocally identified, although synaptophysin has been proposed as a vesicular component. The Ca^{2+}-channel must be closely associated with the fusion pore since highly localized Ca^{2+} triggers release.

membrane fusion. The protein chemistry of membrane traffic associated with the Golgi apparatus and constitutive secretion has been largely elucidated by analysis of yeast mutations and by *in vitro* reconstitution of mammalian Golgi transport. A brief overview is of relevance here since it has recently become apparent that the regulated exocytosis of SSVs utilizes essentially the same mechanism as constitutive secretion, with additional regulatory components.

Membrane is transferred from the endoplasmic reticulum to the Golgi and within Golgi stacks as 70 nm vesicles. This transport can be reconstituted *in vitro* with isolated Golgi stacks together with cytoplasmic factors and ATP, and monitored by the glycosylation state of the vesicles which changes as the membrane is transported through the Golgi. Vesicle formation begins with the generation of a coated bud in the donor membrane. The coat contains two proteins: (i) the *coatomer*, consisting of seven polypeptides, and (ii) *ADP-ribosylation factor*, arf — a small

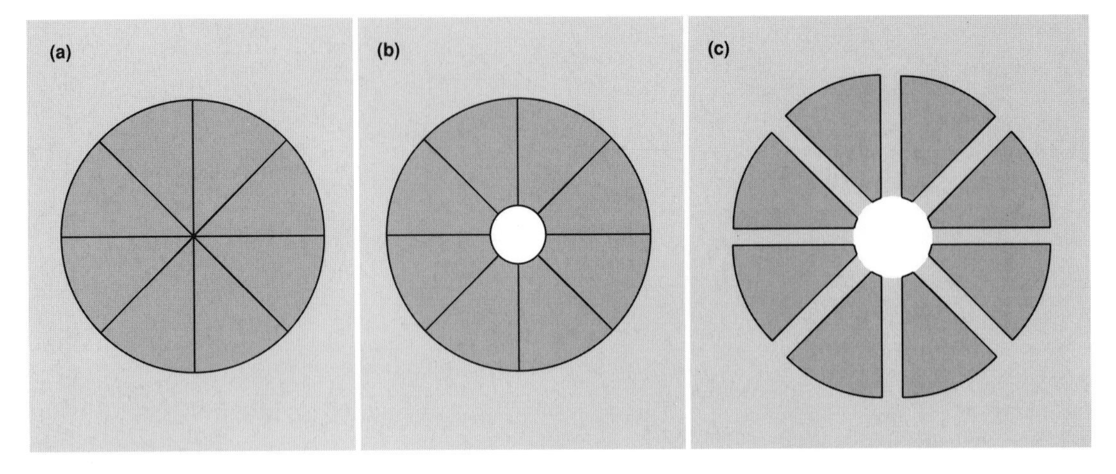

Fig. 7.12 A model for a fusion pore. In the dispersing subunit model a specific oligomeric fusion protein in the synaptic vesicle membrane would insert itself into the presynaptic membrane forming a protein bridge between the membranes in an analogous way to the connexion of a gap junction bridging two cells. It is also possible that proteins from the synaptic vesicle and presynaptic membrane both contribute to this complex. In (a) the oligomer is closed. In (b) Ca^{2+} has induced a conformational change opening a central fusion pore allowing the contents of the synaptic vesicle to exocytose. In (c) phospholipids infiltrate between the monomers, allowing the fusion pore to dilate rapidly. Adapted from Almers & Tse (1990).

G-protein related to ras. Arf-GTP is membrane associated via an apparently extended myristoyl anchor while arf-GDP is cytoplasmic, suggesting that the fatty acid anchor is retracted. Attachment of the coat to the budding membrane occurs via this myristoyl anchor, and a GDP/GTP exchange enzyme converts arf-GDP to arf-GTP. *Brefeldin A* blocks the GDP/GTP exchange by inhibiting the nucleotide exchange enzyme. For fission to occur *acylCoA* is additionally required.

Fusion of these vesicles with the acceptor membrane requires removal of the coat. This in turn involves hydrolysis of the GTP bound to arf, allowing dissociation of the G-protein and uncoating. Thus the non-hydrolysable analogue *GTP-γ-S* causes the accumulation of coated vesicles and inhibits fusion. The arf-GTP hydrolysis factor is not currently characterized.

N-ethylmaleimide, NEM, prevents fusion of the uncoated vesicle with the acceptor membrane by inhibiting a soluble *NEM-sensitive factor* (NSF). NSF is a tetramer of 76 kDa subunits. (Fig. 7.13). In cell-free systems NSF, which in yeast is coded for by the *SEC18* gene, appears to be essential for a variety of intracellular fusion processes since in its absence vesicles accumulate at the acceptor membrane without fusing. *Soluble NSF attachment proteins* or SNAPs with M_rs from 35 to 39 kDa associate with NSF and allow the complex to recognize and bind to *SNAP receptors*, SNAREs, present on specific membranes. NSF is a Mg-ATPase, and ATP addition to detergent-solubilized 20S NSF/SNAP/SNARE complexes causes a dissociation of NSF, a property which can be exploited to identify SNAREs in whole brain homogenates (Fig. 7.13). The three SNAREs which were identified in this way from the whole brain homogenate were each previously characterized presynaptic proteins: *synaptobrevin*, *syntaxin-B* and *SNAP-25* (synaptosomal-associated protein 25).

Synaptobrevin is a 18 kDa, 116 residue, integral synaptic vesicle membrane protein. The protein exists as two isoforms having a single transmembrane segment and cytoplasmic N-terminal domain rich in proline. Synaptobrevin is restricted to small synaptic vesicles and small vesicles of peptide-secreting endocrine cells but is not found in LDCVs. Its presence throughout different brain regions suggests that it is not transmitter specific. Closely homologous proteins are found in *Drosophila* and in *Torpedo* synaptic vesicles, named VAMP-1 and VAMP-2. Synaptobrevin is a target for the potent *Clostridial* botulinum neurotoxin B and tetanus toxin (Box 7.1 and Fig. 7.14).

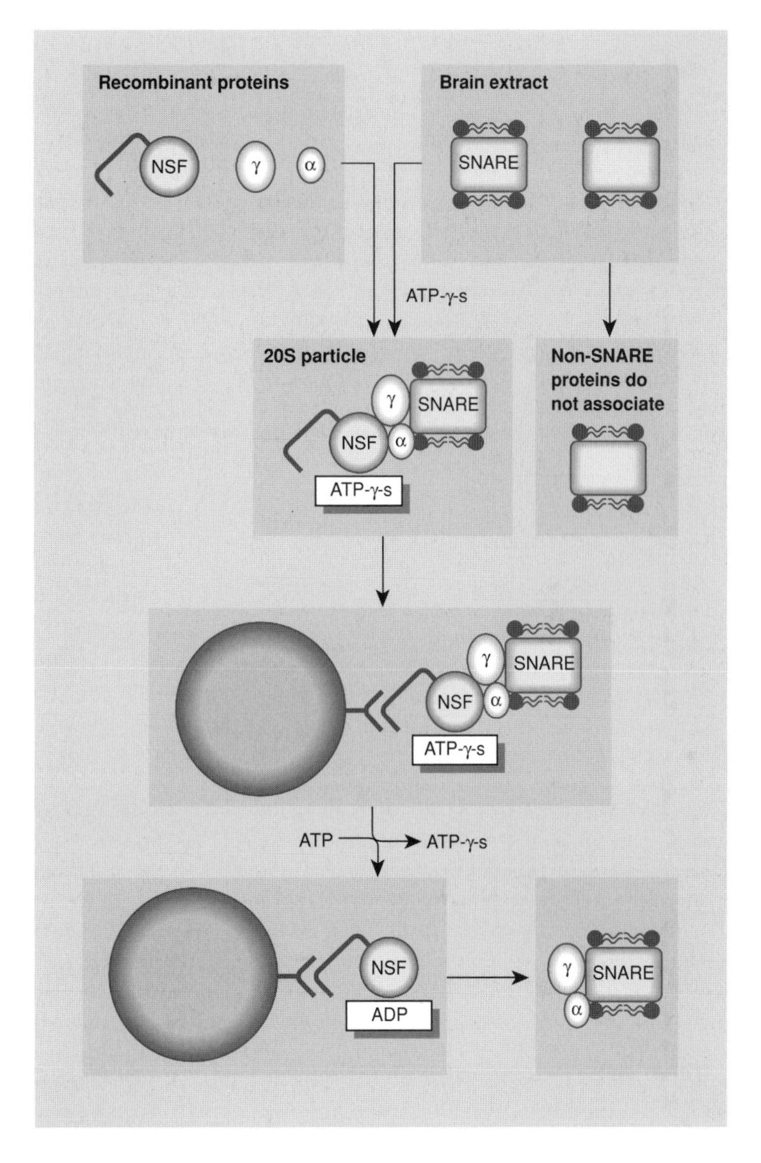

Fig. 7.13 The identification of SNAP receptors (SNAREs). A Triton X-100 extract of bovine brain was mixed with recombinant α- and γ-SNAPs together with *N*-ethylmaleimide-sensitive fusion protein (NSF). In the presence of a non-hydrolysable ATP analogue, ATP-γ-S, a 20S soluble complex was formed. NSF possessed a C-terminal extension allowing its attachment to immunobeads, which were formed into a column. Non-specific proteins were eluted, after which MgATP was substituted for ATP-γ-S, causing SNAP/SNARE complexes to dissociate from the bound NSF. Gel electrophoresis of this specific eluate revealed, in addition to the SNAPs, three major bands which were identified by microsequencing as the plasma membrane located syntaxin-B, the vesicular synaptobrevin-2 and synaptosomal associated protein-25 (coincidentally called SNAP-25). Data from Söllner *et al.* (1993).

The second protein identified as a SNARE is *syntaxin-B*, which was previously identified as a 35 kDa carboxy-terminal anchored protein also called *HPC-1* and associated with active zones. This is of particular interest since its location suggests that it might be a target or t-SNARE, and that the NSF/SNAP complex could serve to cross-link the synaptic vesicle to the presynaptic membrane. A second botulinum neurotoxin serotype, C1, has been shown to cleave syntaxin.

The localization of *SNAP-25* within the presynaptic terminal is currently unclear. Although it lacks obvious hydrophobic regions, it behaves as an integral protein, probably due to multiple palmitoylation of C-terminal cysteine residues. Botulinum neurotoxin A cleaves SNAP-25.

NSF, SNAPs and SNAREs have homologues in yeast where they catalyse constitutive membrane transport and fusion. This suggests that the default condition may be membrane fusion, and that in regulated secretion additional proteins are required to inhibit the final fusion event until the appropriate signal is received in the form of Ca^{2+} entry via the voltage-activated channel. The key protein in this regulation may be *synaptotagmin*.

Box 7.1 Clostridial neurotoxins

Tetanus toxin (TeTx) and the family of botulinum neurotoxins (BoNT types A–G) are synthesized as 150 kDa single-chain protoxins containing one intrachain disulphide bridge. Proteolysis ('nicking') by trypsin or a protease endogenous to the bacterium gives the active toxin containing 50 and 100 kDa chains linked by the disulphide (Fig. 7.14). The toxins are taken up *in vivo* into peripheral terminals via initial binding to high-affinity receptors associated with (but probably not purely composed of) gangliosides and are internalized by subsequent receptor-mediated endocytosis. Although BoNT serotypes A and B have different sites of action, both inhibit ACh release at the neuromuscular junction and other peripheral cholinergic terminals. TeTx can inhibit at the neuromuscular junction, where it has a site of action in common with BoNT type B (see below) but primarily blocks inhibitory synapses of the spinal cord, which it accesses by retrograde transport up the axon of the peripheral neurone, at a rate of 3–30 $mm \cdot h^{-1}$, apparently utilizing the same retrograde transport system as nerve growth factor. At the cell body, TeTx crosses a synaptic cleft to the inhibitory terminal. Thus the convulsant effects of TeTx are due to unopposed glutamatergic excitation while the flaccid paralysis of BoNT is due to cholinergic block.

In vitro the toxins are less specific and will inhibit most exocytotic systems, including glutamatergic terminals and chromaffin cells. The 100 kDa heavy chains are responsible for targetting the toxin to the presynaptic terminals and delivering the neurotoxin light chains into the terminals where they block exocytosis. Ion channels are not affected by the toxins, but the light chains of TeTx and BoNt type B function as Zn^{2+} metalloproteases after reduction of the interchain disulphide and will cleave synaptobrevin-2 on purified synaptic vesicles. Synaptobrevin-1 which is found together with synaptobrevin-2 on the rat vesicles differs at the cleavage site and remains intact. In *Aplysia* neurones, injection of a synthetic peptide corresponding to the cleavage site appears to protect by competing for the toxin. It is not yet clear whether synaptobrevin-2 is the sole site of action of the toxins, nor whether they act by a different mechanism to inhibit exocytosis in cells whose vesicle apparently lack synaptobrevin-2.

7.4.3 Synaptotagmin

Synaptotagmin (p65 or 65 kDa calmodulin-binding protein) is a major calmodulin-binding protein which is present on all synaptic vesicles and chromaffin granules. The protein is highly conserved with a single putative transmembrane span and a large cytoplasmic C-terminal domain which binds calmodulin and also has a repeated sequence which is homologous to the regulatory domain of PKC and, like the kinase, interacts with acidic phospholipids such as phosphatydylserine. As in the case of synaptobrevin, specific neurotoxins have provided an important clue to the action of synaptotagmin. The effect of the toxin will first be described, followed by the evidence linking the toxin action to synaptotagmin.

The venom from the black widow spider, *Lactrodectus mactans*, contains a 130 kDa polypeptide toxin, *α-latrotoxin*, which has a dramatic effect on vesicle exocytosis (Fig. 7.15). This is best characterized at the NMJ, where nanomolar toxin causes a massive exocytosis of synaptic vesicles even in the absence of external Ca^{2+}. Under these conditions, when endocytosis appears to be blocked, the vesicular membrane incorporates into the presynaptic membrane. The toxin also causes an extensive Ca^{2+} uptake (not via voltage-activated Ca^{2+}-channels) and depolarization. In some systems, e.g. cerebrocortical synaptosomes, this permeabilization so dominates that ATP levels collapse and block exocytosis — the massive release of transmitter being of cytoplasmic origin.

α-Latrotoxin binds with subnanomolar affinity to a presynaptic acceptor protein. When presynaptic proteins are subjected to affinity chromatography on an immobilized α-latrotoxin column, and after allowing for less specific interactions, an 'α-latrotoxin acceptor' was retained consisting of at least five high molecular weight components (160–22 000 kDa) together with low molecular weight, 29 kDa, peptides. Importantly, synaptotagmin was also associated with the acceptor, binding to the toxin–acceptor complex but not to the toxin itself. This suggests a trimeric association of toxin/acceptor/synaptotagmin.

As further evidence for an interaction between the α-latrotoxin acceptor and synaptotagmin, the acceptor was found to inhibit phosphorylation of the latter protein, probably mediated by

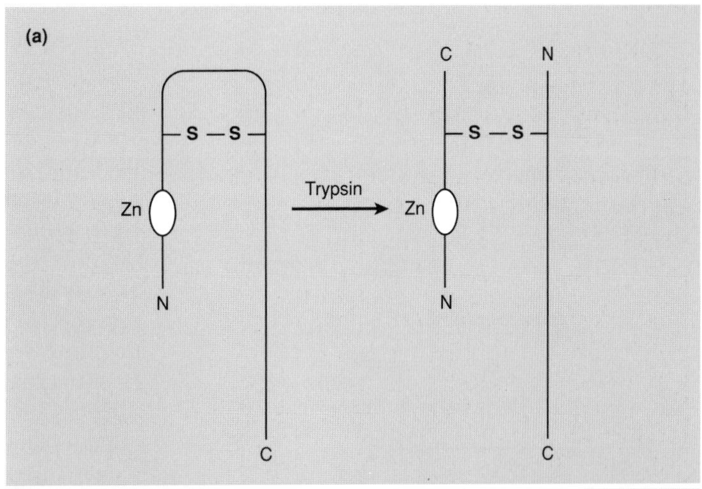

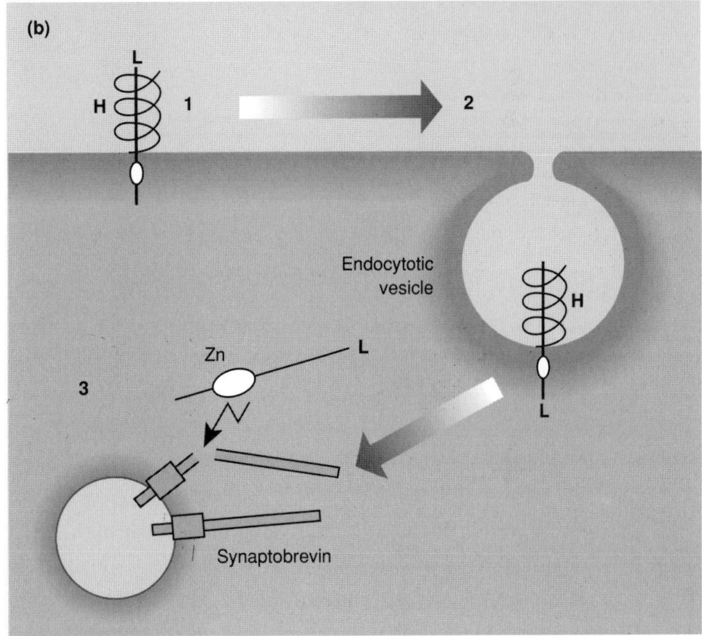

Fig. 7.14 Botulinum and tetanus neurotoxins. (a) Tetanus and botulinum neurotoxins are synthesized as single polypeptides of approximately 150 kDa. Cleavage by trypsin, or a trypsin-like protease generate the active toxin with two chains linked by a disulphide bridge. (b) The H-chain binds to a surface acceptor (1) and, probably after internalization into an endocytotic vesicle, injects the L-chain into the cytoplasm (2). The L-chain has Zn-endopeptidase activity. The essential target for the inhibition of exocytosis by tetanus and botulin neurotoxin type B has been proposed to be the vesicle-specific protein synaptobrevin, which can be cleaved by the peptidases (3).

CaMKII. The implication is that the latrotoxin acceptor may form part of the plasma membrane half of a docking complex involving vesicular synaptotagmin. It has been suggested that synaptotagmin may form the Ca^{2+} sensor in exocytosis although it must be borne in mind that the protein has been reported not to interact with Ba^{2+} up to 1 mM, even though this cation is capable of triggering exocytosis (*see* Section 7.4.6). An association has also been reported between synaptotagmin and N-type Ca^{2+}-channels (or rather ω-conotoxin-GVIA binding sites) mediated by the SNARE syntaxin.

There is controversy as to whether synaptotagmin is obligatory for exocytosis. In one report a PC12 cell line largely deficient in synaptotagmin could still secrete normally, whereas other reports indicate that injection of antisynaptotagmin antibodies or of peptides mimicking the cytoplasmic C2 domain of the protein diminish exocytosis (Fig. 7.16). It should be noted that PC12 secretion of noradrenaline is more closely related to LDCV exocytosis than to SSVs. Finally, synaptotagmin has been implicated in the autoimmune disease of the NMJ, Lambert–Eaton myasthenic syndrome (Box 7.2).

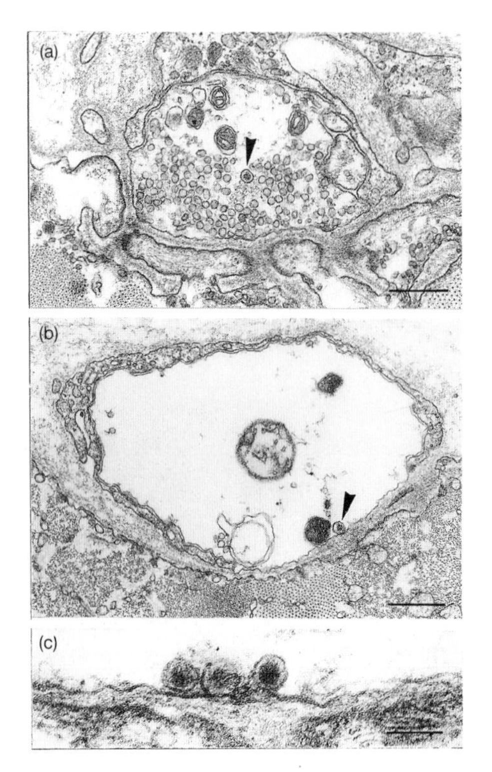

Fig. 7.15 α-Latrotoxin releases SSVs but not LDCVs at the neuromuscular junction (NMJ). α-Latrotoxin causes a massive Ca^{2+}-independent exocytosis at the NMJ: (a) cross-section of a control terminal, bar = 340 μm; (b) terminal exposed to toxin for 1 h — the terminal is completely depleted of SSVs although LDCVs are still present, bar = 530 μm (shown at higher magnification in (c), bar = 180 μm). From Matteoli *et al.* (1988), reproduced with permission from the National Academy of Sciences.

Box 7.2 Lambert–Eaton myasthenic syndrome

Lambert–Eaton myasthenic syndrome (LEMS) is an autoimmune disease of the neuromuscular junction. The pathological autoantibodies appear to downregulate presynaptic Ca^{2+}-channels at the neuromuscular junction, causing muscle weakness. *In vitro* the antibodies will precipitate ω-conotoxin-GVIA labelled N-channels from solubilized neuroblastoma cells. The antigen however is not the Ca^{2+}-channel itself but rather a 58 kDa peptide closely related or identical to synaptotagmin, providing further evidence for a close relation between synaptotagmin and the Ca^{2+} N-channel.

In vitro association experiments indicate that brain SNAREs interact directly with each other without requiring either NSF or SNAPs. It therefore appears that vesicle docking might occur by SNARE interaction, and provide the mechanism by which the vesicle membrane recognizes and fuses with the plasma membrane at the active zone, rather than with another synaptic vesicle or other inappropriate membrane. Association experiments also show that synaptotagmin prevents the binding of the soluble SNAPs. It has therefore been proposed that fusion is prevented by synaptotagmin. Ca^{2+} binding to the exocytotic trigger (which, as discussed above, may be synaptotagmin itself or an additional component) would allow the SNAPs to displace synaptotagmin and allow NSF to bind. Hydrolysis of the ATP bound to NSF would disrupt the SNARE/NSF/SNAP complex allowing fusion to proceed.

7.4.4 Synaptophysin

Synaptophysin (also called p38) is the major integral protein of the membrane of small synaptic vesicles from central and peripheral neurones as well as neuroendocrine cells, but is only present at very low concentration in LDCVs and chromaffin vesicles. This differential location may underlie some of the differences in the release mechanisms for small and LDCVs.

Synaptophysin is a characteristic integral protein with four predicted transmembrane domains and a large cytoplasmic proline-rich C-terminal domain. A single *N*-glycosylation site is located within the vesicle. The 307 amino acid protein has a molecular weight of 38 kDa, which is reduced to 34 kDa on deglycosylation. It binds Ca^{2+} *in vitro* although the physiological relevance of this has been disputed. Synaptophysin can be phosphorylated in intact synaptosomes on serine residues by CaMKII and on tyrosine residues by the proto-oncogene product $pp60^{c\text{-}src}$ which is enriched in vesicle fractions. Unlike tyrosine receptor kinases, $pp60^{c\text{-}src}$ lacks transmembrane domains but is associated with the bilayer via N-terminal myristoylation.

The structure of synaptophysin is similar to that of the hexameric *connexin* proteins which form gap junctions between cells, and the relationship is strengthened by the finding that synaptophysin readily forms hexameric structures in the synaptic vesicle membrane with a

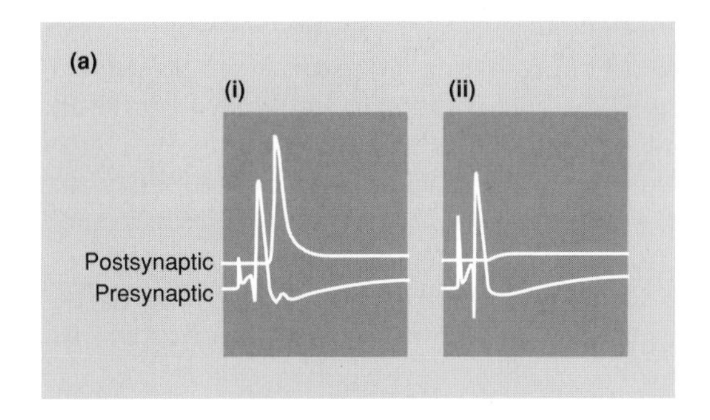

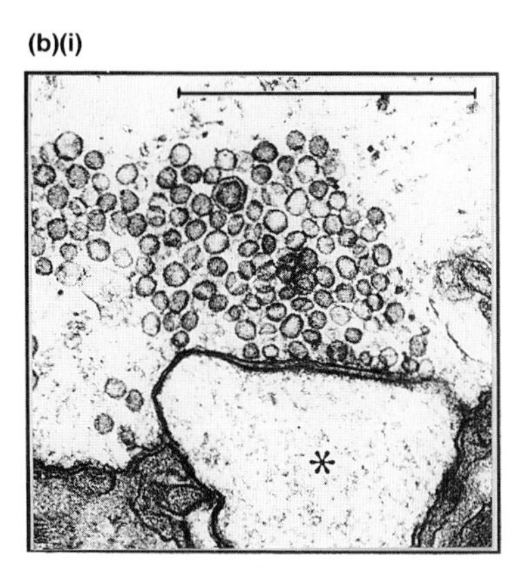

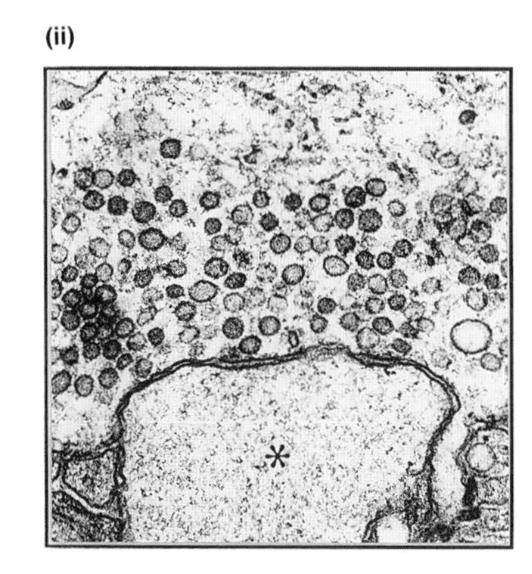

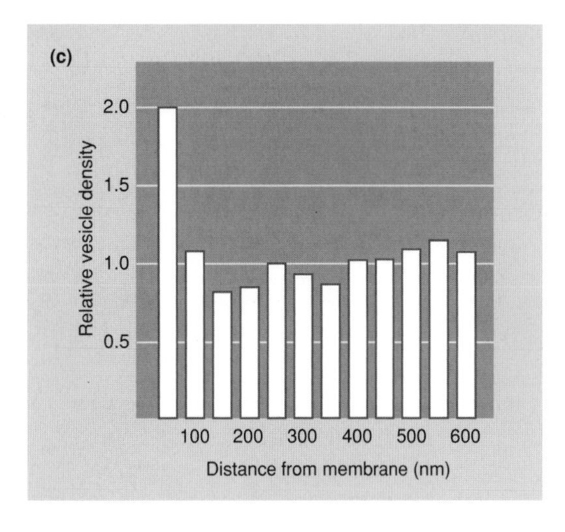

Fig. 7.16 The involvement of synaptotagmin in exocytosis from squid giant terminals. (a) Injection of a 20 amino acid peptide based on the C2-domain sequence of synaptotagmin blocks exocytosis: (i) control, (ii) after peptide injection, note the disappearance of the evoked postsynaptic action potential. (b) Electron micrographs of peptide-injected terminals show an accumulation of vesicles at the active zone (i) relative to controls (ii). Bar = 1 μm. (c) Consistent with the block of exocytosis, the peptide causes a doubling of vesicle density immediately adjacent to the active zone membrane. From Bommert *et al.* (1993), reproduced with permission from *Nature*. © 1993 Macmillan Magazines Ltd.

molecular weight of 230 kDa. Negatively stained preparations of Triton X-100-solubilized and purified synaptophysin reveal a largely homogeneous preparation of 8 nm particles with suggestions of a 1–2 nm central pit. Purified synaptophysin reconstituted into planar lipid bilayers gives channels with a 150 pS conductance, whose open probability increases with positive potential, while no activity is seen at negative potentials. Antibodies to synaptophysin injected into the NMJ inhibit neurotransmission, although this does not distinguish between an integral role for synaptophysin and a non-specific block of exocytosis due to binding of the bulky antibody molecule. *P29* shows a very similar distribution to synaptophysin, being present on small but not large synaptic vesicles. It is an integral protein largely exposed on the cytoplasmic face. It has sufficient homology to synaptophysin for cross-reaction of monoclonal antibodies.

Obviously any channel in the synaptic vesicle membrane could not be constitutively active or the vesicle contents would leak out. Instead the channel should be activated only when the vesicle is docked at the release site. A possible presynaptic plasma membrane protein which could trigger this channel has been termed *physophilin*. This is a 36 kDa integral protein which binds to synaptic vesicles probably by interacting with synaptophysin. It is thus a candidate for a presynaptic plasma membrane component of a vesicle docking complex.

7.4.5 Small G-proteins

Multiple 20–25 kDa monomeric *small G-proteins*, many of which resemble the *ras* gene product, are implicated in the membrane traffic associated with both constitutive and regulated secretion. The C-termini of small G-proteins contain Cys residues which are frequently polyisoprenylated or otherwise modified to provide membrane anchors. As with large trimeric G-proteins (*see* Section 5.3.1) GDP/GTP exchange and GTP hydrolysis are essential for function. *GTPase-activating protein* (GAP, not to be confused with GAP-43, *see* Section 7.5.1) interacts with the G-proteins to increase the low intrinsic GTPase activity of the small G-proteins, while in yeast a number of proteins have been identified which control the GDP/GTP state of the proteins.

In mammalian cells at least 12 members of the Rab family of small G-proteins are known. Both yeast and mammalian cells undergo constitutive secretion and a number of the mammalian proteins are homologues of yeast G-proteins. One Rab protein of particular relevance to regulated secretion is rab3A, which is found to be specifically associated with SSVs via a polyisoprenylated C-terminal anchor. Cross-linking studies have identified a possible rab3A receptor, which

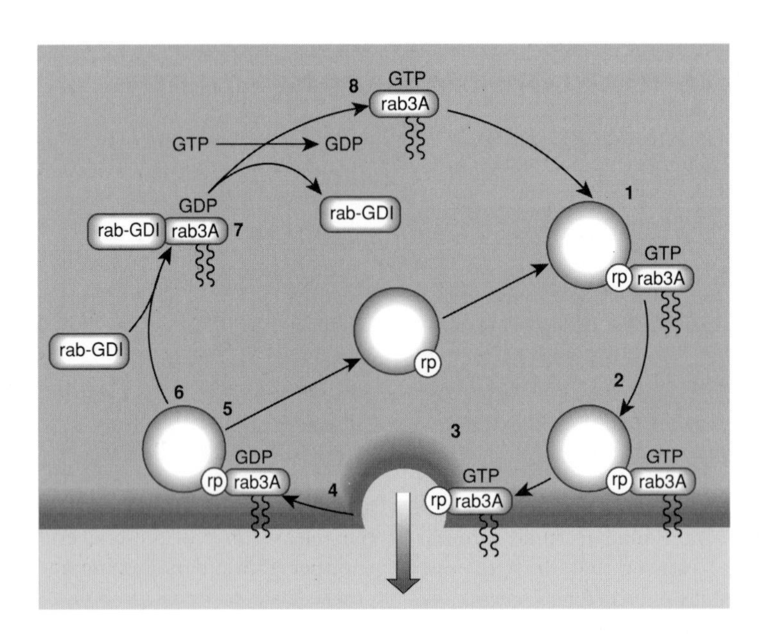

Fig. 7.17 rab3A and small synaptic vesicle (SSV) traffic to and from the membrane. A plausible scheme of rab3A association with SSVs involves a cycle of (1) GTP-rab3A association with a mature synaptic vesicle (perhaps via a receptor *rabphilin*, rp), (2) translocation to the membrane, (3) exocytosis, (4) hydrolysis of the bound GTP facilitated by a GTPase-activating protein, (5) dissociation of the GDP form of rab3A from the membrane, (6) release from the depleted synaptic vesicle, (7) reversible association of a GDP dissociation inhibitor (rab-GDI) protein with rab3A, (8) regeneration of GTP-rab3A, allowing a new cycle to commence.

has been named *rabphilin*, which is a 78 kDa protein with homology to synaptotagmin. When synaptosomes are depolarized and exocytosis occurs, rab3A dissociates from the synaptic vesicles and it is possible that this reversible association may play a role in synaptic vesicle targetting (Fig. 7.17). The role of rab3A has also been investigated in mast cells and chromaffin cells where injected oligopeptides based on the protein can stimulate large vesicle exocytosis.

By analogy to the yeast constitutive system, it has been hypothesized that vesicle-associated rab3A in its GTP-bound form might bind to a presynaptic receptor molecule, ensuring that only synaptic vesicles associate with the exocytotic machinery. After exocytosis and membrane retrieval a triggered hydrolysis of GTP to GDP would prevent the empty vesicle from reassociating with the release site, rab3A would then dissociate and ultimately reassociate with a mature synaptic vesicle (Fig. 7.17). However, in common with the other small G-proteins there is as yet limited information on the receptor mechanism which would be responsible for triggering GTP hydrolysis and activating signalling.

7.4.6 The synapsins

The synapsins (*synapsins Ia/b and synapsins IIa/b*) are a set of four homologous proteins derived from two genes by alternative splicing and differing in their C-terminal sequences (Fig. 7.18). All synapsins contain an N-terminal ('head domain') Ser^{9} phosphorylation site for both cAMP-dependent protein kinase and CaM-kinase I. The synapsins Ia and Ib have additionally two C-terminal ('tail domain') Ser^{566} and Ser^{603} phosphorylation sites for CaMKII, this proline-rich domain is absent from synapsins IIa/b. The synapsins are specific for presynaptic nerve terminals, although the relative proportions of the isoforms differ between terminals.

In vitro synapsins Ia/b bind to microtubules and microfilaments, causing bundling of the latter. This actin bundling is inhibited when synapsins Ia/b are phosphorylated at the CaMKII sites. Synapsins Ia/b are also found to be associated as peripheral proteins with small (but not large) synaptic vesicles when these are isolated under low ionic strength conditions and this appears to mediate their association with actin filaments *in vitro* (Fig. 7.19) and *in vivo* where synapsin-like filaments are seen connecting synaptic vesicles to the cytoskeleton. The synaptic vesicle protein responsible for this specific attachment may be Ca^{2+}/CaMKII itself.

Phosphorylation of the tail region by CaMKII reduces the affinity of the synaptic vesicle/synapsin Ia/b interaction. This has led to the theory that synapsin I phosphorylation regulates the availability of synaptic vesicles for exocytosis by controlling their attachment to the cytoskeleton. There is evidence for and against this hypothesis. It would be predicted that a maintained dephosphorylation of synapsin Ia/b would inhibit release by making synaptic vesicles 'sticky' whereas enhanced tail-domain phosphorylation would have the reverse effect. Microinjection of dephospho-synapsins Ia/b into the squid giant synapse terminal was shown to inhibit neurotransmission, whereas the introduction of exogenous CaMKII to enhance release. Less ambigously, the movement of synaptic vesicles through squid axoplasm can be inhibited by dephospho-synapsin I. However the criticism has been raised that synapsins may not be normal constituents of invertebrate terminals.

Depolarization of synaptosomes with KCl causes a rapid cycle of phosphorylation/dephosphorylation of synapsins Ia/b at the CaMKII sites. The causal relationship between this cycle and the release of transmitter has been the matter of some debate. Although the endogenous kinase appears to be adequate for stoichiometric phosphorylation, some enhancement of glutamate release can be seen when additional CaMKII is introduced by freeze-thawing into synaptosomes although total release is greatly decreased by this procedure.

A further test of this hypothesis is to consider the release of transmitters evoked by Ba^{2+}. Ba^{2+} enters through the same VACCs as Ca^{2+}, indeed in electrophysiological experiments it is frequently used in place of Ca^{2+}. Ba^{2+} can trigger the release of transmitters, either when added together with a depolarizing agent such as KCl, or alone, since it also inhibits K^{+}-channels and can induce spontaneous action potentials in terminals. However Ba^{2+} only interacts very weakly with CaM and hence does not induce the CaMKII-dependent phosphorylation of synapsin I. Nevertheless the ion is as efficient as Ca^{2+} in releasing glutamate from synaptosomes, suggest-

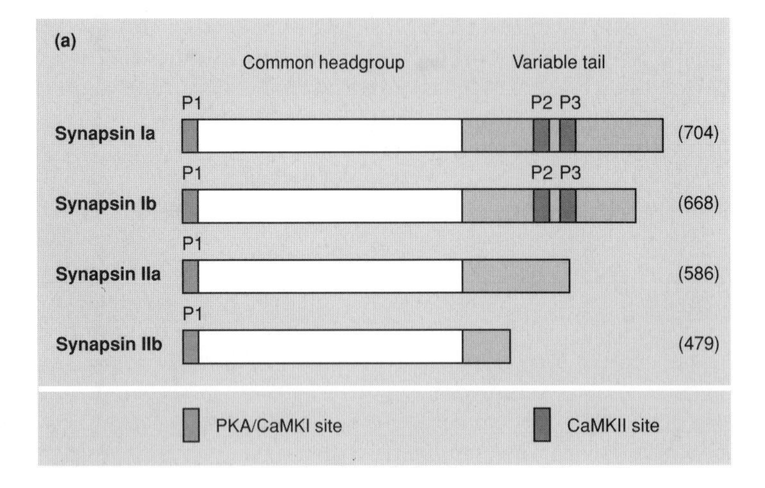

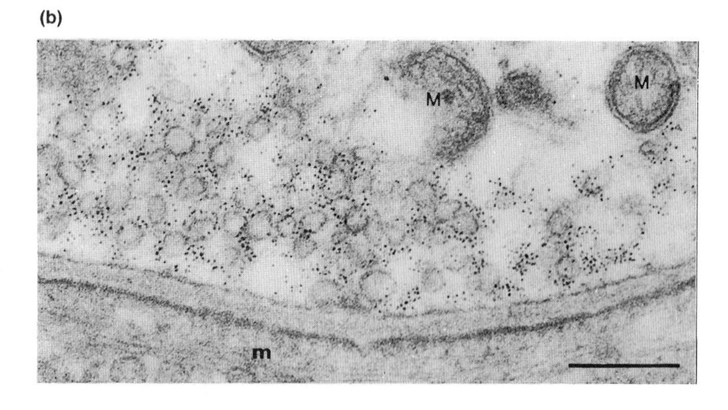

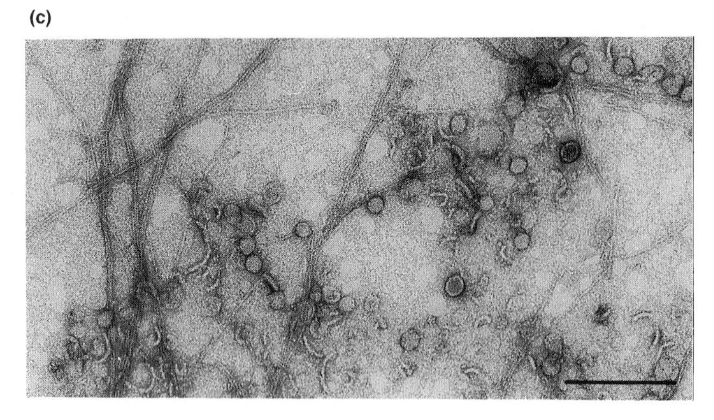

Fig. 7.18 Synapsins and synaptic vesicle associations with the cytoskeleton. (a) The synapsin family: synapsins Ia and Ib are coded by a single gene by alternative splicing, as are synapsins IIa and IIb. All synapsins are substrates for PKA and CaMKI, but synapsins IIa and IIb have abbreviated C-termini which lack a CaMKII site. (b) Synapsins are found presynaptically where they are peripherally associated with small synaptic vesicles, here at the neuromuscular junction: M, mitochondria; m, muscle fibre. Synapsin I was visualized by immunoferritin labelling. Bar = 0.25 μm. (c) Synaptic vesicles associate *in vitro* with actin filaments. Bar = 0.25 μm. From Greengard *et al.* (1993), reproduced with permission from *Science*. © 1993 by the AAAS.

ing that calmodulin-dependent phosphorylation is not an obligatory step in the release of glutamate from this preparation.

In the retina, specific populations of terminals have been identified with distinct distributions of synapsin isoforms. Colocalization with transmitter-specific enzymes suggested that synapsin I^-/II^+ terminals were cholinergic, while synapsin II^- terminals may be GABAergic. The presence of terminals lacking synapsin I (and hence the CaMKII-dependent phosphorylation site) argues against an obligatory role for this kinase in all small synaptic vesicle release.

The primary role of the synapsins may be in

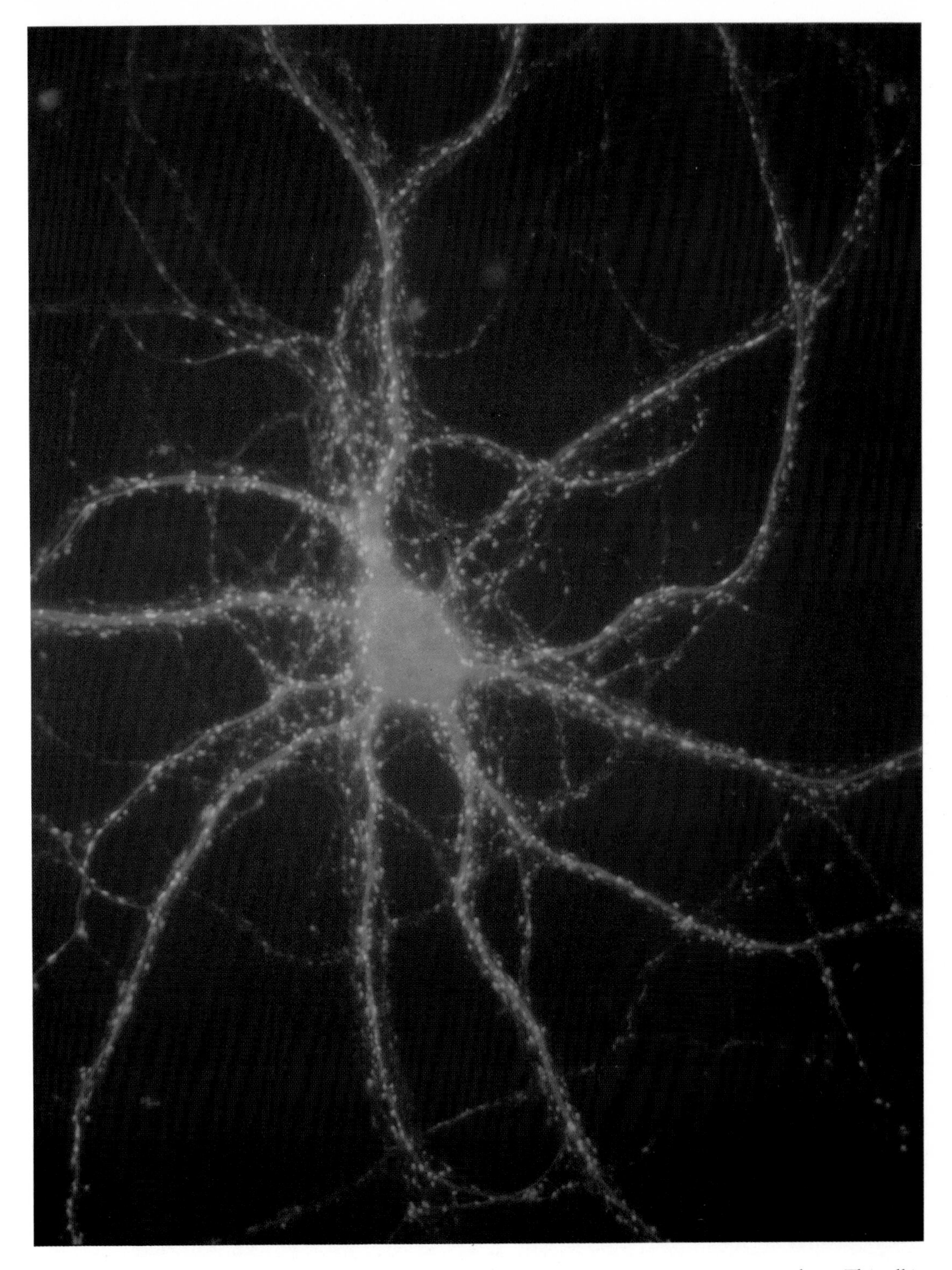

Plate 1 The synaptic input to a neurone. A neurone from the hippocampus is grown in primary culture. The cell is labelled with a dendritic marker (microtubule-associated protein, MAP2) giving a green fluorescence. Terminals from other cells synapsing onto the dendrites are visualized by antibody against the presynaptic protein synaptotagmin (orange fluorescence). The axons from these other cells are invisible since they contain no MAP2. Courtesy of Matteoli *et al.* (1993), front cover of *Neuron* **10** (Suppl.), reproduced with permission from *Neuron*.

[*facing page 132*]

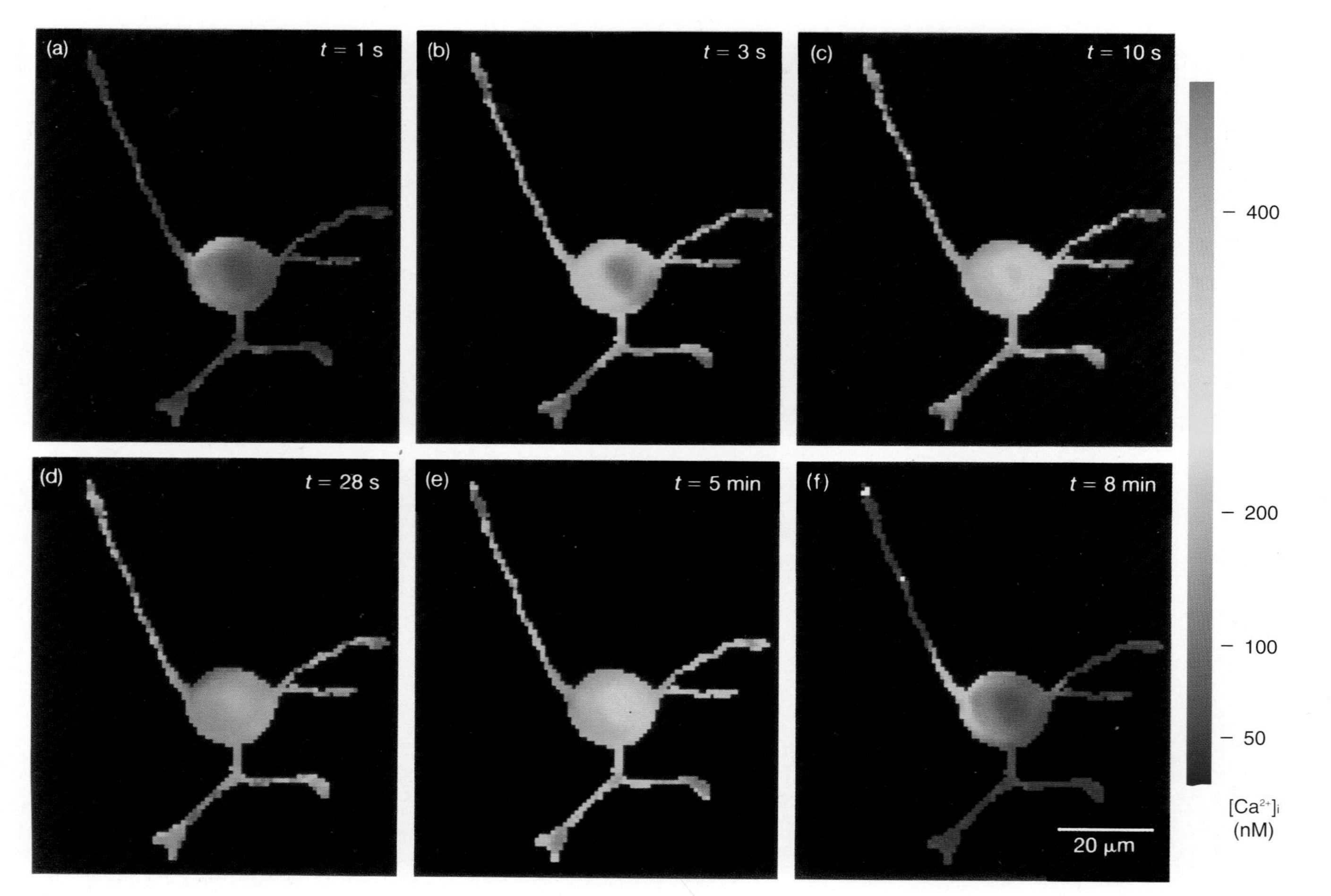

Plate 2 Fura-2 Ca^{2+} imaging of an electrically stimulated rat sympathetic neurone. A cultured superior cervical ganglion neurone was stimulated at 10 Hz by a cell-attached patch electrode (not visible). $[Ca^{2+}]_c$ (see bar at right for calibration) increases first in the developing growth cones and then in the soma. Stimulation was stopped at 5 min and the cell recovers its resting Ca^{2+} by 8 min. From Garyantes & Regehr (1992), by permission of the Society for Neuroscience.

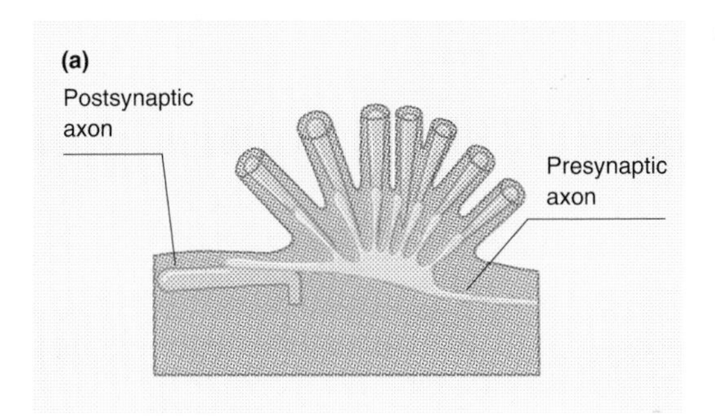

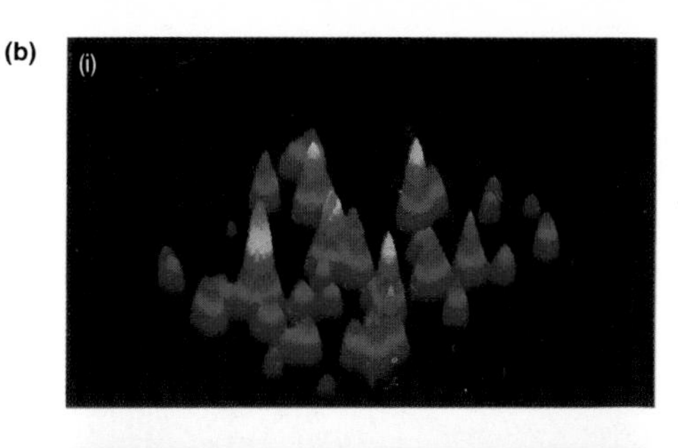

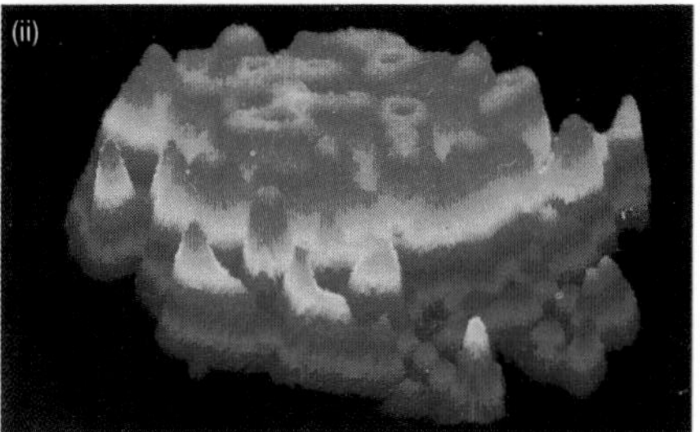

Plate 3 Localized Ca^{2+} and exocytosis at the squid giant synapse. The presynaptic terminal of the squid giant synapse (a) was injected with a modified aequorin which responded to Ca^{2+} concentrations in excess of about 100 μM with photon emission. Signals were obtained from discrete microdomains on the presynaptic membrane which could be integrated as a pseudo-colour three-dimensional projection (b) sampled before (i) and during (ii) electrical stimulation of the axon. Data from Llinas *et al.* (1992), reproduced with permission from *Science*. © 1992 by the AAAS.

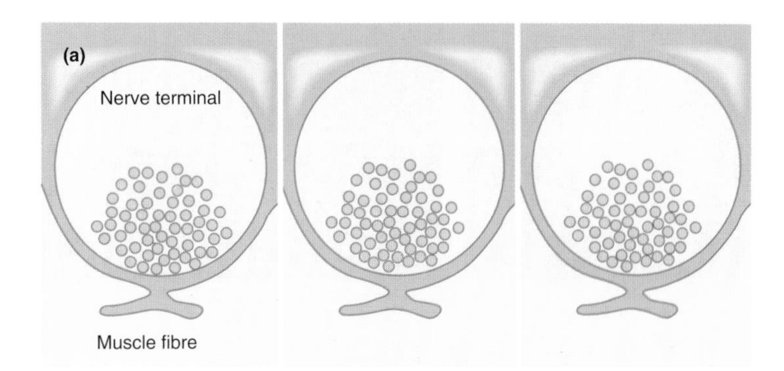

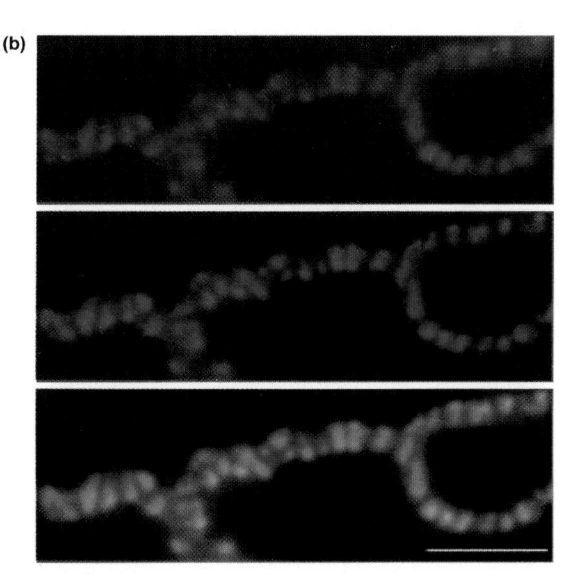

Plate 4 Fluorescent monitoring of endocytosis at the frog neuromuscular junction. (a) Three models by which recycling vesicles (red) might re-enter the vesicle pool: at the back of the queue (left), at the front of the queue (centre) or randomly (right). (b) The synaptic vesicle population within two motor nerve terminals were loaded sequentially with the impermeant styryl dye RH414 (red fluorescence) followed by the green FMI-43. The dyes were incorporated by endocytosis following electrical stimulation. The combined image (bottom) shows that no spatial separation can be detected between the two population of vesicles. Furthermore they were released randomly during subsequent stimulation. From Betz & Bewick (1992), reproduced with permission from *Science*. © 1992 by the AAAS.

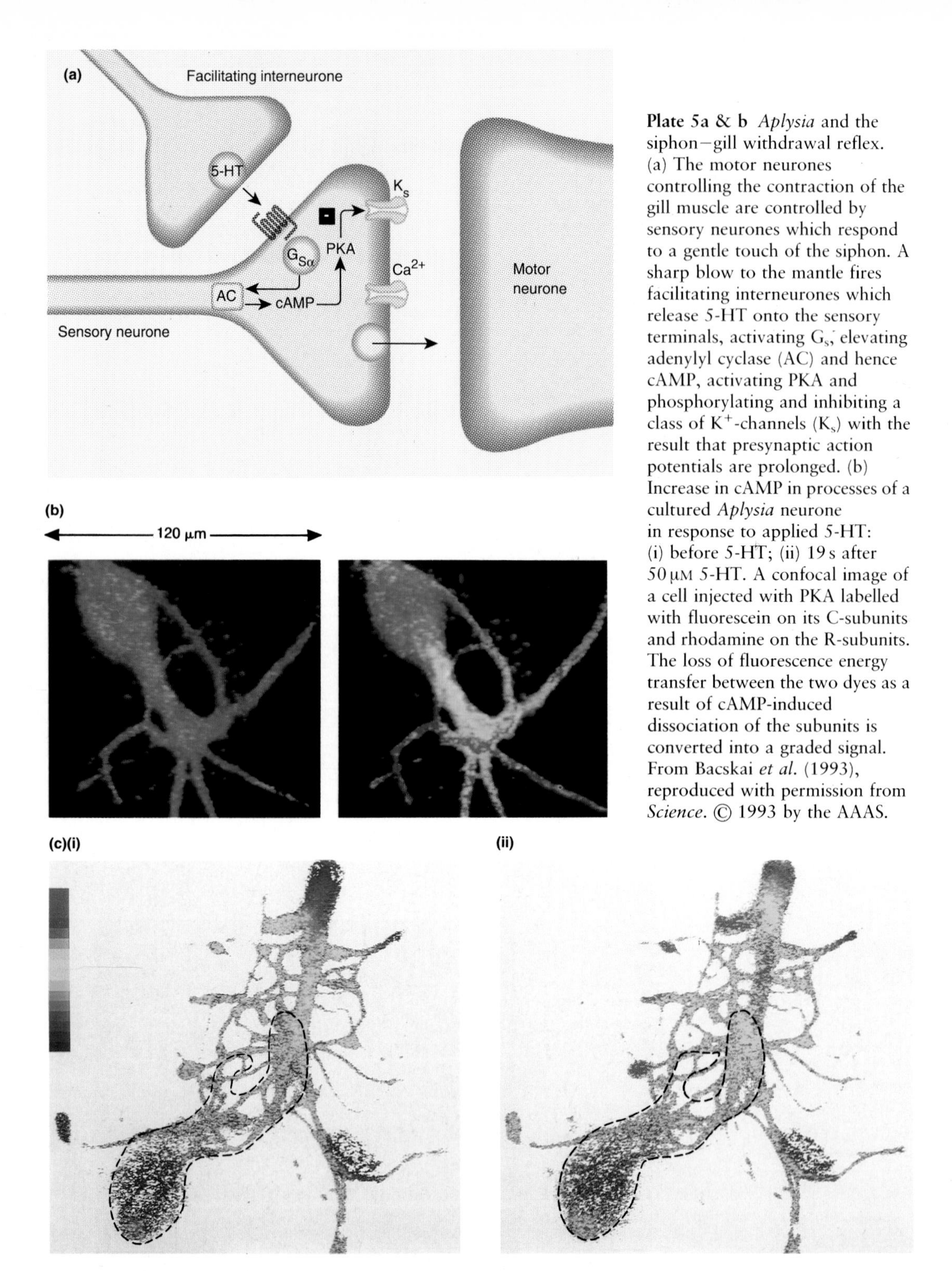

Plate 5a & b *Aplysia* and the siphon–gill withdrawal reflex. (a) The motor neurones controlling the contraction of the gill muscle are controlled by sensory neurones which respond to a gentle touch of the siphon. A sharp blow to the mantle fires facilitating interneurones which release 5-HT onto the sensory terminals, activating G_s, elevating adenylyl cyclase (AC) and hence cAMP, activating PKA and phosphorylating and inhibiting a class of K^+-channels (K_s) with the result that presynaptic action potentials are prolonged. (b) Increase in cAMP in processes of a cultured *Aplysia* neurone in response to applied 5-HT: (i) before 5-HT; (ii) 19 s after 50 μM 5-HT. A confocal image of a cell injected with PKA labelled with fluorescein on its C-subunits and rhodamine on the R-subunits. The loss of fluorescence energy transfer between the two dyes as a result of cAMP-induced dissociation of the subunits is converted into a graded signal. From Bacskai *et al.* (1993), reproduced with permission from *Science*. © 1993 by the AAAS.

Plate 5c $[Ca^{2+}]_c$ within a fura-2 injected cultured *Aplysia* sensory neurone in the region of synaptic contact with a motor neurone (not labelled but indicated by the dashed lines). In the control (i) the cell was stimulated with a 10 Hz train of action potentials; the same procedure was repeated but in the presence of 15 μM 5-HT (ii), showing the enhanced presynaptic Ca^{2+} elevation. From Eliot *et al.* (1993), reproduced with permission from *Nature*. © 1993 Macmillan Magazines Ltd.

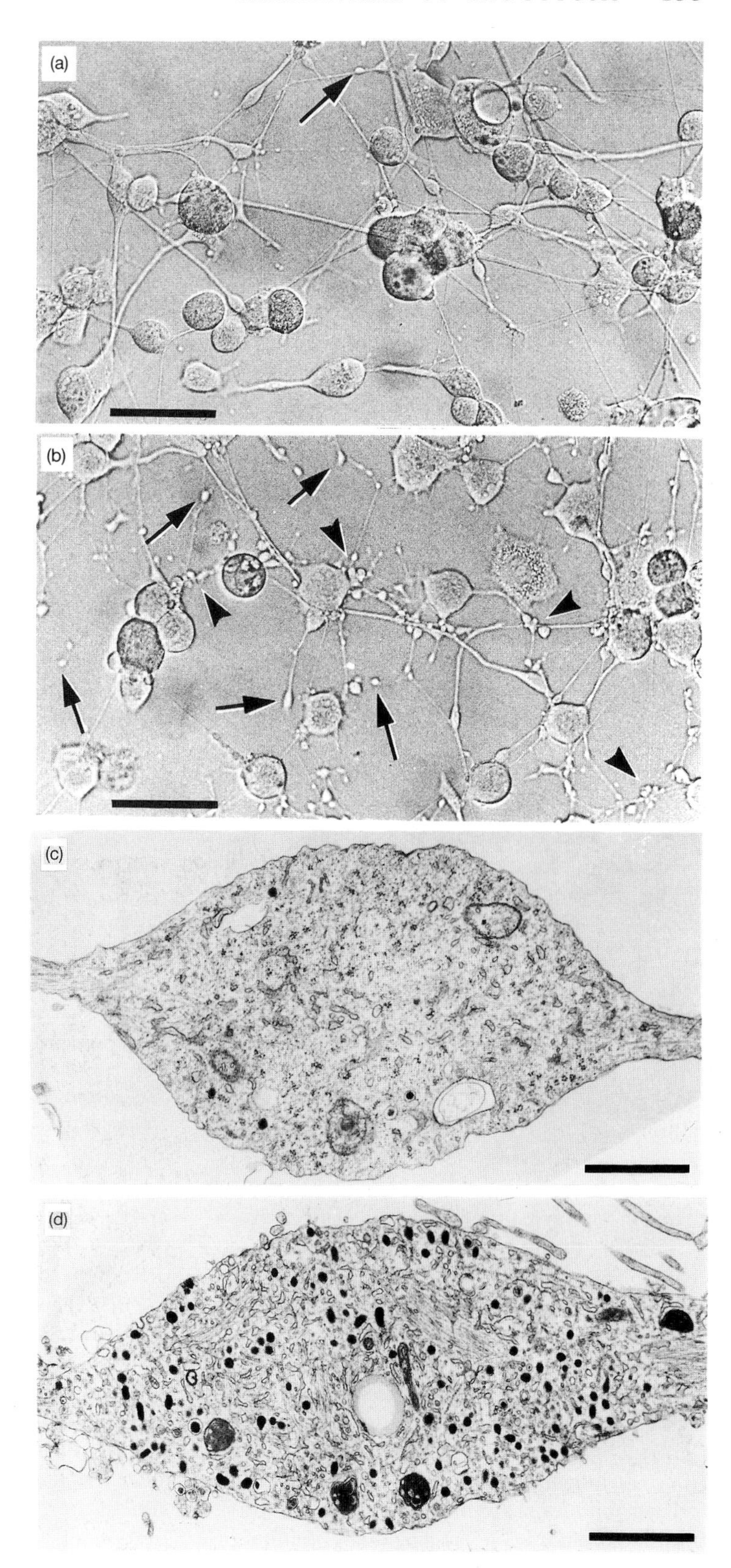

Fig. 7.19 Synapsin IIb induces presynaptic terminal formation in neuroblastoma cells. (a) Light micrograph of control NG108–15 neuroblastoma cells grown in monolayer culture after 3 days of differentiation. (b) Parallel cells from a synapsin IIb transfected clone, note the increased number of varicosities (arrows). Bars = 100 μm. (c) Electron micrograph of a varicosity from a control cell and (d) from a transfected cell showing increased numbers of synaptic vesicles. Bars = 1 μm. From Han *et al.* (1991), reproduced with permission from *Nature*. © 1991 Macmillan Magazines Ltd.

presynaptic development and synaptic vesicle recruitment. Transfection of NG108−15 (a neuroblastoma × glioma hybrid clonal cell line, *see* Section 2.4) with the cDNA for rat synapsin IIb, which caused over expression of the protein, produced a phenotype in which there was a large increase in varicosity formation and in synaptic vesicle content within each varicosity (Fig. 7.19).

7.4.7 Dephosphin, endocytosis and membrane retrieval

There are two possible fates for the vesicle membrane immediately after exocytosis (*see* Fig. 7.10); firstly the vesicles may transiently 'kiss' the membrane, release their contents but are then immediately retrieved before their lipids and proteins have a chance to mingle with the synaptic membrane. The second possibility is a complete flattening and randomization of the vesicle membrane into the synaptic membrane with a subsequent loss of the vesicle's identity. In this case endocytosis would occur, probably as clathrin-coated vesicles, followed by transport to an endosomal compartment in the terminal and reformation of mature synaptic vesicles.

Under conditions of moderate activity, it appears that the contents of SSVs can be exocytosed without the necessity for the vesicle membrane to completely flatten into the synaptic membrane. All that may be required is a transient fusion followed by a direct retrieval of the emptied vesicle, without involving clathrin or an intermediate 'endosome' compartment. This would be consistent with the failure to see SSV integral proteins in the synaptic membrane during quantal secretion. In contrast, extreme stimulation leads to a depletion of SSV numbers, an increase in exocytotic ω-profiles and clathrin-coated vesicles and the appearance of SSV proteins in the synaptic membrane. Under these conditions the retrieved membrane appears to pass to an 'endosomal' compartment where SSVs are reformed.

After exocytosis, vesicle membrane must be retrieved, reformed into synaptic vesicles, refilled and returned to the releasable pool of vesicles near the active zone. Endocytosis is more difficult to monitor than exocytosis because there is no obvious signal corresponding to the release of transmitter. One technique is to include horse radish peroxidase in the medium; fixation of the neuronal preparation allows vesicles to be detected which have endocytosed and incorporated the electron-dense label. A more recent method, which allows endocytosis to be monitored in living tissue, substitutes the impermeant styryl fluorescent indicators *RH414* or *FM1-43* (Plate 4, between pages 132 and 133). Frog motor nerve terminals were stimulated in the presence of the indicators to induce exocytosis. The subsequent endocytosis caused some of the indicator to become trapped inside the recycling vesicles and beads of fluorescence were seen corresponding to clusters of synaptic vesicles at release sites along the length of the terminal.

An important question is whether newly recycled vesicles mix and randomize with the older population of vesicles or whether they enter a queuing system. In *Torpedo* there is evidence that far from queuing, newly recycled synaptic vesicles jump the queue and are preferentially released by subsequent stimulation (see below). However, at the frog NMJ, vesicles labelled with one styryl indicator appeared to randomize with those labelled previously with the other indicator. Furthermore, the subsequent loss of fluorescence from pre-labelled terminals due to exocytosis of the styryl dyes showed a random release of recycled and endogenous vesicles.

Recycling of cholinergic vesicles in *Torpedo* has been examined in considerable detail (Fig. 7.20). After synaptic vesicles fuse with the plasma membrane there is a rapid retrieval of membrane as endocytosis occurs. The endocytotic vesicle entraps some of the extracellular medium, and if this contains an identifying marker the newly endocytosed vesicles can be identified and their fate monitored. Using either stimulated perfused electric organs or isolated groups of electrocytes suspended in dextran-containing medium, more and more vesicles in the electrocytes are found in electron micrographs to contain incorporated dextran as exocytosis proceeds.

These recycling vesicles are about 25% smaller in diameter than before exocytosis, indicating a >50% decrease in volume. The presence of dense dextran within the freshly endocytosed vesicle population is sufficient to enable them to be isolated as a distinct fraction, termed VP2, which can be separated from 'mature' VP1 vesicles on zonal density centrifugation. The proportion of VP2 vesicles increases from a few

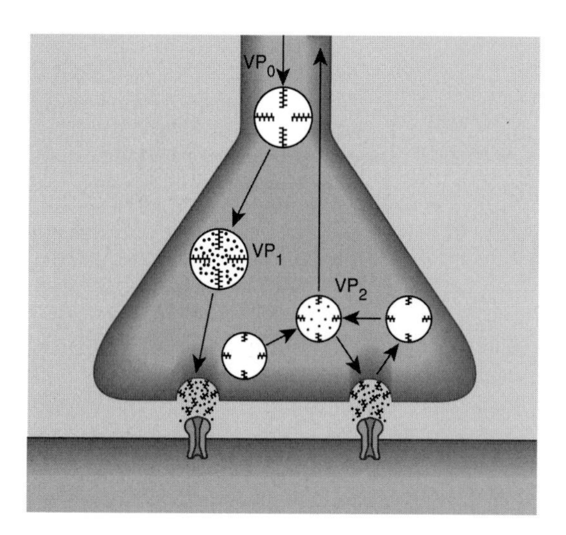

Fig. 7.20 Membrane recycling in *Torpedo* terminals. New vesicles (VP0) arriving from the axon accumulate ACh and ATP to form the VP1 population, rich in ACh and proteoglycan. After exocytosis, vesicle membrane is retrieved into a rapidly recycling VP2 population which is smaller and contains less ACh but probably accounts for most of the quantal ACh release from this terminal.

percent in resting *Torpedo* electric organ to 70% after extensive stimulation.

The endogenous content of ACh and ATP in the VP2 fraction is much lower (five- to ten-fold) than for VP1 vesicles. However, if [^{3}H]-acetate or the choline analogue homocholine is additionally included in the medium perfusing the electric organ, it is found that the VP2 fraction is very heavily labelled with [^{3}H]-acetylcholine or [^{3}H]-acetylhomocholine when compared with the VP1 synaptic vesicles. It thus appears that there is rapid uptake of ACh into freshly endocytosed synaptic vesicles to aid their refilling. Furthermore, a second electrical stimulation releases label characteristic of the recycling VP2 vesicle fraction, rather than the VP1 fraction which appears to be held in reserve, to replace any loss from the metabolically active VP2 pool. The proteoglycan content of VP2 vesicles is also much lower than for VP1. The proteoglycan appears to be processed in the VP1 to VP2 transition from a membrane anchored to a free, secretable, form. One suggestion for its function is that its acts in the extracellular matrix of the synaptic cleft to aid clustering of the ACh receptors, since it is similar (or identical) to a heparin sulphate proteoglycan which has been reported to cluster receptors. Mammalian synaptic vesicles may contain a similar secretable proteoglycan.

Independent support for the concept that VP2 vesicles are the ones active in transmission comes from electrophysiological analysis of the size of ACh quanta in the electric organ. Estimates of 7000–10 000 molecules per quanta are in much better agreement with the content of VP2 vesicles than with the much more heavily loaded VP1 vesicles. Exocytosis of VP1 vesicles may be a rare event. VP1 vesicles are readily available from *Torpedo* electric organ, and most studies on the mechanism of ACh storage have been performed with this fraction, although recently a comparative study of VP1 and VP2 vesicles has been performed with the New World *Torpedo californica*.

Only VP1 (but not VP0 or VP2) vesicles are capable of accumulating Ca^{2+}. It has been suggested that Ca^{2+} may activate a protease which splits the membrane anchored proteoglycan to create its secretory form. The specific vesicle heparin sulphate proteoglycan can be labelled *in vivo* with ^{35}S and exploited to follow the synthesis and transport of new vesicles down the axon. The axonal (VP0) vesicles have a protein composition identical to that of mature terminal vesicles, but are empty of transmitter and ATP. VP0 vesicles share with VP2 (but not VP1 vesicles which are already full), the ability to accumulate ACh and ATP. 'Recycling' and 'reserve' cholinergic synaptic vesicles are not unique to *Torpedo* but have also been detected in mammalian peripheral and central cholinergic terminals.

By analogy with other endocytotic sytems, it might be expected that clathrin is involved in the retrieval of synaptic vesicle membrane from the plasma membrane. Clathrin coats (Fig. 7.21) are formed from the self-assembly of clathrin *triskelions*, three-legged structures which are in turn made from three polypeptides of 180, 36 and 33 kDa. Virtually all coated vesicles purified from synaptosomes contain characteristic SSV proteins including synaptophysin, synaptotagmin and V-ATPase subunits. Interestingly, the small G-protein rab3A is absent, consistent with its dissociation during endocytosis (*see* Fig. 7.17). This indicates that clathrin in the nerve terminal is primarily involved in SSV endocytosis.

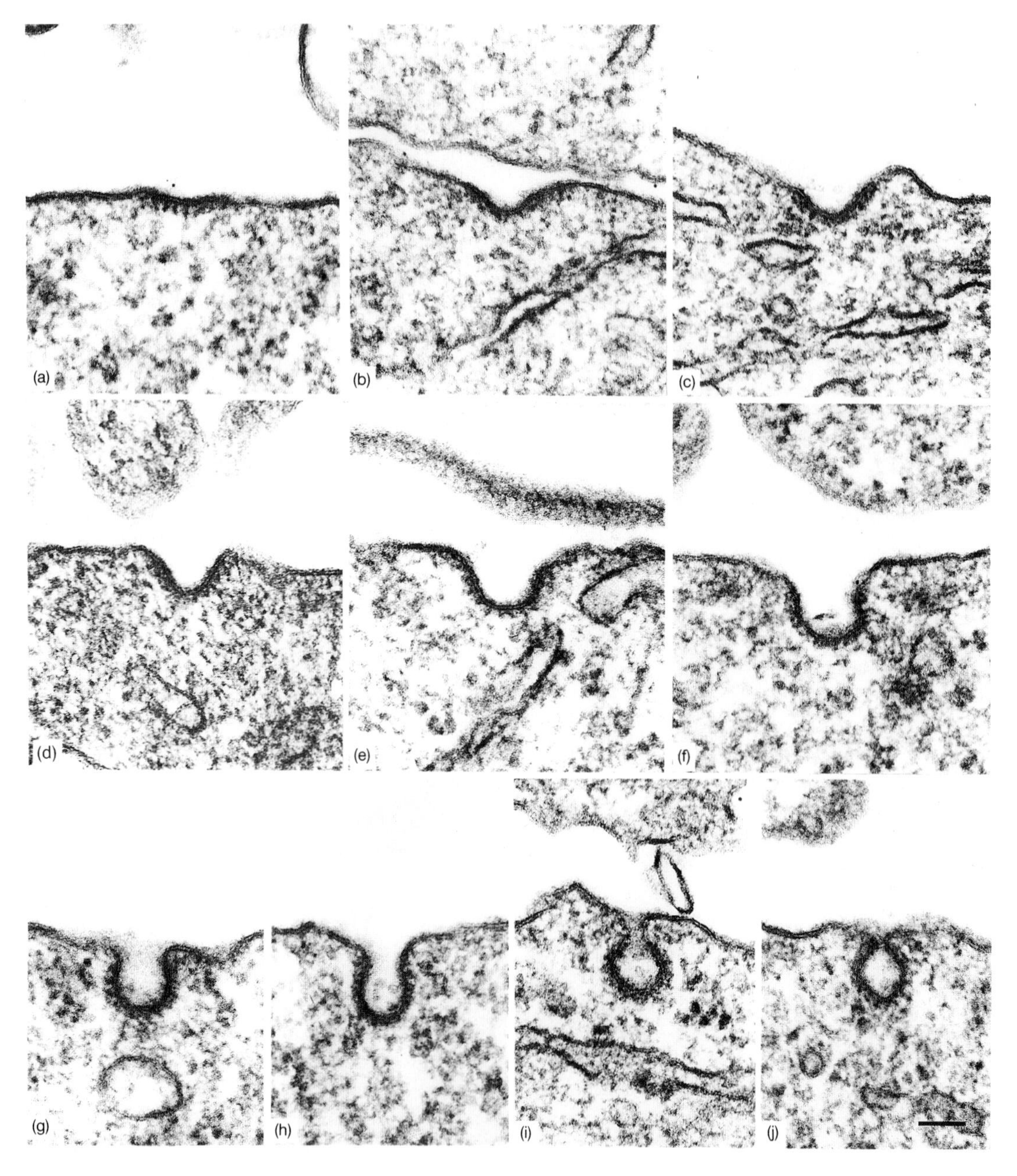

Fig. 7.21 Formation of clathrin-coated vesicles. A series of electron micrographs arranged in order to show sequential stages in the formation of a clathrin-coated vesicle. Data from Pypaert *et al.* (1987).

A temperature-sensitive mutant of *Drosophila*, *shibire ts-1* has normal neuromuscular transmission at 19°C, but on increasing the temperature to 29°C endocytosis becomes reversibly blocked, by a single mutation in a protein involved in endocytosis (see below). There is no effect on exocytosis with the result that the terminals become rapidly depleted of synaptic vesicles. Returning the temperature to 19°C results in the formation, over a period of 2–3 min, of uncoated invaginations of the plasma membrane formed which enlarge and then pinch off to form large cisternae (Fig. 7.22). Newly formed synaptic vesicles are found to be associated with

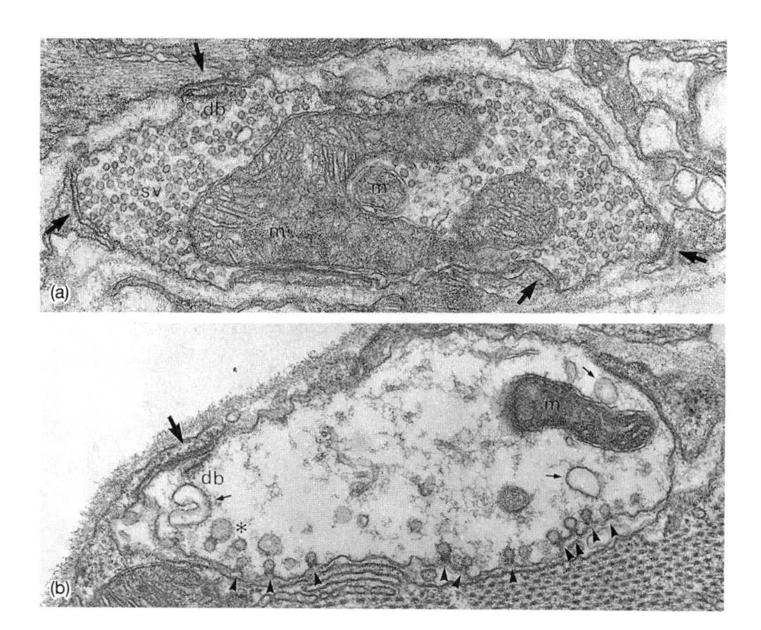

Fig. 7.22 Temperature-sensitive mutations of endocytosis in *Drosophila*. The temperature-sensitive *Drosophila* mutant *shibere*$^{ts-1}$ is normal at 19°C but endocytosis is reversibly blocked at 29°C. (a) Typical terminal of fly at 19°C. Thick arrows represent active zones; db, dense body; m, mitochondria; sv, synaptic vesicle. (b) Terminal from fly exposed to 29°C for 8 min. Note the loss of synaptic vesicles and appearance of collared pits at the plasma membrane (arrowheads). Invaginations whose necks are not in the plane of section are indicated by small arrows. An active zone is shown by the thick arrow. From Koenig & Ikeda (1989), by permission of the Society for Neuroscience.

these cisternae and gradually increase in number as cisternae decrease. By 30 min at 19°C the full complement of synaptic vesicles return.

The mutation in *shibere* is in a microtubule-binding protein, *dynamin*, which possesses microtubule-activated GTPase activity. Dynamin is identical to a presynaptic protein called *P96* or *dephosphin* which exists as a 96 kDa and 93 kDa doublet and is phosphorylated by PKC under resting conditions. Phosphorylation greatly enhances the GTPase activity of the protein. Dephosphorylation of dephosphin occurs when synaptosomes are depolarized; this requires Ca^{2+} although Ca^{2+}-ionophores are relatively ineffective, indicating an involvement of the high localized Ca^{2+} in the region of the Ca^{2+}-channel and the release apparatus (*see* Section 7.3). Dephosphin is thus a component of the endocytotic pathway.

7.5 Other presynaptic proteins implicated in SSV exocytosis

In cell-free systems, PKC can phosphorylate a wide variety of neuronal proteins (Fig. 7.23), including tyrosine hydroxylase (*see* Section 11.1.1), the nicotinic acetylcholine receptor (*see* Section 10.3.1), voltage-dependent Na^+- and K^+-channels, microtubule-associated proteins MAP2 and tau, and two presynaptic substrates *GAP-43* (which has many pseudonyms, see below) and *MARCKS* (*myristoylated, alanine-rich C-kinase substrate*), although suggestions for the involvement of these proteins in SSV exocytosis must be balanced against the insensitivity of SSV exocytosis to PKC inhibition.

7.5.1 GAP-43

GAP-43, *B50*, *neuromodulin*, *F1*, *pp46* and *p57* all refer to the same highly acidic, rod-shaped calmodulin-binding protein with an apparent molecular weight of 57 000 but an actual molecular mass of 25 kDa. The protein is growth-related and is targetted to axonal rather than dendritic growth cones. GAP-43 has a widespread distribution in embryonic neurones and its expression correlates with the period of neurite outgrowth both during development and during nerve regeneration following injury. In the initial stages of outgrowth, before differentiation into axonal and dendritic neurites, all neurites contain GAP-43. However as soon as one neurite starts to develop an axonal morphology the others lose the protein and differentiate into dendritic neurites, bearing the microtubule-specific marker, MAP2.

The ability of dendritic neurites to grow in the absence of GAP-43 shows that the protein is not necessary for neurite growth *per se*. In the adult, the mRNA only continues to be expressed in

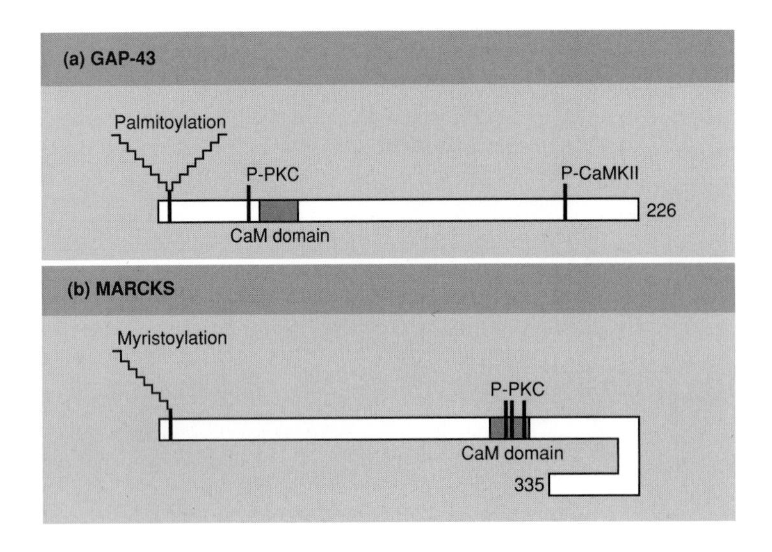

Fig. 7.23 Neuronal substrates for protein kinase C (PKC). The domain organization of some major PKC substrates in neurones. P-CaMKII, CaMKII-dependent phosphorylation; P-PKC, PKC-dependent phosphorylation.

nerve terminals in regions such as the hippocampus where plasticity may occur. Although GAP-43 is a highly acidic hydrophilic protein lacking obvious membrane-spanning regions, it associates with the growth cone plasma membrane, via two covalently attached palmitate residues attached via cysteines 3 and 4 close to the N-terminus (Fig. 7.23). It also copurifies with the plasma membrane skeleton (as opposed to the cytoskeleton), suggesting that it may be involved in the shape determination of the presynaptic membrane.

GAP-43 can be phosphorylated by PKC at a single serine 41 residue. Antibodies that block the phosphorylation inhibit the release of NA from permeabilized synaptosomes, while transfection of PC12 cells with a vector coding for antisense GAP-43 cRNA inhibit dopamine release from these cells. The latter system is an example of LDVC exocytosis, which differs in many respects from SSV exocytosis in that amino acid exocytosis from synaptosomes can proceed when PKC is inhibited. GAP-43 is dephosphorylated by the calmodulin-activated calcineurin.

7.5.2 MARCKS

MARCKS is a 32 kDa protein which migrates anomalously on SDS gells giving an apparent molecular weight of 87 000 due to its highly extended rod-like structure (about 36 nm in length). MARCKS has an unusual amino acid composition, with no less than 30% alanine residues, and is also highly acidic. The protein is expressed in both neuronal and non-neuronal cells, and is a substrate for PKC (although the isoenzyme specificity is not established); a 25 amino acid phosphorylation site domain contains three or four serines which can be phosphorylated *in vivo* (Fig. 7.23). N-terminal myristoylation (whether co- or posttranslationally is unclear) provides a hydrophobic anchor to attach MARCKS to the membrane — probably directly to the lipid bilayer rather than to a membrane protein.

PKC-dependent phosphorylation causes an apparent translocation to the cytoplasm, where it remains associated with actin filaments. Ca^{2+}/calmodulin can bind to dephosphorylated MARCKS, and either phosphorylation or Ca^{2+}/calmodulin binding abolishes the actin cross-linking activity of the protein. The dual regulation by Ca^{2+}, via calmodulin and by PKC, can be interpreted as providing a more plastic cytoskeleton by inhibiting actin cross-linking at the plasma membrane and releasing elements of the actin cytoskeleton from membrane attachment respectively. MARCKS has been implicated in membrane trafficking, cell motility, the regulation of the cell cycle and cell transformation. As will be discussed in Chapter 8, cytoskeletal disruption appears to play a role in the regulation of LDCV exocytosis. However, although MARCKS is the major PKC substrate in SSV-secreting nerve terminals and is phosphorylated, for example, by activation of the phospholipase C-coupled presynaptic metabotropic glutamate receptor, PKC has no obligatory involvement in SSV release.

The binding of CaM to both dephospho-GAP-43 and dephospho-MARCKS is abolished by phosphorylation. In the case of MARCKS this is because the calmodulin binding site is identical to the phosphorylation site domain, phosphorylation decreasing the affinity for calmodulin by 200-fold and causing the disruption of MARCKS/CaM complexes. It has been suggested that both GAP-43 and MARCKS may act as reservoirs for CaM, which would be released as a consequence of PKC-dependent phosphorylation; in brain the proteins are present at comparable concentrations to CaM. There is some evidence in support of this, since activation of PKC in cultured neurones increases cytoplasmic CaM and decreases the pool associated with the membrane and presumably associated with one or both of these proteins.

Further reading

The neuromuscular junction

Ceccarelli, B. & Hurlbut, W.P. (1980) Vesicle hypothesis of the release of quanta of acetylcholine. *Physiol. Rev.* **60**, 396–441.

Heuser, J.E. (1989a) Review of electron microscopic evidence favouring vesicle exocytosis as the structural basis for quantal release during synaptic transmission. *Q. J. Exp. Physiol.* **74**, 1051–1069.

Heuser, J.E. (1989b) The role of coated vesicles in recycling of synaptic vesicle membrane. *Cell Biol. Int. Rep.* **13**, 1063–1076.

Heuser, J.E. & Reese, T.S. (1981) Structural changes after transmitter release at the frog neuromuscular junction. *J. Cell Biol.* **88**, 564–580.

Torri-Tarelli, F., Grohovaz, F., Fesce, R. & Ceccarelli, B. (1985) Temporal coincidence between synaptic vesicle fusion and quantal secretion of acetylcholine. *J. Cell Biol.* **101**, 1386–1399.

The CNS synapse

Landis, D.M., Hall, A.K., Weinstein, L.A. & Reese, T.S. (1988) The organization of cytoplasm at the presynaptic active zone of a central nervous system synapse. *Neuron* **1**, 201–209.

Electrophysiological techniques for monitoring exocytosis

Korn, H. & Faber, D.S. (1991) Quantal analysis and synaptic efficiency in the CNS. *Trends Pharmacol. Sci.* **12**, 439–445.

The neurochemical approach

Nicholls, D.G. (1993) The glutamatergic nerve terminal. *Eur. J. Biochem.* **212**, 613–631.

Nicholls, D.G. & Attwell, D.A. (1990) The release and uptake of excitatory amino acids. *Trends Pharmacol. Sci.* **11**, 462–468.

Tibbs, G.R., Barrie, A.P., Van-Mieghem, F., McMahon, H.T. & Nicholls, D.G. (1989) Repetitive action potentials in isolated nerve terminals in the presence of 4-aminopyridine: effects on cytosolic free Ca^{2+} and glutamate release. *J. Neurochem.* **53**, 1693–1699.

Localized Ca^{2+} and the trigger for SSV exocytosis

Fujiwara, T. & Nagakuro, C. (1989) Three-dimensional structure of the presynaptic nerve ending of the ciliary ganglion of the chick embryo: a scanning electron microscopic study. *Neurosci. Lett.* **98**, 125–128.

Llinas, R., Sugimori, M. & Silver, R.B. (1992) Microdomains of high calcium concentration in a presynaptic terminal. *Science* **256**, 677–679.

Smith, S.J. & Augustine, G.J. (1988) Calcium ions, active zones and synaptic transmitter release. *Trends Neurosci* **11**, 458–464.

Stanley, E.F. (1991) Single calcium channels on a cholinergic presynaptic nerve terminal. *Neuron* **7**, 585–591.

Verhage, M., McMahon, H.T., Ghijsen, W.E.J.M., Boomsma, F., Wiegant, V. & Nicholls, D.G. (1991) Differential release of amino acids, neuropeptides and catecholamines from nerve terminals. *Neuron* **6**, 517–524.

Protein chemistry of SSV exocytosis

Almers, W. & Tse, F.W. (1990) Transmitter release from synapses: does a preassembled fusion pore initiate exocytosis? (Review). *Neuron* **4**, 813–818.

Bennett, M.K., Calakos, N., Kreiner, T. & Scheller, R.H. (1992) Synaptic vesicle membrane proteins interact to form a multimeric complex. *J. Cell Biol.* **116**, 761–775.

De Camilli, P. & Jahn, R. (1990) Pathways to regulated exocytosis in neurons. *Annu. Rev. Physiol.* **52**, 625–645.

Greengard, P., Valtorta, F., Czernik, A.J. & Benfenati, F. (1993) Synaptic vesicle phosphoproteins and regulation of synaptic function. *Science* **259**, 780–785.

Jahn, R. & De Camilli, P. (1990) Membrane proteins of synaptic vesicles: markers for neurons and neuroendocrine cells and tools for the study of neurosecretion. In *Markers for Neural and Endocrine Cells*, M. Gratz & K. Langley (eds). Weinheim: VCH.

Pypaert, M., Lucocq, J.M. & Warren, G. (1987) Coated pits in mitotic and interphase A341 cells. *Eur. J. Cell Biol.* **45**, 23–29.

Sudhof, T.C. & Jahn, R. (1991) Proteins of synaptic vesicles involved in exocytosis and membrane recycling (Review). *Neuron* **6**, 665–677.

Synaptic vesicle transporters

Bajjalieh, S.M., Peterson, K., Shinghal, R. & Scheller, R.H. (1992) SV2, a brain synaptic vesicle protein homologous to bacterial transporters. *Science* **257**, 1271–1273.

Cain, C.C., Trimble, W.S. & Lienhard, G.E. (1992) Members of the VAMP family of synaptic vesicle proteins are components of glucose transporter-containing vesicles from rat adipocytes. *J. Biol. Chem.* **267**, 11681–11684.

Feany, M.B., Lee, S., Edwards, R.H. & Buckley, K.M. (1992) The synaptic vesicle protein SV2 is a novel type of transmembrane transporter. *Cell* **70**, 861–867.

SNAPs and SNAREs

Bagetta, G., Nistico, G. & Bowery, N.G. (1991) Characteristics of tetanus toxin and its exploitation in neurodegenerative studies. *Trends Pharmacol. Sci.* **12**, 285–289.

Bennett, M.K., Calakos, N. & Scheller, R.H. (1992) Syntaxin: a synaptic protein implicated in docking of synaptic vesicles at presynaptic active zones. *Science* **257**, 255–259.

Dolly, J.O., Ashton, A.C., Mcinnes, C., Wadsworth, J.D.F., Poulain, B., Tauc, L., Shone, C.C. & Melling, J. (1990) Clues to the multi-phasic inhibitory action of botulinum neurotoxins on release of transmitters. *J. Physiol. (Paris)* **84**, 237–246.

Schiavo, G., Benfenati, F., Poulain, B., Rossetto, O., Polverino-de-Laureto, P., Dasgupta, B.R. & Montecucco, C. (1992b) Tetanus and botulinum-B neurotoxins block neurotransmitter release by proteolytic cleavage of synaptobrevin. *Nature* **359**, 832–835.

Schiavo, G., Poulain, B., Rossetto, O., Benfenati, F., Tauc, L. & Montecucco, C. (1992a) Tetanus toxin is a zinc protein and its inhibition of neurotransmitter release and protease activity depend on zinc. *EMBO J.* **11**, 3577–3583.

Simpson, L.L. (1986) Molecular pharmacology of botulinum toxin and tetanus toxin. *Annu Rev. Pharmac. Toxicol.* **26**, 427–453.

Söllner, T., Whiteheart, S.W., Brunner, M., Erdjument-Bromage, H., Geronamos, S., Tempts, P. & Rothman, J.E. (1993) SNAP receptors implicated in vesicle targeting and fusion. *Nature* **362**, 318–324.

Synaptotagmin

Bommert, K., Charlton, M.P., DeBello, W.M., Chin, G.J., Betz, H. & Augustine, G.J. (1993) Inhibition of neurotransmitter release by C2-domain peptides implicates synaptotagmin in exocytosis. *Nature* **363**, 163–165.

Brose, N., Petrenko, A.G., Südhof, T.C. & Jahn, R. (1992) Synaptotagmin: a calcium sensor on the synaptic vesicle surface. *Science* **256**, 1021–1025.

Elferink, L.A., Peterson, M.R. & Scheller, R.H. (1993) A role for synaptotagmin (p65) in regulated exocytosis. *Cell* **72**, 153–159.

Leveque, C., Hoshino, T., David P., Shoji-Kasai, Y., Leys, K., Omori, A., Lang, B., El Far, O., Sato, K., Martin-Moutot, N., Newsom-Davis, J., Takahashi, M. & Seagar, M.J. (1992) The synaptic vesicle protein synaptotagmin associates with calcium channels and is a putative Lambert–Eaton myasthenic syndrome antigen. *Proc. Natl. Acad. Sci. USA* **89**, 3625–3629.

Matteoli, M., Haimann, C., Torri-Tarelli, F., Polak, J.M., Ceccarelli, B. & De Camilli, P. (1988) Differential effect of alpha-latrotoxin on exocytosis from small synaptic vesicles and from large dense-core vesicles containing calcitonin gene-related peptide at the frog neuromuscular junction. *Proc. Natl. Acad. Sci. USA* **85**, 7366–7370.

Petrenko, A.G., Lazaryeva, V.D., Geppert, M., Tarasyuk, T.A., Moomaw, C., Khokhlatchev, A.V., Ushkaryov, Y.A., Slaughter, C., Nasimov, I.V. & Südhof, T.C. (1993) Polypeptide composition of the α-latrotoxin receptor. High affinity binding protein consists of a family of related high molecular weight polypeptides complexed to a low molecular weight protein. *J. Biol. Chem.* **268**, 1860–1867.

Popoli, M. (1993) Synaptotagmin is endogenously phosphorylated by Ca^{2+}/calmodulin protein kinase II in synaptic vesicles. *FEBS Lett.* **317**, 85–88.

Rosenthal, L. & Meldolesi, J. (1989) α-Latrotoxin and related toxins. *Pharmac. Ther.* **42**, 115–134.

Shoji-Kasai, Y., Yoshida, A., Sato, K., Hoshino, T., Ogura, A., Kondo, S., Fujimoto, Y., Kuwahara, R., Kato, R. & Takahashi, M. (1992) Neurotransmitter release from synaptotagmin-deficientclonal variants of PC12 cells. *Science* **256**, 1820–1823.

Synaptophysin

Alder, J., Xie, Z.-P., Valtorta, F., Greengard, P. & Poo, M. (1992) Antibodies to synaptophysin interfere with transmitter secretion at neuromuscular synapses. *Neuron* **9**, 759–768.

Lichte, B., Veh, R.W., Meyer, H.E. & Kilimann, M.W. (1992) Amphiphysin, a novel protein associated with synaptic vesicles. *EMBO J.* **11**, 2521–2530.

Thomas, L., Hartung, K., Langosch, D., Rehm, H., Bamberg, E., Franke, W.W. & Betz, H. (1988) Identification of a synaptophysin as a hexameric channel protein of the synaptic vesicle membrane. *Science* **242**, 1050–1053.

Small G-proteins

Oberhauser, A., Monck, J.R., Balch, W.E. & Fernandez, J.M. (1992) Exocytotic fusion is activated by Rab3A proteins. *Nature* **360**, 270–273.

Senyshyn, J., Balch, W.E. & Holz, R.W. (1992) Synthetic peptides of the effector-binding domain of rab enhance secretion from digitonin-permeabilized chromaffin cells. *FEBS Lett.* **309**, 41–46.

Shirataki, H., Kaibuchi, K., Sakoda, T., Kishida, S., Yamaguchi, T., Wada, K., Miyazaki, M. & Takai, Y. (1993) Rabphilin-3A, a putative target protein foe smg p25A/rab3A p25 small GTP-binding protein related to synaptotagmin. *Mol. Cell Biol.* **13**, 2061–2068.

The synapsins

Benfenati, F., Valtorta, F., Bahler, M. & Greengard, P. (1989) Synapsin I, a neuron-specific phosphoprotein interacting with small synaptic vesicles and F-actin. *Cell Biol. Int. Rep.* **13**, 1007–1021.

De Camilli, P., Benfenati, F., Valtorta, F. & Greengard, P. (1990) The synapsins. *Annu. Rev. Cell Biol.* **6**, 433–460.

Han, H.Q., Nichols, R.A., Rubin, M.R., Bahler, M. & Greengard, P. (1991) Induction of formation of presynaptic terminals in neuroblastoma cells by synapsin IIb. *Nature* **349**, 697–700.

Lu, B., Greengard, P. & Poo, M. (1992) Exogenous synapsin I promotes functional maturation of developing neuromuscular synapses. *Neuron* **8**, 521–529.

McGuinness, T.L., Brady, S.T., Gruner, J.A., Sugimori, M., Llinas, R. & Greengard, P. (1989) Phosphorylation-dependent inhibition by synapsin-1 of organelle movement in squid axoplasm. *J. Neurosci.* **9**, 4138–4149.

Mandell, J.W., Czernik, A.J., De Camilli, P., Greengard, P. & Townes-Anderson, E. (1992) Differential expression of synapsins I and II among rat retinal synapses. *J. Neurosci.* **12**, 1736–1749.

Nestler, E.J., & Greengard, P. (1986) Synapsin I: a review of its distribution and biological regulation. *Prog. Brain Res.* **69**, 323–339.

Sudhof, T.C., Czernik, A.J., Kao, H.-T., Takei, K., Johnston, P.A., Horiuchi, A., Kanazir, S.D., Wagner, M.A., Perin, M.S., De Camilli, P. & Greengard, P. (1989) Synapsins: mosaics of shared and individual domains in a family of synaptic vesicle phosphoproteins. *Science* **245**, 1474–1480.

Dephosphin, endocytosis and membrane retrieval

Betz, W.J. & Bewick, G.S. (1992) Optical analysis of synaptic vesicle recycling at the frog neuromuscular junction. *Science* **255**, 200–202.

Koenig, J.H. & Ikeda, K. (1989) Disappearance and reformation of synaptic vesicle membrane upon transmitter release observed under reversible blockage of membrane retrieval. *J. Neurosci.* **9**, 3844–3860.

Maycox, P.R., Link, E., Reetz, A., Morris, S.A. & Jahn, R. (1992) Clathrin-coated vesicles in nervous tissue are involved primarily in synaptic vesicle recycling. *J. Cell Biol.* **118**, 1379–1388.

Robinson, P.J., Sontag, J.M., Liu, J.-P., Fykse, E.M., Slaughter, C., McMahon, H.T. & Sudhof, T.C. (1993) Dynamin GTPase regulated by protein kinase C phosphorylation in nerve terminals. *Nature* **365**, 163–166.

Smythe, E. & Warren, G. (1991) The mechanism of receptor-mediated endocytosis. *Eur. J. Biochem.* **202**, 689–699.

GAP-43, MARCKS

Aderem, A. (1992) Signal transduction and the actin cytoskeleton: the roles of MARCKS and profilin. *Trends Biochem. Sci.* **17**, 438–443.

Blackshear, P.J. (1993) The MARCKS family of cellular protein kinase C substrates. *J. Biol. Chem.* **268**, 1501–1504.

Coggins, P.J. & Zwiers, H. (1991) B-50 (GAP-43): biochemistry and functional neurochemistry of a neuron-specific protein. *J. Neurochem.* **56**, 1095–1106.

Dekker, L.V., DeGraan, P.N.E., Oestreicher, A.B., Versteeg, D.H.G. & Gispen, W.H. (1989) Inhibition of noradrenaline release by antibodies to B-50 (GAP-43). *Nature* **342**, 74–76.

Ivins, K.J., Neve, K.A., Feller, D.J., Fidel, S.A. & Neve, R.L. (1993) Antisense GAP-43 inhibits the evoked release of dopamine from PC12 cells. *J. Neurochem.* **60**, 626–633.

8

Exocytosis of large dense-core vesicles from neurones and secretory cells

8.1 Introduction

Neuropeptide release from CNS terminals may have more features in common with the secretion of hormones from endocrine cells than with the exocytosis of small synaptic vesicles (SSVs) (Fig. 8.1). Firstly neuropeptides are stored in large (>70 nm) dense-core vesicles (LDCVs) which are located away from active zones and are released ectopically, i.e. not directly into the synaptic cleft. Secondly there can be no local recycling of LDCVs in the terminal since neuropeptides are not retrieved but must be synthesized *de novo* by proteolytic cleavage of precursor peptides synthesized in the cell body. Thirdly LDCVs either lack, or have much lower amounts of, many of the specific proteins associated with SSVs. Finally LDCVs are released by a bulk increase in $[Ca^{2+}]_c$ rather than a localized coupling between Ca^{2+}-channels and exocytosis.

8.1.1 Origins of SSVs and LDCVs

The presence of two discrete classes of regulated secretory vesicles in the same terminal demands a mechanism for their segregated synthesis. Current evidence suggests that the trans-Golgi network in the cell body generates mature LDCVs (Fig. 8.1). The fungal metabolite *brefeldin A* inhibits the formation of secretory vesicles from the trans-Golgi network without inhibiting the release of preformed vesicles.

In contrast, *de novo* SSVs may be produced from early endosomal compartments in the terminal itself. The delivery of membrane and SSV-specific proteins to the endosomal compartment may occur via axonally targetted constitutive secretory vesicles which fuse with the terminal membrane and are then retrieved into the endosomal compartment to replenish the proteins required for SSV biogenesis. The mechanism by which integral and intravesicular proteins are sorted into either LDCVs or vesicles destined for the endosomal/SSV compartment is unclear, although a physical aggregation and precipitation of secretory proteins within the Golgi may be involved in the selection of proteins for the contents of LDCVs.

The model in which SSVs originate from the early endosomal compartment, rather than directly from the trans-Golgi, relies in part upon interpretation of the morphology of vesicles travelling down the axon. Ligation of axons does not lead to a build-up of vesicles with the morphology of SSVs before the block, suggesting an absence of mature SSVs travelling down the axon, although a build-up of SV2 and synaptophysin can be detected, and elongated tubular structures accumulate. The actual anterograde (forward) transport of the organelles is driven by *kinesin*. At the terminal cAMP-dependent phosphorylation of kinesin may be responsible for releasing the motor from the transported vesicle. Retrograde transport of vesicles back to the cell body is driven by *dynein* (also called MAP1C).

Further evidence for the origin of the two classes of vesicle has come from studies with chromaffin cell derived PC12 cells (*see* Section 2.3), which in common with other endocrine cells possess not only LDCV-like secretory granules, but also a population of SSV-like '*synaptic-like microvesicles*' (*SLMVs*) which contain many of the integral proteins of SSVs

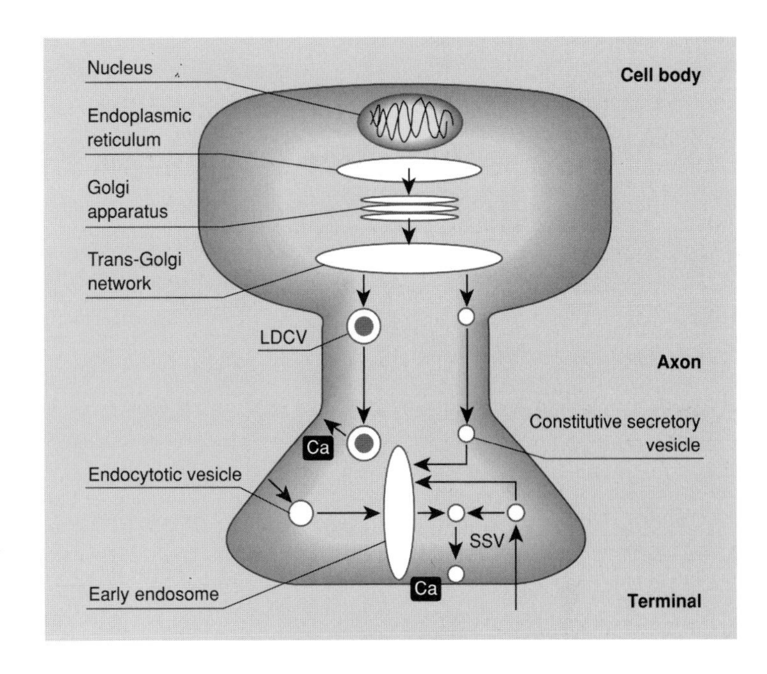

Fig. 8.1 The origin of small synaptic vesicles (SSVs) and large dense-core vesicles (LDCVs). LDCVs originate from the endoplasmic reticulum via the Golgi apparatus and the trans-Golgi network. After transport to the terminal they are released by regulated secretion at predominantly non-synaptic sites. LDCVs do not undergo local recycling at the terminal. *De novo* SSV membrane proteins are packaged into constitutive synaptic vesicles in the trans-Golgi, delivered to the terminal plasma membrane, retrieved either directly or into endocytotic vesicles and delivered to an early endosomal compartment, from which mature SSVs are formed. These then undergo several exocytotic/endocytotic cycles. Adapted from Regnier-Vigouroux & Huttner (1993).

(e.g. synaptophysin, synaptobrevin and p29) and can clearly be distinguished from the large secretory granules (Fig. 8.2). SLMVs appear to undergo a similar local recycling to neuronal SSVs involving endosomes but not the trans-Golgi network. Controversy as to whether LDCVs and chromaffin vesicles possess low concentrations of synaptophysin, synaptobrevin and SV2 is thought by some to be due to contamination of preparations with SLMVs.

The labelled microvesicles are concentrated in the region of the Golgi complex, in contrast to the secretory granules which are concentrated in the peripheral regions of the cell. The function of the endocrine cell microvesicles is unclear, but may be related to the membrane fraction responsible for the recycling of plasmalemmal receptors such as that for transferrin. Thus neuronal SSVs could represent an adaptation of a general mechanism present in non-neuronal cells. The microvesicles of endocrine cells can undergo regulated exocytosis and recycle to endosomal compartments.

Synaptophysin immunobeads pull down an almost homogeneous population of microvesicles from chromaffin cell vesicle fractions, supporting a complete demarkation between the two vesicle populations. This is important since it bears on two fundamental questions; firstly, it indicates that SSVs are derived from a discrete parallel pathway of synthesis and transport to the terminals, and secondly it more precisely defines the roles of proteins such as synaptophysin if exocytosis of LDCVs can occur without their involvement.

Large dense-core vesicles contain neuropeptides, sometimes co-stored with Type I or II transmitters. The appearance under the electron microscope of an electron-dense core is due to the presence of protein within the lumen of the vesicle. Neuronal LDCVs are difficult to prepare in a yield or purity sufficient for biochemical analysis and most studies are performed of chromaffin vesicles which are closely analogous to LDCVs.

8.2 LDCV exocytosis

There are fundamental differences between the release mechanisms for LDCVs and SSVs. Apart from the lack of the integral proteins characteristic of SSVs (Table 8.1), little synapsin I is associated with LDCVs; LDCVs do not cluster at active zones, and electron micrographs only occasionally show LDCV exocytosis at the zones. Instead LDCV exocytosis is most frequently

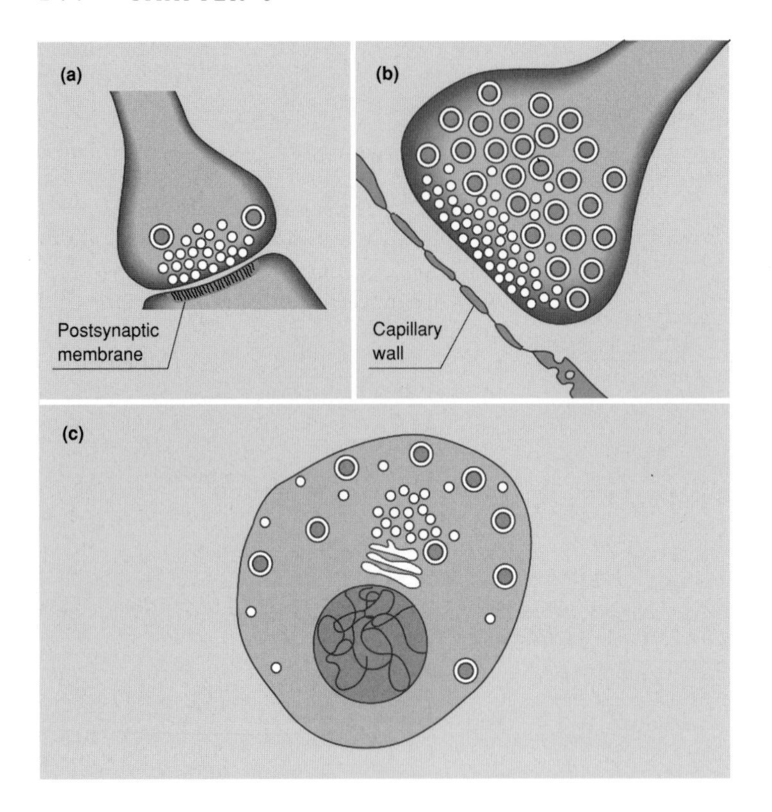

Fig. 8.2 SSVs and LDCVs coexist in terminals and secretory cells. (a) In most CNS terminals SSVs dominate, but LDCVs can also be found. (b) The nerve ending of a hypothalamic neurone in the neurohypophysis only secretes peptide hormones stored in LDCVs, nevertheless the terminals also contain SSVs of unknown function. (c) Neuroendocrine cells contain two populations of vesicles; secretory granules closely related to LDCVs and synaptic-like microvesicles related to SSVs. From De Camilli (1991).

Table 8.1 Constituents of noradrenergic small synaptic vesicles, large dense-core vesicles and chromaffin vesicles.

	Small dense vesicles	Large dense-core vesicles	Chromaffin vesicles
Soluble components			
Noradrenaline	Present	Present	Present
ATP	Present	Present	Present
Chromogranins	Absent	Present	Present
Neuropeptides	Absent	Present	Present
Membrane-associated proteins			
Dopamine β-hydroxylase	Present (?)	Present	Present
Synapsins	Present	Absent	Absent
Membrane proteins			
V-ATPase	Present	Present	Present
Synaptotagmin	Present	Present	Present
Synaptobrevin	Present	Present	Present
Synaptophysin	Present	Present	Present

observed at non-synaptic sites, and hence are not directed immediately onto postsynaptic receptors; this is consistent with a neuromodulator role for neuropeptides, where they might be expected to diffuse a significant distance before interacting with their receptors.

The evidence that fast-acting neurotransmitters are released by high localized $[Ca^{2+}]_c$ at the active zone has been summarized in Chapter 7. Since neuropeptide-containing LDCVs are predominantly released from non-synaptic sites it is likely that their release mechanism might differ, raising the possibility of differential release. In preparations where electrically evoked release of coexisting transmitters can be quantified, it is found that lower frequencies of stimulation are required to release the classical transmitter than the neuropeptide. Thus increasing the frequency of electrical stimulation of terminals containing co-stored classical and neuropeptide transmitters causes a disproportionate increase in neuropeptide release relative to that of classical transmitter, for example vasoactive intestinal peptide (VIP) relative to ACh, neuropeptide Y (NPY) relative to NA or substance P (SP) relative to 5-HT (Fig. 8.3). Since the terminal stores of the classical transmitter may be several orders of magnitude higher than the neuropeptide, and since the latter can only be replenished by axonal transport, the neuropeptide can therefore be used more sparingly than the classical transmitter. This also implies that the mix of transmitters may vary not only with the acute stimulation frequency, but also chronically due to temporary depletion of peptide stores in the terminal.

A mechanism for this differential release of neuropeptides was outlined in Fig. 7.8 in the context of local Ca^{2+} gradients and SSV exocytosis. Hippocampal synaptosomes contain both glutamate and the neuropeptide cholecystokinin. When the average $[Ca^{2+}]_c$ in a population of synaptosomes is increased to a uniform 400 nM by ionomycin, glutamate is not released since the ionophore does not create the high Ca^{2+} concentration in the immediate vicinity of the presynaptic Ca^{2+}-channels required for amino acid exocytosis (*see* Section 9.5.4). However the neuropeptide *is* released by this small elevation in the bulk cytoplasmic $[Ca^{2+}]_c$, indicating the presence of a high-affinity Ca^{2+} trigger responding to bulk Ca^{2+}. Thus paradoxically neuropeptides are released by a high-affinity Ca^{2+} trigger, but require repetitive stimulation of the terminal in order to elevate the bulk $[Ca^{2+}]_c$, whereas amino acids are released by Ca^{2+} interacting with a low-affinity trigger but one which responds immediately to the Ca^{2+} influx accompanying isolated action potentials.

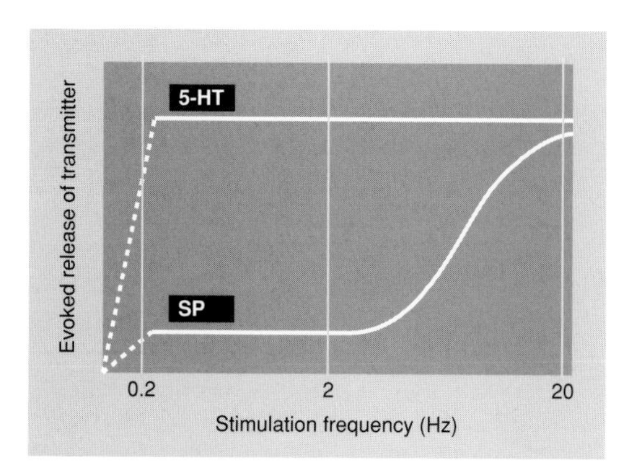

Fig. 8.3 Frequency-dependent release of substance P from spinal cord neurones. The amount of 5-HT released from spinal cord slices by 900 pulses of electrical field stimulation does not depend on the frequency of that stimulation, whereas substance P-like immunoactivity (SP) is only released by high frequency stimulation. Adapted from Iverfledt *et al.* (1989).

LDCVs are present in small numbers in many terminals (*see* Fig. 8.2). However in order to study LDCV exocytosis it is convenient either to use a specialized synaptosomal preparation, the neurosecretosome (*see* Section 2.6.1) from neurohypophyseal terminals which release vasopressin and oxytocin, or to study exocytosis from the chromaffin cell (an 'honorary neurone') or from the *mast cell*, which has a different mechanism for triggering exocytosis, but is a valuable model for direct electrophysiological monitoring of capacitance changes when the vesicle fuses with the membrane. The details of the transmitter-specific biochemistry of these cells will be discussed in later chapters. However these cells have made a significant contribution to our understanding of the exocytotic process itself, which it is relevant to discuss here.

Both L- and N-type Ca^{2+}-channels can be detected by patch clamping of large neurosecretosomes, and these are randomly scattered along the membrane rather than being concentrated in 'hot-spots'. Multiple action potentials at a frequency of >1 Hz are required to release the peptides, in agreement with the need to increase bulk $[Ca^{2+}]_c$ as discussed above. Physiologically the terminals receive bursts of action potentials at 5–15 Hz. Exocytosis can be monitored by the increase in membrane capacitance (*see* Fig. 12.3); a train of action potentials will release up to about 75 vesicles per terminal, the release being a function of the bulk elevation in $[Ca^{2+}]_c$. Interestingly in view of hypotheses concerning the action of tetanus toxin on SSV exocytosis, the toxin is able to block exocytosis in this system, even though the vesicles are believed not to contain synaptobrevin.

8.2.1 Chromaffin cells

Chromaffin cells (*see* Fig. 2.2) and the derived phaeochromocytoma cell line *PC12* provide an invaluable model for LDCV secretion. Chromaffin cells contain about 30 000 vesicles and optimal stimulation *in vitro* can release about 30% of this complement. Exocytosis can be monitored by direct assay of released catecholamine, by continuous voltametric analysis of catecholamine concentration in the medium, or by capacitance changes during whole-cell patch-clamp. An indirect assay for exocytosis makes use of the fact that ATP is co-stored and co-released together with the catecholamine and that this ATP can be detected by its ability to increase $[Ca^{2+}]_c$ in surrounding fibroblasts by acting on ATP receptors. Electron microscopy of stimulated cells allows 'ω-profiles' to be seen, corresponding to vesicles in the act of fusion. If tannic acid is present in the medium, the protein content of the vesicles denatures and remains at the site of exocytosis.

Although elevated cytoplasmic Ca^{2+} is necessary and sufficient for catecholamine release, there is no direct correlation between the maximal $[Ca^{2+}]_c$ and the extent of exocytosis evoked by different means. Depolarization (by activation of nicotinic receptors or by high KCl) followed by Ca^{2+} entry through voltage-activated Ca^{2+}-channels is the most effective way of releasing catecholamine (Fig. 8.4). The cells possess a large number of PLC-coupled metabotropic receptors

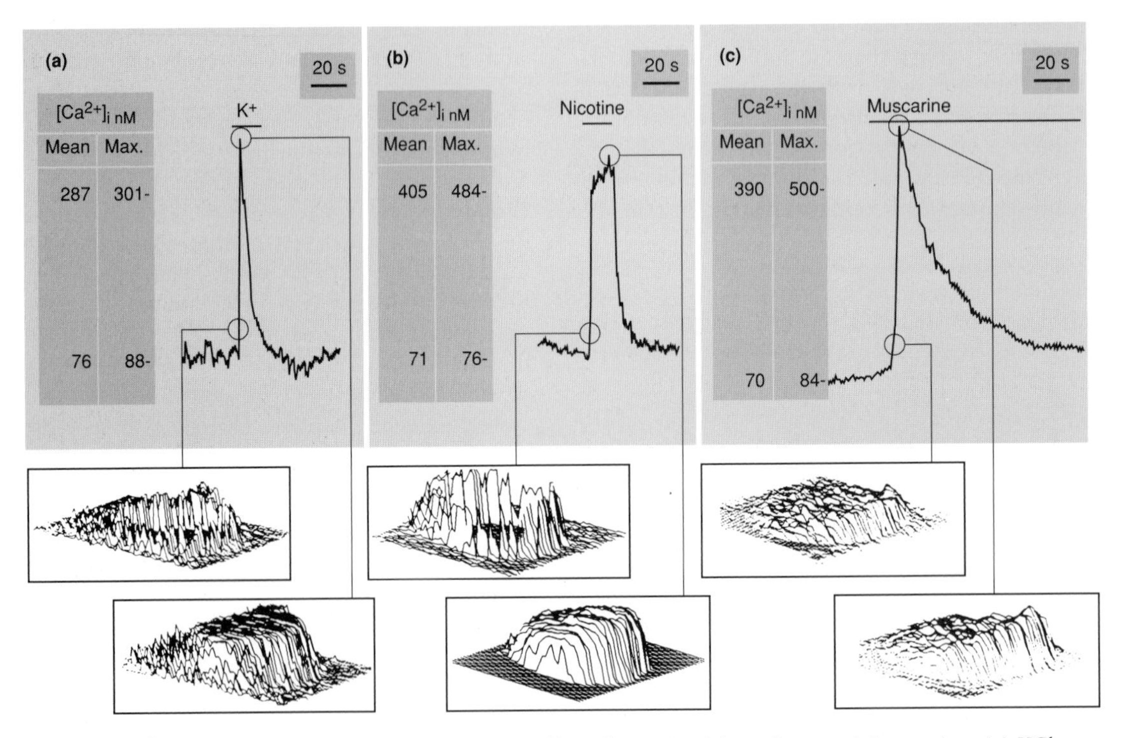

Fig. 8.4 Ca^{2+} imaging profiles of response of chromaffin cells to nicotinic and muscarinic agonists. (a) KCl depolarization produces a sharp 'spike' of $[Ca^{2+}]_c$ in a single cell photodiode trace. Imaging of the cell at the onset of the response (left hand side) shows a 'crown' of high Ca^{2+} immediately under the membrane which has evened out to a 'hat' by the peak of the response. (b) The response with nicotine is similar, whereas muscarine (c) produces a relatively uniform initial increase in $[Ca^{2+}]_c$, which may subsequently be supplemented by local influx at one pole of the cell. Nicotinic stimulation is more effective than muscarinic stimulation in triggering exocytosis. From Cheek *et al.* (1989), by permission of *FEBS Letters*.

(including those for ACh, histamine (H_1), angiotensin II, ATP, bradykinin, prostaglandin E_2 (PGE_2) and VIP) which are each coupled to $(1,4,5)IP_3$ production and consequent release of internal $(1,4,5)IP_3$-sensitive Ca^{2+} stores. Furthermore caffeine will release Ca^{2+} from $(1,4,5)IP_3$-insensitive stores. This internally released Ca^{2+} is however not able to trigger exocytosis directly, as shown by the complete inhibition of release if external Ca^{2+} is withdrawn; rather the increased $[Ca^{2+}]_c$ triggers secondary Ca^{2+} entry across the plasma membrane (*see* Section 5.8.1). The actual Ca^{2+} concentration at the release site has been estimated to be 5 μM, and it has been suggested that concentric shells of decreasing $[Ca^{2+}]_c$ are formed during exocytosis (*see* Fig. 7.9a) intermediate between the fully delocalized Ca^{2+} which triggers peptide exocytosis and the highly localized Ca^{2+} responsible for amino acid exocytosis.

It must be emphasized that exocytosis from chromaffin cells is much slower (capacitance changes indicate typically a 50 ms – 1 s lag) than the submillisecond exocytosis of ACh and amino acids in the CNS (*see* Fig. 7.1). Thus the chain of events between Ca^{2+} entry and release will not necessarily be the same in both cases. It is safest to assume that the two processes are distinct until proven otherwise. The kinetics of exocytosis from chromaffin cells can be monitored by following capacitance changes of the cells in whole-cell patch mode (described in more detail in the next section) when $[Ca^{2+}]_c$ is suddenly increased by photolysis of caged Ca^{2+} compounds such as *nitr-5* or *DM-nitrophen*, whose affinity for Ca^{2+} decreases dramatically on exposure to UV radiation. Flash photolysis is thus an extremely rapid means of increasing $[Ca^{2+}]_c$ uniformly within a cell. Four types of Ca^{2+}-evoked capacitance changes are observed; ultrafast, fast and slow secretion with respective time constants of <0.5, 3 and 10–30 s and a Ca^{2+}-evoked capacitance *decrease* which may reflect Ca^{2+}-triggered endocytosis. The differing release kinetics may reflect differentially located pools of vesicles, as has been suggested for glutamate release from synaptosomes (*see* Section 9.5).

Catecholamine secretion can still occur from PC12 cells which lack protein kinase A and are downregulated in PKC by prolonged exposure to phorbol ester. Secretion from these cells can also be evoked by Ba^{2+} which does not activate calmodulin and hence CaMKII. Thus as in the case of small synaptic vesicle exocytosis (*see* Section 7.4.6) protein phosphorylation does not appear to play an obligatory role in the basic release mechanism.

PKC may regulate exocytosis. In polarized chromaffin cells a cortical (peripheral) actin network prevents the chromaffin vesicles from reaching the plasma membrane. Chromaffin vesicles interact with actin filaments via actin-binding proteins such as α-actinin which is present on the cytoplasmic face of the vesicle membrane. Exocytosis must therefore be preceded by actin disassembly, and is accompanied by a clearing of the Ca^{2+}-dependent actin-binding proteins *caldesmon* and *fodrin* from the plasma membrane, perhaps to open up the structure of the cytoplasm to allow the granules to reach their exocytotic site on the plasma membrane. Ca^{2+} and probably PKC are involved in this disassembly, but the exact mechanism is unclear (Fig. 8.5).

Chromaffin cells can be permeabilized to allow direct access to the site of exocytosis (*see* Fig. 2.6). *Electro-permeabilized cells* (which have small holes which do not leak proteins), release catecholamine when the Ca^{2+} concentration is increased to about 10 μM as long as MgATP is present. The clostridial toxins botulinum neurotoxins type A, B and D and tetanus toxin inhibit exocytosis in permeabilized cells as in CNS terminals although this must be reconciled with reports that the BoNT type B acts on the small synaptic vesicle-specific protein synaptobrevin.

Digitonin-permeabilized chromaffin cells slowly lose the ability to release catecholamines even in the presence of Ca^{2+} and ATP. This appears to be due to the loss of cytoplasmic enzymes, and the process can be slowed or reversed by the re-addition of a number of proteins. PKC will increase the Ca^{2+} sensitivity of secretion as will GTP-γ-S, suggesting some involvement of small GTP-binding proteins (*see* Section 7.4.5). Another factor which can slow the rundown of exocytosis is *annexin II* (also known as *calpactin*). Annexins are an extensive group of related proteins which have unique Ca^{2+}-binding domains which are distinct from the 'EF-hands' found in calmodulin and related proteins (Fig. 8.6). They share a related 70 residue sequence which may be repeated four or eight times. The proteins associate with phospholipids

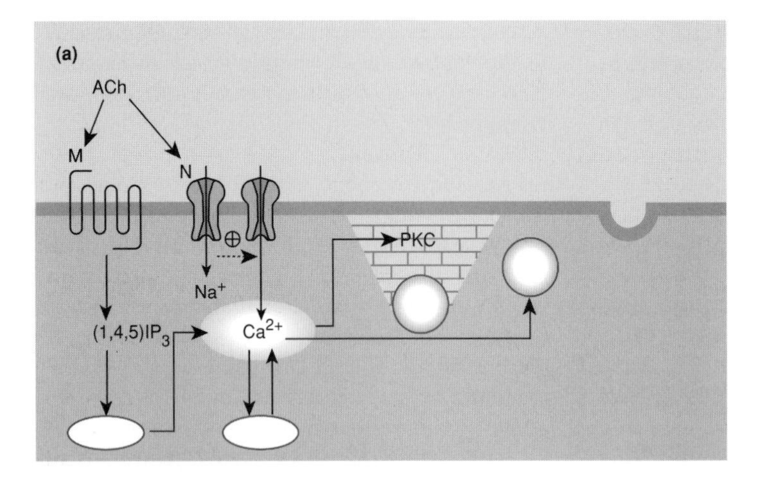

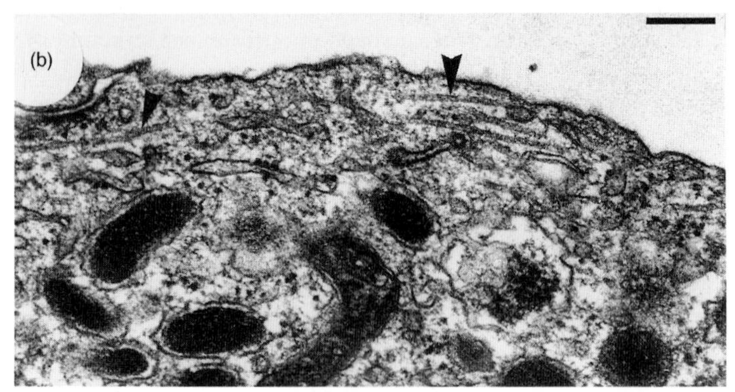

Fig. 8.5 The cytoskeleton and chromaffin cell exocytosis. (a) ACh, acting on muscarinic (M) receptors activates phospholipase C generating $(1,4,5)IP_3$ and releasing internal Ca^{2+} stores. Ca^{2+} can also be elevated by entry through voltage-activated Ca^{2+}-channels in response to nicotinic (N) receptor-mediated depolarization. Ca^{2+} is envisaged as having two main effects; firstly a PKC-mediated disruption of the cytoskeleton immediately below the cytoskeleton, and secondly modulation of the triggering of exocytosis itself. It should be noted that as in the case of amino acid exocytosis Ca^{2+} entering through voltage-activated channels may be more effective in triggering exocytosis. (b) The submembrane cytoskeleton of the chromaffin cell and the exclusion of vesicles from this region. Bar = 200 nm. Data from Burgoyne (1991), by permission of Elsevier Science Publishers.

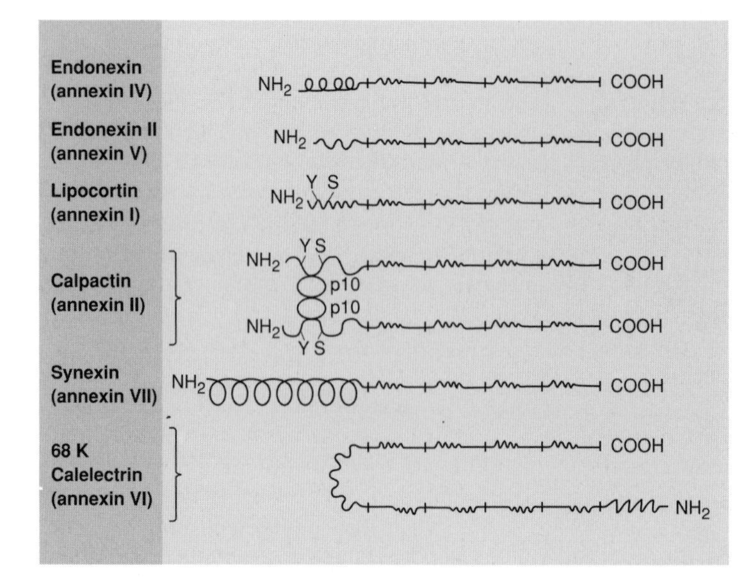

Fig. 8.6 Annexins. Schematic structures of six annexins. The four (or eight) homologous domains in each protein are 70 amino acids in length. The N-termini differ and contain tyrosine (Y) and serine (S) phosphorylation sites. Calpactin may function as a dimer linked by p10 dimers. Adapted from Creutz (1992).

when complexed with Ca^{2+}, although their affinity for Ca^{2+} is low, with K_ds in the region of 200 μM. Annexin II will aggregate isolated chromaffin vesicles in the presence of micromolar $[Ca^{2+}]_c$. A 20 amino acid annexin II consensus peptide injected in the cells inhibits release, and re-addition of annexin II to chronically permeabilized cells can partially restore the extent of exocytosis. Finally the protein can be seen in cryo-sections to anchor chromaffin vesicles to the sites of exocytosis in stimulated cells. *Synexin* (also referred to as annexin VII) is a related member of the annexin family which will also aggregate isolated chromaffin vesicles, but in the presence of high (100 μM) Ca^{2+}.

Prolonged digitonin-permeabilization of chromaffin cells reveals a requirement for two further leaked protein fractions, named *Exo I* and *Exo II*. Little is known about Exo II, but Exo I belongs to the '14–3–3' gene family which is widely distributed and highly conserved. Exo I and PKC are synergistic in reconstituting exocytosis. The relevant substrate for PKC in PC12 cells is a 145 kDa protein termed *p145*. It should however be noted in passing that Ca^{2+}-secretion coupling of SSV exocytosis does not appear to involve PKC-dependent phosphorylation, other than indirectly via channel modulation (*see* Section 7.5).

Exocytosis in PC12 cells can be resolved into two sequential stages (Fig. 8.7). The first is, reversible, ATP (but not GTP)-dependent, can be inhibited by protein kinase inhibitors, including those active against PKC, and appears to be a Ca^{2+}-independent priming step. The second step is exocytosis itself, which is Ca^{2+}-dependent but does not require ATP. Whether the first stage relates to the cytoskeletal disruption discussed above remains to be established.

8.2.2 Mast cells

Mast cells are characterized by the presence of large (700 nm diameter) secretory vesicles containing histamine in a complex with heparin and proteins. The role of histamine as a neurotransmitter in the CNS will be discussed later (*see* Section 11.1), here we shall just discuss the contribution the histamine secreting mast cell has made to our understanding of LDCV exocytosis.

Whole-cell mode patch clamping (*see* Section 2.8) of secretory cells allows the plasma membrane capacitance to be determined (Fig. 8.8). As individual vesicles fuse with the membrane, step increases in capacitance can be detected. While most work with this technique has been performed with mast cells, it is also possible to study capacitance changes in chromaffin cells and PC12 cells during exocytosis. Their secretion can be studied in detail, although it must be borne in mind that firstly mast cell secretion is peculiar particularly in terms of its coupling to Ca^{2+}, and secondly the slow secretion of histamine from mast cells may have limited relevance to events in the CNS. Mast cells from *beige mice* possess 10–20 giant granules (with mean diameters of 2.5 μm) and are particularly suited to the study of capacitance changes. The technique works best when many large vesicles are released in a short time since there must be an imbalance between exocytosis and endocytosis during the experiment. Since capacitance changes are detected by the current flow in

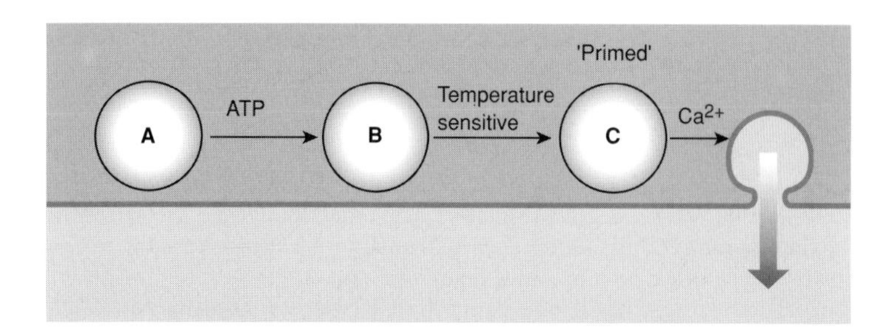

Fig. 8.7 Chromaffin cell exocytosis can be resolved into two stages. In the absence of ATP (A) there is a limited release from permeabilized cells of chromaffin vesicles which are apparently 'primed' for release (C). Priming involves two separable steps, the first (A–B) requires ATP while the second (B–C) is blocked at low temperature. Different soluble proteins, so far uncharacterized, potentiate the priming and release stages.

response to imposed voltage fluctuations it is also necessary to chose a potential range in which voltage-activated channels do not become activated.

The initial fusion event is the formation of a 'neck' between the vesicle and the plasma membrane resembling a 200–300 pS ion-conducting channel which can be quantified from the transient current which flows when the positively charged vesicle lumen first fuses with the plasma membrane. This conductance would be consistent with a *fusion pore* some 2 nm in diameter and 15 nm long (sufficient to span both the vesicle and plasma membrane bilayers). In some cases the capacitance is seen to flicker between open and closed states before the increase in surface area becomes permanent (Fig. 8.8). There is debate as to whether release of the vesicle contents can occur during this 'flickering' phase or only after the final dilation of the pore. Beige mouse mast cell granules contain 5-HT co-stored with histamine; release of the former can be detected by a carbon fibre microelectrode situated close to the surface of the cell and it is possible to observe some release of 5-HT during the flickering phase, although much more release occurs after the presumed pore dilation. 'Stand-alone flickering' can also be seen which does not lead to permanent fusion, and again a partial exocytosis is seen. Extrapolation from these

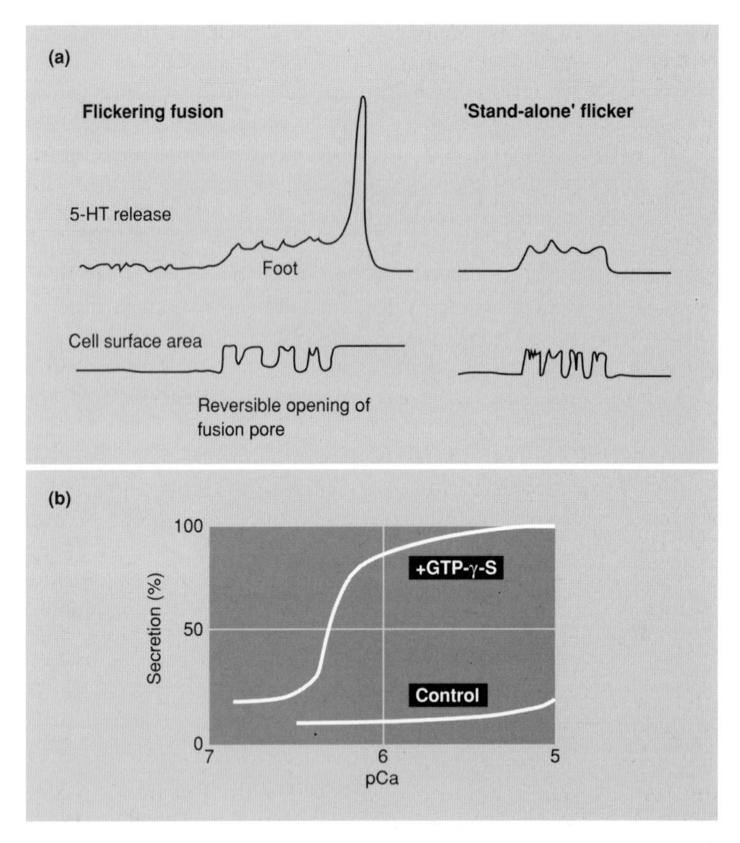

Fig. 8.8 The monitoring of exocytosis from mast cells. (a) Release of 5-HT from a beige mouse mast cell measured with carbon fibre amperometry. Whole-cell recording of capacitance changes in the cell show flickering fusion consistent with a reversible opening of a fusion pore. Note that some 5-HT is released in this phase although most is released when the increase in surface area becomes permanent, due to irreversible fusion. In some cases 'stand-alone' flicker occurs associated with partial release of the vesicle contents. From Neher (1993), reproduced with permission from *Science*. © 1993 by the AAAS. (Adapted from Alvarez de Toledo *et al.* (1993).) (b) Mast cells permeabilized by streptolysin-O secrete hexosaminidase (contained with their secretory vesicles) in response to elevated Ca^{2+} and activation of G-protein. Cells were exposed for 10 min prior to sampling. Note that neither stimulus is sufficient in itself in permeabilized cells although the role of Ca^{2+} may be modulatory rather than essential. Data from Lindau & Gomperts (1991).

enormous secretory vesicles (with volumes 100 000 times greater than SSVs!) suggests that SSVs might be able to release most or all of their contents with a time constant of *c*. 0.2 ms through a reversible fusion pore.

Mast cells do not secrete in response to an increased $[Ca^{2+}]_c$ but instead (or additionally, the point is contentious) require activation (by antigens, synthetic polycations or SP) of cell-surface receptors coupled to pertussis-sensitive G-protein(s) (Fig. 8.8). The non-hydrolysable GTP analogue GTP-γ-S added to permeabilized mast cells, causes complete exocytosis (measured by capacitance increase) after a delay of about 2 min. A specific G-protein termed 'G_e' (for exocytosis), appears to be involved, rather than a G-protein coupled through PI-PLC or adenylyl cyclase activation, since GTP-γ-S can still initiate secretion in the presence of neomycin, which inhibits PI breakdown, while an increase in cAMP blocks rather than activates secretion.

GTP-γ-S interacts with a membrane-associated G-protein, since it is still effective in extensively dialysed cells which have lost their cytoplasmic contents. Prolonged activation of the G-protein appears to be necessary since inclusion of GTP itself in the patch pipette is insufficient, suggesting that the lifetime of the GTP-activated form of the G-protein is too short to trigger exocytosis. Physiological mast cell exocytosis could perhaps be evoked by pathways which prolong the lifetime of the GTP-form, although molecular details are unclear. The involvement of small molecular weight G-proteins in SSV exocytosis and membrane trafficking is discussed elsewhere (*see* Section 7.4.5). In the present system, intracellular perfusion of the cells with peptides derived from the portion of the sequence of Rab3A which is believed to interact with effectors causes exocytosis in the absence of GTP-γ-S. The peptides, which lack GTP-binding domains apparently substitute for the intact GTP-γ-S-activated G-protein in binding to an unknown component of the exocytotic machinery and triggering release.

Further reading

Origins of SSVs and LDCVs

De Camilli, P. (1991) Co-secretion of multiple signal molecules from endocrine cells via distinct exocytotic pathways. *Trends Pharmacol. Sic.* **12**, 446–448.

Regnier-Vigouroux, A. & Huttner, W.B. (1993) Biogenesis of small synaptic vesicles and synaptic-like microvesicles. *Neurochem. Res.* **18**, 59–64.

Rosa, P., Barr, F.A., Stinchcombe, J.C., Binacchi, C. & Huttner, W.B. (1992) Brefeldin A inhibits the formation of constitutive secretory vesicles and immature secretory granules from the trans-Golgi network. *Eur. J. Cell Biol.* **59**, 265–274.

LDCV exocytosis

Bartfai, T., Iverfeldt, K., Fisone, G. & Serfözö, P. (1988) Regulation of the release of coexisting neurotransmitters. *Annu. Rev. Pharmac. Toxicol.* **28**, 285–310.

Boarder, M.R. (1989) Presynaptic aspects of cotransmission: relationship between vesicles and neurotransmitters (Review). *J. Neurochem.* **53**, 1–11.

Campbell, G. (1987) Cotransmission. *Ann. Rev. Pharmac. Toxicol.* **27**, 51–70.

Creutz, C.E. (1992) The annexins and exocytosis. *Science* **258**, 924–931.

Fatatis, A., Holtzclaw, L., Payza, K. & Russell, J.T. (1992) Secretion from rat neurohypophysial nerve terminals (neurosecretosomes) rapidly inactivates despite continued elevation of intracellular Ca^{2+}. *Brain Res.* **574**, 33–41.

Iverfeldt, K., Serfozo, P., Diaz-Arnesto, L. & Bartfai, T. (1989) Differential release of coexisting neurotransmitters: frequency dependence of the efflux of substance P, thyrotropin releasing hormone and [3H] serotonin from tissue slices of rat ventral spinal cord. *Acta Physiol. Scand.* **137**, 63–71.

Chromaffin cells

Bittner, M.A. & Holz, R.W. (1992) Kinetic analysis of secretion from permeabilized adrenal chromaffin cells reveals distinct components. *J. Biol. Chem.* **267**, 16 219–16 225.

Burgoyne, R.D. (1990) Secretory vesicle-associated proteins and their role in exocytosis. *Annu. Rev. Physiol.* **52**, 647–659.

Burgoyne, R.D. (1991) Control of exocytosis in adrenal chromaffin cells. *Biochim. Biophys. Acta* **1071**, 174–202.

Cheek, T.R., O'Sullivan, A.J., Moreton, R.B., Berridge, M.J. & Burgoyne, R.D. (1989) Spatial localization of the stimulus-induced rise in cytosolic Ca in bovine adrenal chromaffin cells. *FEBS Lett.* **247**, 429–434.

Hay, J.C. & Martin, T.F.J. (1992) Resolution of regulated secretion into sequential MgATP-dependent and calcium-dependent stages mediated by distinct cytosolic proteins. *J. Cell Biol.* **119**, 139–151.

Isobe, T., Hiyane, Y., Ichimura, T., Okuyama, T., Takahashi, N., Nakajo, S. & Nakaya, K. (1992) Activation of protein kinase C by the 14-3-3 proteins homologous with Exo1 protein that stimulates calcium-dependent exocytosis. *FEBS Lett.* **308**, 121–124.

Matthies, H.J.G., Palfrey, H.C. & Miller, R.J. (1988) Calmodulin and protein phosphorylation-independent release of catecholamines from PC-12 cells. *FEBS Lett.* **229**, 238–242.

Neher, E. & Zucker, R.S. (1993) Multiple calcium-dependent processes related to secretion in bovine chromaffin cells. *Neuron* **10**, 21–30.

Nishizaki, T., Walent, J.H., Kowalchyk, J.A. & Martin, T.F.J. (1992) A key role for a 145-kDa cystosolic protein in the stimulation of Ca^{2+}-dependent secretion by protein kinase C. *J. Biol. Chem.* **267**, 23 972–23 981.

Trifaro, J.M., Vitale, M.L. & Del Castillo, A.R. (1992) Cytoskeleton and molecular mechanisms in neurotransmitter release by neurosecretory cells. *Eur. J. Pharmacol.* **225**, 83–104.

Winkler, H. & Fischer-Colbrie, R. (1990) Common membrane proteins of chromaffin granules, endocrine and synaptic vesicles: properties, tissue distribution, membrane topography and regulation of synthesis. *Neurochem. Int.* **17**, 245–262.

Mast cells

Alvarez de Toledo, G., Fernandez-Chacon, R. & Fernandez, J.M. (1993) Release of secretory products during transient vesicle fusion. *Nature* **363**, 554–558.

Lindau, M. & Gomperts, B.D. (1991) Techniques and concepts in exocytosis: focus on mast cells. *Biochim. Biophys. Acta* **1071**, 429–471.

Neher, E. (1992) Ion channels for communication between and within cells. *Science* **256**, 498–502.

Neher, E. (1993) Secretion without full fusion. *Nature* **363**, 497–498.

Penner, R. & Neher, E. (1989) The patch-clamp technique in the study of secretion. *Trends Neurosci.* **12**, 159–163.

Part 3
Transmitters and Synapses

9

Amino acids as neurotransmitters

9.1 Introduction

Amino acids account for the large majority of fast synaptic transmission in the mammalian CNS, and a striking feature is the extremely high concentration of the transmitters within brain, even after taking account of the metabolic roles which glutamate, aspartate and glycine perform; thus GABA and glutamate are present at concentrations two to three orders of magnitude higher than Type II transmitters such as ACh and NA. Indeed it was initially difficult to believe that such universal metabolites could act as specific transmitters. As will be seen, specific transport and consequent precise compartmentation are the keys to the action of the amino acid transmitters.

The confirmed and putative amino acid neutrotransmitters are listed in Table 1.1. While the roles of glutamate, glycine, GABA and to a lesser extent aspartate are well documented, a number of candidate amino acids remain in the 'not proven' category. It is therefore instructive to consider the criteria which the established amino acid transmitters satisfy:

1 *A transmitter amino acid should be localized presynaptically within synaptic vesicles in specific neurones.* GABA can most easily be localized in histochemical sections by the specific presence of the enzyme glutamic acid decarboxylase, glutamate decarboxylase (GAD), which is responsible for the synthesis of the amino acid. The amino acid itself can be made immunoreactive by *in situ* cross-linking to proteins with glutaraldehyde. Antibodies against cross-linked GABA can identify GABA-rich brain regions, or by immunogold electron microscopy (Fig. 9.1) individual GABAergic terminals. While the latter technique gives some indication of punctate labelling of vesicle clusters, better evidence for vesicular accumulation of the amino acid comes from the ability to purify synaptic vesicles containing a subpopulation capable of accumulating the amino acid (*see* Section 9.4). While the plasma membrane of GABAergic terminals contains a high-affinity Na^+-coupled GABA co-transporter (*see* Section 9.3), a related carrier is also present on glial cells and so the high-affinity uptake of labelled GABA cannot be taken as being diagnostic of the presence of GABAergic terminals.

Although glycine has metabolic roles, immunohistochemistry reveals that it is concentrated in specific terminals, particularly in the brain stem and spinal cord. Since the synaptic vesicle GABA transporter also accumulates glycine, terminals must discriminate between these transmitters by the presence of discrete selective plasma membrane transporters.

The universal metabolic role of glutamate and aspartate limits the validity of presynaptic localization as a sufficient criterion for identifying the glutamatergic and aspartatergic terminals. In the case of glutamate the high background due to metabolic pools means that it is unusual to find regions where the amino acid is elevated more than two-fold above the general background signal (*see* Fig. 9.1a). Furthermore there are no specific enzymes to allow the histochemical mapping of glutamatergic pathways. The respective roles of glutamate and aspartate are difficult to resolve, since the two acidic amino acids are accumulated into terminals on the same plasma membrane transporter (*see* Section 9.3) and are in any case freely interconvertable in the terminal cytoplasm (Fig. 9.2). However acidic amino acid uptake into isolated synaptic vesicles shows a strong selectivity for glutamate over aspartate (*see* Section 9.4), consistent with the dominant role of the former.

2 *The amino acid should be released exocytotically in response to a physiologically relevant stimulus.* The key word here is 'exocytotically': KCl depolarization of synaptosomes, cultured neurones or brain slices will each lead to the release of amino acids. However the amino acids are not only present in synaptic vesicles, but are also present in the cytoplasm of neurones and glia at high concentrations; these cytoplasmic pools can be released by stimuli which decrease the Na^+ electrochemical gradient across the plasma membrane allowing reversal of the plasma membrane carriers (*see* Section 3.3.1). It is therefore essential to distinguish exocytotic release from cytoplasmic release. Ca^{2+}-dependency is a necessary but not sufficient criterion, since the existence of Ca^{2+}-permeant ionotropic receptors (e.g. the NMDA receptor and some forms of the AMPA receptor (*see* Section 9.6) means that *in vitro* Ca^{2+} can potentiate plasma membrane depolarization and hence cytoplasmic release. The extent of release is also important: some sulphur amino acids can be released in a Ca^{2+}-dependent manner, but the amount of release is very small in relation to that of glutamate or GABA, limiting their role (Fig. 9.3).

3 *An amino acid applied to the postsynaptic membrane should mimic the physiological*

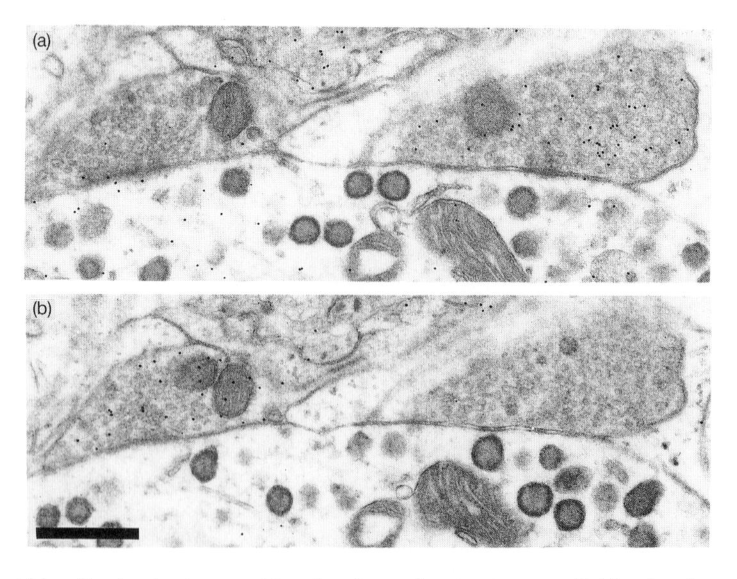

Fig. 9.1 Amino acid localization by immunohistochemistry after *in situ* cross-linking by glutaraldehyde. Glutaraldehyde-fixed tissue was sectioned and embedded and then reacted with primary antibody raised in the rabbit against protein-conjugated glutamate or GABA. These antibodies were then visualized by secondary goat anti-rabbit IgG antibody adsorbed to 10 nm colloidal gold particles. Electron microscopy was carried out after counter-staining with lead citrate. Two serial sections are shown through adjacent hypothalamic terminals. (a) Glutamate immunoreactivity is seen in the terminal to the right. (b) GABA-immunoreactivity is exlusively in the left-hand terminal. Bar = 500 nm. From Meeker *et al.* (1993), by permission of Elsevier Science Publishers.

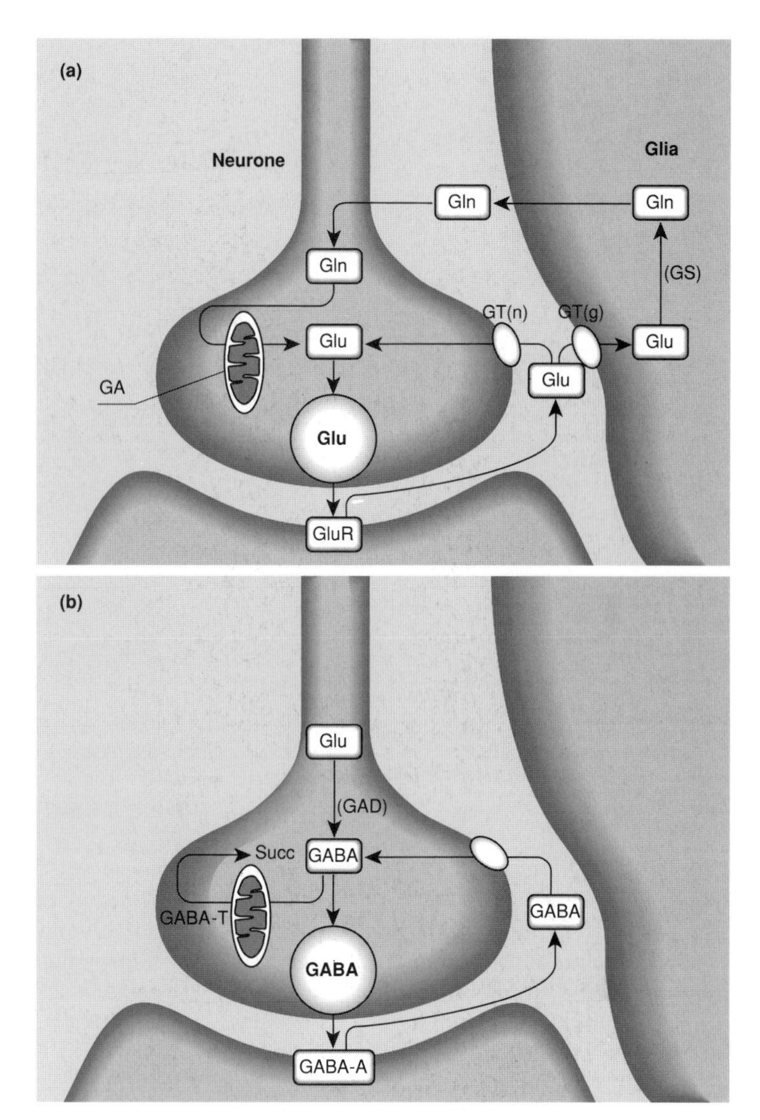

Fig. 9.2 Pathways of glutamate (Glu) and GABA metabolism. (a) Glu released into the synaptic cleft is retrieved by neuronal-type (GT(n)) and glial-type (GT(g)) Na^+-coupled glutamate transporters. Glial glutamate is converted to glutamine (Gln) by the enzyme glutamine synthetase (GS). Gln is present at high concentration in the CSF (*c.* 0.5 mM) and can enter the neurone to help replenish glutamate after hydrolysis by the mitochondrial glutaminase (GA). (b) GABA is synthesized in the cytoplasm of GABAergic neurones from glutamate. Most released GABA is retrieved into terminals, although GABA can also be taken up into glia. GABA is degraded by the mitochondrial GABA-transaminase (GABA-T) yielding succinic semialdehyde which is then oxidized to succinate (Succ).

response. This is perhaps the least discriminating criterion, since the specificity of most amino acid receptors for putative agonists is limited. For example glutamate, aspartate and minor sulphur amino acids (Fig. 9.3) are frequently equipotent at a given receptor.

4 *A high-affinity re-uptake or inactivation mechanism should exist to terminate transmission.* Amino acid transmission is terminated by re-uptake into terminals and surrounding glia (*see* Section 9.3). Again this criterion fails to distinguish between glutamate and aspartate, which are substrates for both the terminal and glial forms of the Na^+-coupled acidic amino acid carrier (*see* Section 9.2).

While it is not easy to quantify the relative numbers of transmitter-specific synapses in brain, it is evident that amino acid synapses substantially exceed those of all other transmitters combined, and that amino acids are responsible for almost all the fast ionotropic signalling between neurones, leaving predominantly modulatory roles via G-protein-coupled metabotropic receptors for the other transmitters. An individual transmitter is not inherently excitatory or inhibitory since its action depends on the nature of the receptor. Nevertheless the postsynaptic actions of glutamate, and to a much more restricted extent aspartate, are overwhelmingly excitatory. Figure 9.4 shows some of the major projecting

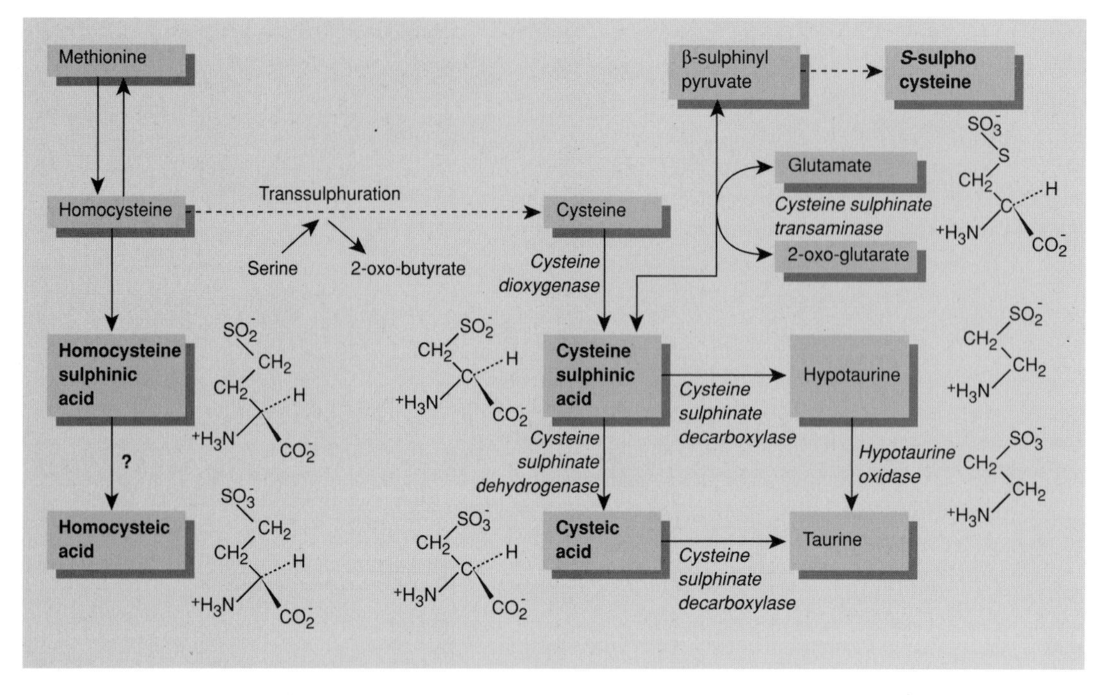

Fig. 9.3 Sulphur amino acids. Methionine and cysteine are precursors for a range of sulphur amino acids. Some, such as homocysteic acid, are ligands for excitatory amino acid receptors, and their possible role as minor neurotransmitters is actively debated. Courtesy of J. Griffiths.

pathways which utilize glutamate. Almost all the pathways originate from cell bodies within the neocortex and the hippocampus. In addition glutamate is the presumed transmitter for an enormous number of intrinsic neurones, particularly in the neocortex, hippocampus and cerebellum. A crude estimate would be that >50% of all CNS synapses may be glutamatergic.

GABA is the main inhibitory transmitter in higher brain areas and is present uniformly in short-axoned interneurones in different brain regions. Rough estimates suggest that 25–45% of all nerve terminals may be GABAergic, judged by the occurrence of the specific re-uptake pathway. *Glycine* performs an inhibitory role in the brain stem and spinal cord. Glycinergic transmission is important for the control of most motor and sensory functions. In evolutionary terms, therefore, glycine may have been the older inhibitory transmitter, which was subsequently complemented by GABA. There are striking similarities in the biochemistry of the two systems, for example, in the nature of their postsynaptic ionotropic receptors (*see* Section 9.8).

A number of sulphur amino acids are in the 'not-proven' category of candidate transmitters. *Taurine* is present in whole brain at a concentration which is only exceeded among amino acids by glutamate (Fig. 9.3). Its synthesis from cysteine involves oxidation of the sulphydryl group to a sulphinate (SO_2^-), decarboxylation to hypotaurine and finally oxidation to taurine. Taurine's function is still unclear; the amine is present at high concentration outside the nervous system. *Homocysteic acid* (HCA) is a naturally occurring amino acid which has an excitatory action. HCA acts on NMDA receptors; indeed, in the absence of glycine, HCA appears to be more effective at NMDA receptors in slice preparations than glutamate itself. However the amount of HCA which is released by slice preparations is some 100-fold less than for glutamate, limiting the scope of the former as a significant neurotransmitter.

9.2 Pathways of amino acid synthesis and degradation

Glutamate and aspartate do not cross the blood–brain barrier and must therefore be synthesized

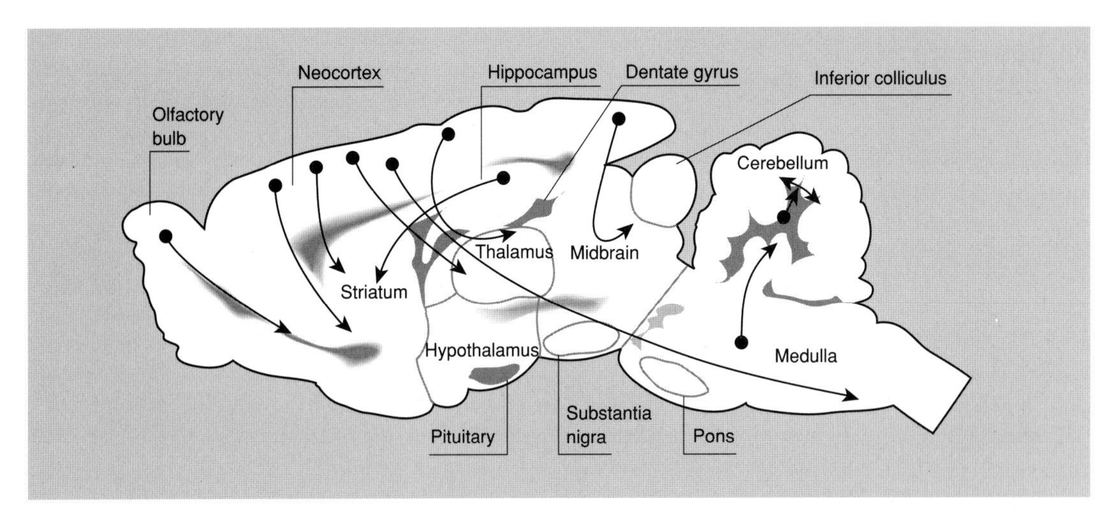

Fig. 9.4 Overview of the major projecting glutamatergic pathways in the rat brain. Nearly all the projecting glutamatergic pathways originate from the neocortex and the hippocampus. In addition there are enormous numbers of glutamatergic interneurones contained within a single anatomical region (GABAergic cells function overwhelmingly as interneurones).

within the brain. Potential pathways for glutamate synthesis include transmination, reduction of 2-oxoglutarate by glutamate dehydrogenase and deamination of glutamine by glutaminase (*see* Fig. 9.2) in addition to minor pathways from ornithine and proline.

Glutamine is synthesized in astrocytes by glutamine synthetase. The reverse process of glutamine hydrolysis by glutaminase occurs in neurones, where the enzyme is localized on the inner face of the inner mitochondrial membrane. Since both glia and neurones possess active Na^+-coupled glutamate re-uptake pathways, an attractive possibility for the replenishment of transmitter glutamate has been the '*glutamine cycle*' (*see* Fig. 9.2). This proposes that glutamate released into the synaptic cleft is primarily taken up into glia and converted to glutamine by the glial glutamine synthetase. The glutamine would then leave the glia and diffuse into the terminals (there is no concentrative Na^+-coupled uptake pathway for glutamine) being finally hydrolysed to glutamate to refill the transmitter pool.

However, since the terminals themselves possess an avid Na^+-coupled glutamate re-uptake pathway (*see* Section 9.3), it is probable that the glial uptake and glutamine synthetase are there primarily to assist in the rapid removal of glutamate from the synaptic cleft and its inactivation by conversion to glutamine. Thus glutamine synthetase is concentrated in fine astrocytic processes associated with defined glutamatergic synapses. The cerebrospinal fluid (CSF) contains about 0.5 mM glutamine and only about 1 μM glutamate. Glutamine can serve as a precursor for neuronal glutamate, but may not label a specific 'transmitter' glutamate pool as has sometimes been claimed.

GABA is synthesized and degraded by a shunt from the citric acid cycle (*see* Fig. 9.2). *GAD* synthesizes GABA from glutamate. The 85 kDa pyridoxal phosphate cofactor enzyme is specifically localized in the terminal cytoplasm, which accounts for the high concentrations of GABA in terminals rather than cell bodies or dendrites. GABA is degraded by *GABA-2-oxoglutarate transminase (GABA-T)* to form succinic semialdehyde which is then oxidized to succinate by *succinate semialdehyde dehydrogenase* and returned to the citric acid cycle. GABA-T is present in both terminals and in glia. Both the degradative enzymes are mitochondrial.

Glycine, being a common amino acid, requires no specialized metabolic pathways. It is however important to realize that in addition to its role in the activation of the inhibitory glycine receptor, glycine is an activator of the excitatory NMDA-selective glutamate receptor which has a wide distribution throughout the CNS (*see* Section 9.6.3).

9.3 Uptake of amino acid neurotransmitters into terminals and glia

The exchange and net accumulation of GABA across the plasma membrane of glia and GABA-ergic neurones is catalysed by two closely related Na^+-cotransporters which can be distinguished from one another by their inhibitor specificities (Fig. 9.5); the neuronal carrier can be competitively inhibited by the transportable analogues *diaminobutyrate* (DABA) or *nipecotic acid*. The glial transporter is sensitive to *β-alanine*.

The neuronal GABA transporter has been purified to apparent homogeneity and reconstituted into liposomes. Determination of partial sequences of cyanogen bromide fragments of the peptide has allowed the design of oligonucleotide probes for screening a rat brain cDNA library. mRNA from a positive clone when injected into *Xenopus* oocytes resulted in the expression of a Na^+-dependent DABA-sensitive [^{3}H]-GABA uptake pathway which bound polyclonal antibodies against the conventionally purified transporter. The cDNA predicts a 599 amino acid protein of 67 kDa (for the deglycosylated carrier) and a probable 12-transmembrane region with both C- and N-termini in the cytoplasm. The transporter shares a common structural motif with the main family of Na^+-coupled plasma membrane transporters which cotransport Na^+, Cl^- and substrate, including 12 putative transmembrane domains, cytoplasmic N- and C-termini and a large glycosylated extracellular loop between transmembrane domains 3 and 4 (Fig. 9.5).

The kinetics of the GABA transporter have been determined using resealed synaptosomal plasma membrane vesicles (not to be confused with synaptic vesicles) which are synaptosomes depleted of cytoplasmic contents by osmotic shock. Using this preparation it is possible to drive GABA accumulation with an artificially imposed ion gradient as the sole driving force. The vesicle interior can be equilibrated with one ionic medium and resuspended into another to create defined ion gradients across the membrane. Uptake is dependent on an inwardly directed Na^+ gradient, and is electrogenic, i.e. accompanied by the net inward movement of positive charge, since uptake is stimulated when valinomycin allows K^+ to flow out of the plasma membrane vesicle generating an interior negative membrane potential. There is an absolute requirement for external Cl^- for GABA uptake, and a Cl^- gradient out→in can drive GABA uptake in the absence of a Na^+ gradient, indicating that it is cotransported together with GABA and Na^+.

Distinct plasma membrane Na^+-glutamate cotransporters responsible for scavenging glutamate from the synaptic cleft are present on neurones and surrounding glia. It has been estimated that the glutamate concentration in the cleft can reach 1 mM following the exocytosis of a single synaptic vesicle and that this can fall with a time constant of little more than 1 ms due to diffusion and re-uptake. The glutamate transporters will also transport D- and L-aspartate and can be inhibited by kainate, dihydrokainate and threo-3-hydroxyaspartate.

The glutamate transporters belong to a family distinct from that comprising the GABA, glycine, choline, proline and taurine transporters, all of which cotransport Na^+, Cl^- and substrate. Instead Na^+ and glutamate are cotransported in exchange for K^+. This stoichiometry was first investigated using synaptosomal plasma membrane vesicles as for the GABA transporter. More recently the kinetics of the glial transporter have been monitored electrophysiologically by exploiting the ability of a whole-cell patch electrode (Fig. 9.6) to monitor the net charge transfer due to glutamate transport as a function of ion gradient and membrane potential. Glutamate uptake is decreased by a K^+ gradient out→in, confirming that K^+ is transported in the opposite direction to glutamate. The stoichiometry indicated by this technique is a 2 Na^+ plus monoanionic glutamate symport coupled to the counter transport of one K^+ and one OH^-, giving a typical equilibrium accumulation ratio of about 20 000 : 1, sufficient to maintain submicromolar concentrations of glutamate in the external medium even in the presence of millimolar cytoplasmic glutamate.

The transporter appears to operate close to thermodynamic equilibrium. Thus depolarization, an increase in internal Na^+ or a decrease in internal K^+ (all of which can occur under conditions of energy deprivation) will cause a net efflux of glutamate from the cytoplasm. This Ca^{2+}-independent cytoplasmic glutamate release is a major contributor to the excitotoxic gluta-

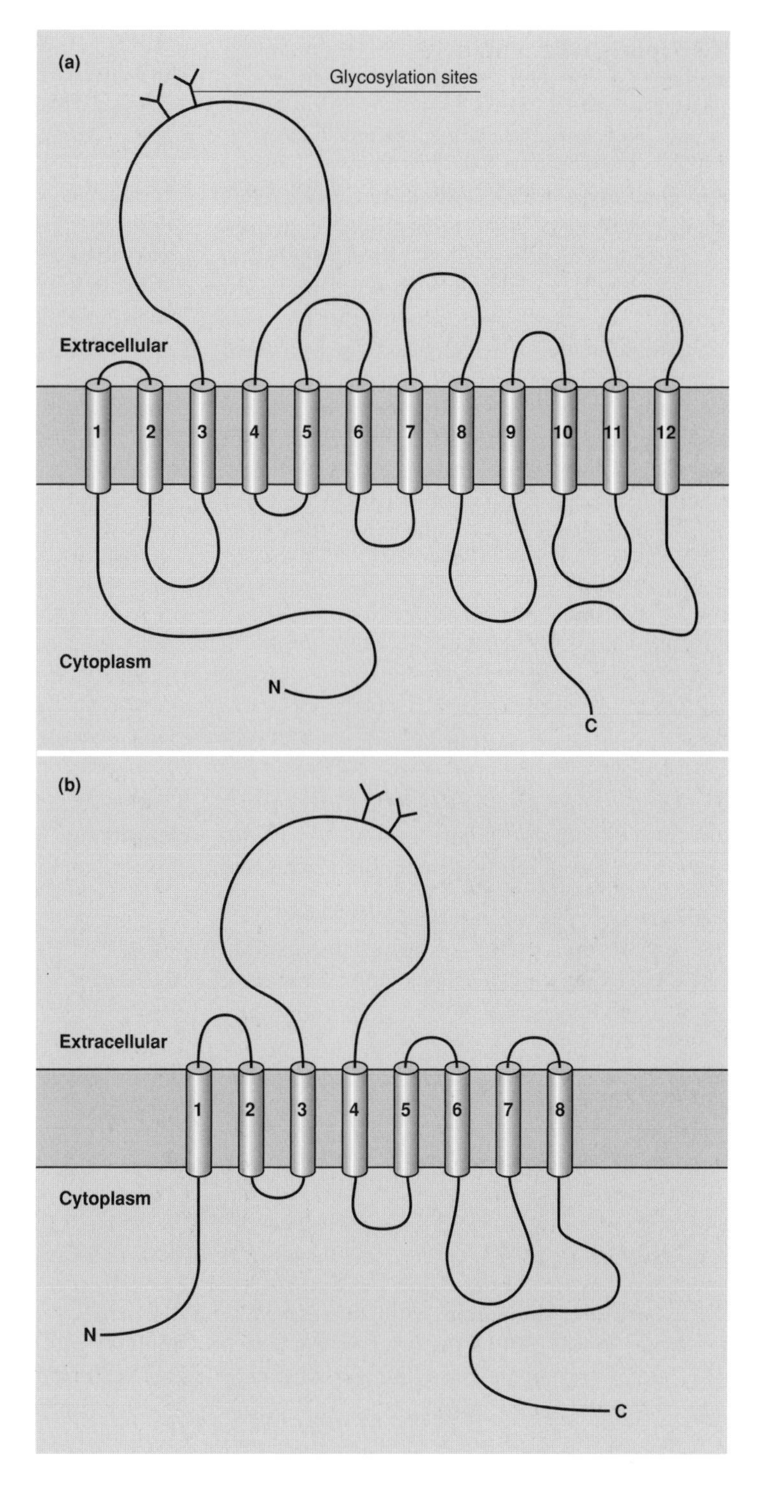

Fig. 9.5 Plasma membrane Na^+-coupled GABA and glutamate carriers. (a) Deduced orientation of the presynaptic GABA transporter. (b) Deduced orientation of a cloned glial plasma membrane glutamate transporter, GLT-1. Note that the exact number of transmembrane segments is undecided due to the presence of hydrophobic regions in the putative cytoplasmic terminus. Adapted from Pines *et al.* (1992).

mate efflux which occurs under these conditions (*see* Section 9.5.5).

Recently three closely homologous glutamate transporters have been cloned, termed GLAST 1, GLT-1 and EAAC1. The first two genes code for glial transporters, while EAAC1 is largely neuronal in the CNS but is also found in peripheral tissue. The transporters possess 500–600 amino acids (approximate molecular weight 60 000), show no homology with the 12-transmembrane spanning GABA class of transporters, but do have some homology with an *Escherichia coli* proton-coupled glutamate transporter. There is currently uncertainty as to the number of putative transmembrane spans due to the presence of a large hydrophobic region near the C-terminus although six to ten spans have been suggested (*see* Fig. 9.5).

9.4 Uptake of amino acids into synaptic vesicles

The early characterization of acetylcholine and catecholamine containing subpopulations of synaptic vesicles (*see* Sections 10.2.1 and 11.2.2) contrasted with the failure to find significant concentrations of intravesicular amino acids and was a stumbling block to the acceptance of an exocytotic model for amino acid transmitter release. This, together with the ease with which cytoplasmic pools of amino acids could be released from synaptosomes and neurones upon prolonged depolarization led to alternative theories proposing a cytoplasmic origin for physiologically released amino acid transmitters.

Even if glutamatergic synaptic vesicles leak their contents during preparation, they should still retain the capacity to reaccumulate the amino acid when incubated under appropriate conditions. ATP-dependent glutamate uptake into extensively purified synaptic vesicles was observed using pure synaptic vesicles immunoprecipitated from crude fractions with anti-synapsin I IgG. More recently controlled-pore glass bead columns have been used to separate vesicles. Using synaptophysin as a vesicular marker (*see* Section 7.4.4) a purification of some 20-fold over the homogenate can be obtained.

The uptake of glutamate is driven by the proton electrochemical gradient (or proton-motive force, pmf) across the vesicular membrane generated by the inwardly directed vesicular proton-ATPase (*see* Section 4.3). The pmf is composed of both membrane potential ($\Delta\psi$) and pH gradient (ΔpH) components:

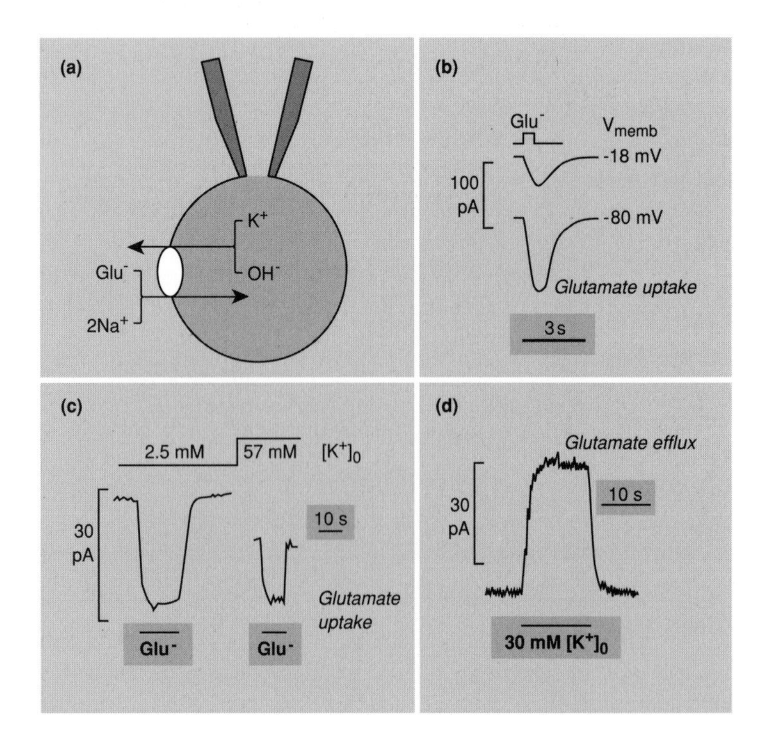

Fig. 9.6 The plasma membrane acidic amino acid carrier. (a) Glutamate (Glu) uptake into salamander retinal glial cells via the plasma membrane transporter can be monitored electrophysiologically by a whole-cell patch since the transporter is electrogenic (one positive charge is transported inward per cycle). (b) Lowering the membrane potential from −80 to −18 mV decreases the driving force for uptake. (c) The inward current is dependent on glutamate and is decreased by an opposing gradient of external K^+. (d) *Efflux* from cells loaded with glutamate and Na^+ via the patch electrode is dependent on external K^+. Data of D. Attwell from Nicholls & Attwell (1990).

$$\text{pmf} = \Delta\psi - 60\Delta\text{pH}$$

Net proton translocation ceases when the free energy available from ATP hydrolysis equals that which is required to pump the next proton across the membrane against the pmf. In the absence of any other ion movements, the low electrical capacity of the vesicle membrane means that a high membrane potential builds up when only a few protons have been translocated, and before a significant ΔpH can be established, i.e. the pmf is largely in the form of a membrane potential capable of driving the electrophoretic uptake of anions into the vesicle (Fig. 9.7). Increasing the concentration of permeant anion, such as Cl^-, dissipates the charge built up by the inward proton translocation, allowing more protons to be pumped into the vesicle, increasing ΔpH at the expense of a lowered Δψ.

By manipulating the balance of Δψ and ΔpH experimentally, it is possible to determine whether glutamate is accumulated as an anion (utilizing Δψ), as a neutral species in exchange for protons (i.e. utilizing ΔpH) or as some intermediate form utilizing the total pmf. It is found that optimal glutamate accumulation occurs in the presence of low (5 mM) Cl^-, when ATP can cause a 30-fold stimulation of uptake (Fig. 9.7). This Cl^- requirement suggests some direct requirement of the transporter itself for the anion.

No effective inhibitors of the vesicular glutamate transporter are known. Significantly, L-aspartate neither competes with L-glutamate for uptake (indicating that aspartate and glutamate are not transported into the same vesicles) nor is accumulated itself (showing the absence of a separate population of aspartatergic vesicles). This is consistent with the low Ca^{2+}-dependent release of aspartate from synaptosomes.

Abolition of the pmf by making the vesicle membrane permeable to protons with a protonophore such as FCCP leads to a release of accumulated labelled glutamate, indicating that a maintained pmf is essential to retain the amino acid within the vesicle. This is different from Type II transmitters such as ACh and NA, which are held in a semistable complex within the vesicle even in the absence of a continued ATP supply. However, inclusion of high concentrations of *N*-ethylmaleimide to block the synaptic vesicle glutamate transporter has allowed vesicles to be prepared containing an estimated 50–100 mM glutamate. It must of course be remembered that these preparations of synaptic vesicles are not homogeneous with respect to neurotransmitter but reflect the heterogeneity of brain area from which they were prepared. Interestingly, these vesicles contain no detectable aspartate.

Glutamatergic synaptic vesicles exist in a milieu of millimolar glutamate in the cytoplasm; it is therefore reasonable that they should possess a carrier with a relatively low affinity for the amino acid, and K_m values in the region of 1–5 mM have been reported.

The Ca^{2+}-dependent GABA released from cortical synaptosomes on depolarization is only about 20% that of glutamate, and this may account for the greater difficulty which was experienced in finding intravesicular GABA or ATP-dependent GABA uptake into vesicle preparations. However both GABA and glycine vesicular transport systems have now been demonstrated. GABA uptake into synaptic vesicles has been demonstrated and a reconstitution of an unfractional carrier in liposomes has been reported, although the rates are less impressive than for glutamate. The uptake mechanism probably involves both Δψ and ΔpH, which may suggest a GABA/H^+ exchange stoichiometry for the carrier. Recent evidence suggests that GABA and glycine share a common synaptic vesicular carrier.

9.5 Pathways of amino acid release from synaptosomes and cells in culture

Before discussing amino acid release in detail, it is important to appreciate the limitations of neurochemical preparations such as the synaptosome, brain slice or cultured neurone for monitoring release. If a direct postsynaptic response is not measured electrophysiologically, it is not possible to determine the release from a single terminal in response to a single action potential. Instead, a population of synaptosomes or cells must be depolarized by one of the methods discussed in Section 3.3.1 and the cumulative release of transmitter detected over a finite time interval, which can vary from as little as 50 ms if rapid mixing and sampling techniques are available to several minutes. If depolarization

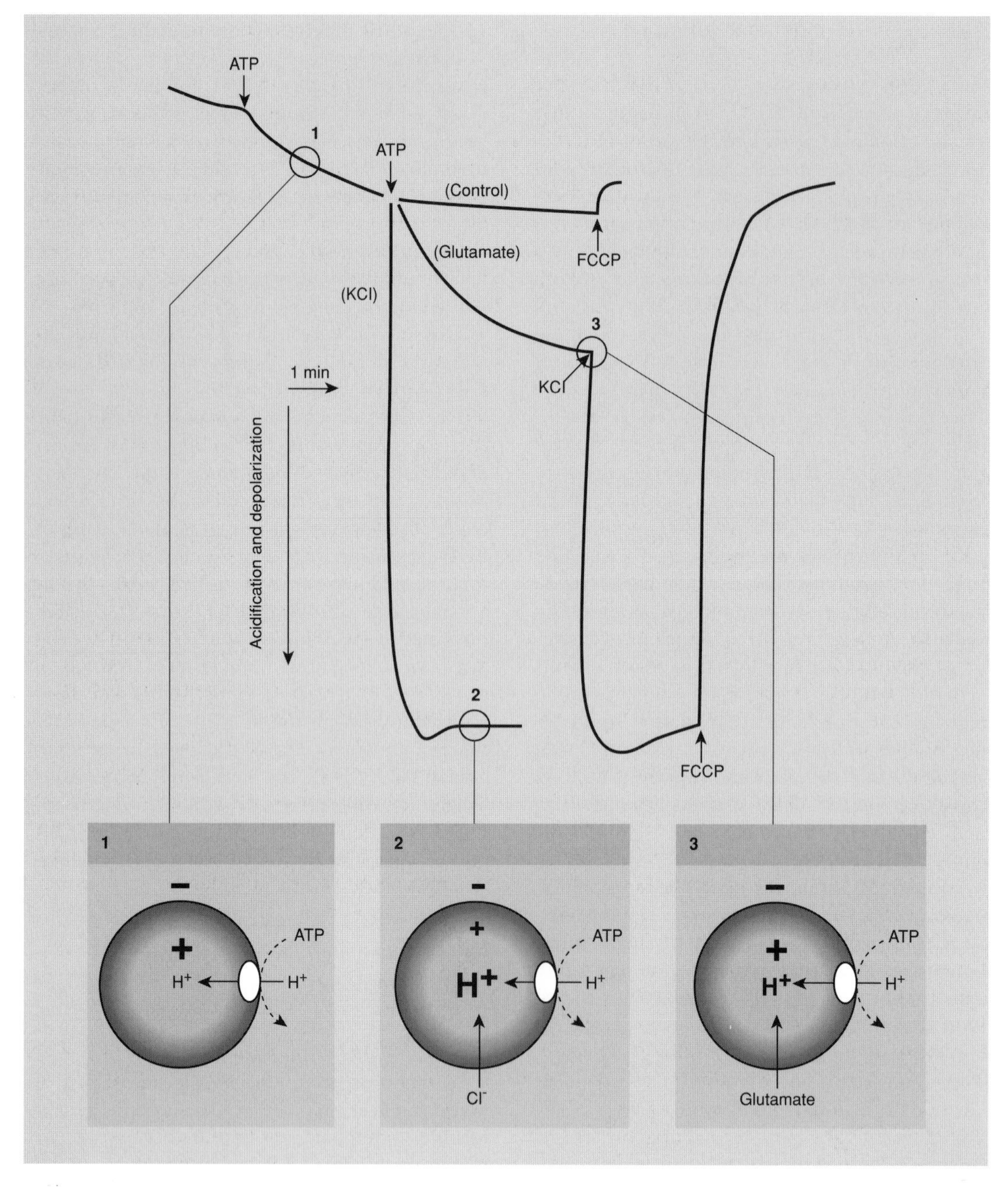

Fig. 9.7 Glutamatergic synaptic vesicles accumulate the amino acid as the anion. Synaptic vesicles were suspended in a sucrose-based buffer containing low (4 mM) KCl and equilibrated with acridine orange to monitor internal pH. (1) Addition of ATP to activate the vesicular H^+-ATPase caused little acidification (downward deflection) due to the rapid build-up of a membrane potential (positive inside). (2) 100 mM KCl caused a large acidification as Cl^- entry into both glutamatergic and non-glutamatergic vesicles lowered the membrane potential allowing further proton injection into the vesicle. (3) 1 mM glutamate caused a similar acidification in the glutamatergic subpopulation. FCCP is a protonophore which collapses both membrane potential and ΔpH. Data from Maycox *et al.* (1988)

is prolonged for more than a fraction of a second, the likelihood is that any pool of synaptic vesicles which were docked at the presynaptic membrane ready for release would become depleted and require replenishment from a 'reserve pool' of vesicles. Since this second process would be much slower than exocytosis itself, prolonged depolarization might be expected to result in a strongly biphasic kinetic of release which must be distinguished from that due to any inactivation of presynaptic Ca^{2+}-channels.

The monitoring of transmitter release is central to much of the neurochemical approach, and the available techniques will now be discussed in detail.

9.5.1 Isotopic detection

Isotopic prelabelling of preparations with amino acids or precursors is convenient and sensitive but is only valid if two criteria are satisfied. Firstly it must be possible either to prevent metabolism of the added radioisotope or to re-isolate the transmitter from a mixture of metabolites. Secondly it must be clear that isotopic equilibration has occurred, particularly into the synaptic vesicles.

Since GABA is not a general metabolite, it is possible to inhibit the metabolism of added [^{3}H]-GABA without killing the terminal. The main pathway for GABA metabolism is via GABA-T (*see* Fig. 9.2), which can be inhibited by transaminase inhibitors such as aminooxyacetate (AOAA). However, since the terminals rely on transaminases to allow reoxidation of cytoplasmic NADH via the malate–aspartate shuttle, inclusion of AOAA in the incubation induces a pseudo-hypoglycaemic condition, where the pyruvate generated by glycolysis is reduced to lactate to allow reoxidation of cytoplasmic NADH, rather than supplying the mitochondrion with its main substrate (*see* Section 3.5). The resulting energy deficit can inhibit release in a way that can be confused with a direct role of transaminases in transmitter metabolism.

Unlike GABA it is not possible to prevent the metabolism of glutamate or its precursor glutamine without killing the preparation. For incubation periods exceeding 1–2 min, an increasing proportion of the radioactivity is no longer present as glutamate but as metabolites (largely aspartate and CO_2) and authentic glutamate must be separated from the released mixture.

In situ synaptic vesicles incorporate label slowly from the cytoplasm. Therefore incomplete equilibration can occur with the result that vesicular release is underestimated while any release from the cytoplasm is exaggerated. This can however be combined with analysis of endogenous amino acid release to obtain information of the amino acid compartmentation within synaptosomes and to support an exocytotic mode of release for glutamate and GABA (Fig. 9.8). The non-metabolizable amino acid D-aspartate, although transported on the acidic amino acid carrier, is not a valid analogue for endogenous glutamate, since it is not accumulated into synaptic vesicles (*see* Section 9.4).

9.5.2 Endogenous amino acids

Chromatographic techniques for monitoring the release of endogenous amino acids allow the full spectrum of amino acids to be determined and avoid errors due to incomplete isotopic equilibration. However the technique is time consuming which does not encourage the acquisition of sufficient data for detailed time courses.

The utility of continuous fluorometric assays for $[Ca^{2+}]_c$, membrane potential, pH and other parameters have been discussed previously (e.g. Sections 3.4 and 5.7.1). A comparable assay for monitoring the release of glutamate from synaptosomes by following the increase in fluorescence of NADPH was introduced in Fig. 7.7.

9.5.3 Batch vs perifusion techniques

'Batch' experiments, in which the cells or synaptosomes are in contact with a fixed volume of incubation, are more versatile than 'superfusion' studies (see below), because it is possible to make parallel determinations of amino acid uptake or release with bioenergetic parameters such as plasma and mitochondrial membrane potentials, Ca^{2+}-fluxes, cytoplasmic free Ca^{2+}, ATP levels and respiratory rates. Supernatants and pellets may be separated by filtration or by centrifugation through silicone oil when it is necessary to determine synaptosome and supernatant contents separately (e.g. for isotopic or HPLC studies). However, substances may accumulate in the restricted volume of the incubation which

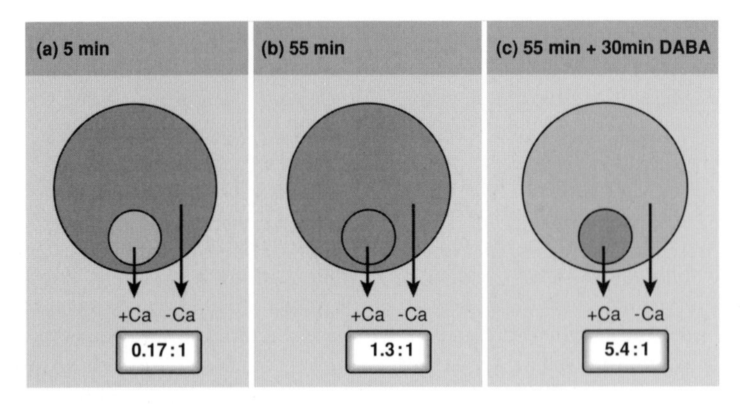

Fig. 9.8 Ca^{2+}-dependent release of GABA from synaptosomes occurs from a non-cytoplasmic compartment. (a) Brief (5 min) exposure of synaptosomes to [^{3}H]-GABA labels the cytoplasm but not the synaptic vesicles. The Ca^{2+}-independent release of GABA evoked by a 15 s exposure to KCl is labelled whereas the Ca^{2+}-dependent release is unlabelled (ratio of Ca^{2+}-dependent to Ca^{2+}-independent release shown). (b) After prolonged equilibration both Ca^{2+}-independent and Ca^{2+}-dependent pools are labelled. (c) Diaminobutyrate (DABA) exchanges out the cytoplasmic pool of labelled GABA but not the vesicular pool. Subsequent KCl depolarization allows Ca^{2+} release of labelled GABA but not Ca^{2+}-independent release. Data from Sihra & Nicholls (1987).

may influence the activity of presynaptic receptors.

The superfusion technique (strictly a 'perifusion') in which synaptosomes are layered on a filter which is then perifused with media has several advantages (Fig. 9.9). Firstly the incubation medium can be changed several times during an experimental allowing, for example, multiple KCl-depolarizations; secondly the continuous removal of released transmitters limits the possibility of re-uptake into the terminal or of uncontrolled activation of presynaptic receptors. It is however necessary to compromise between sensitivity and time resolution in selecting the flow rate, and consequently the technique has been used more extensively for isotopic studies. The time resolution of the superfusion technique has been greatly improved by the use of high pressure, computerized rapid switching of media and high-speed fraction collecting; using [^{3}H]-GABA loaded synaptosomes it is possible to detect release within 30 ms of depolarization (Fig. 9.9).

9.5.4 Ca^{2+} coupling to amino acid transmitter release

The general principles of Ca^{2+} coupling to the release of fast-acting neurotransmitters have been covered in previous chapters. Here we will concentrate on those aspects of particular relevance to the release of amino acids.

The amount of glutamate which can be released by Ca^{2+}-dependent exocytosis from cerebrocortical synaptosomes greatly exceeds that for ACh, catecholamines or neuropeptides, and is substantially greater than for GABA. This reflects the dominant role of glutamate as the major excitatory transmitter in the mammalian brain. For example, the KCl-evoked, Ca^{2+}-dependent release of glutamate from cortical synaptosomes amounts to 5 nmol glutamate per milligram of protein, or 15% of the total content, which it must be remembered includes metabolic glutamate in both glutamatergic and non-glutamatergic terminals.

The clamped depolarization evoked by sustained high KCl is not physiological but it is instructive to follow the kinetics of release of glutamate from cerebrocortical synaptosomes. This occurs biphasically, with 20% released within 2 s, and the remainder released much more slowly, with a half-life of 60 s (Fig. 9.10). ^{45}Ca entry into synaptosomes in response to clamped KCl depolarization is also biphasic (*see* Fig. 7.5), however this does not explain the kinetics of glutamate release since it is the slow, non-inactivating phase of Ca^{2+} entry which is coupled to glutamate release. The biphasic release kinetics seen with KCl may instead reflect a dual localization of vesicles, with those already docked at the active zone ready for immediate

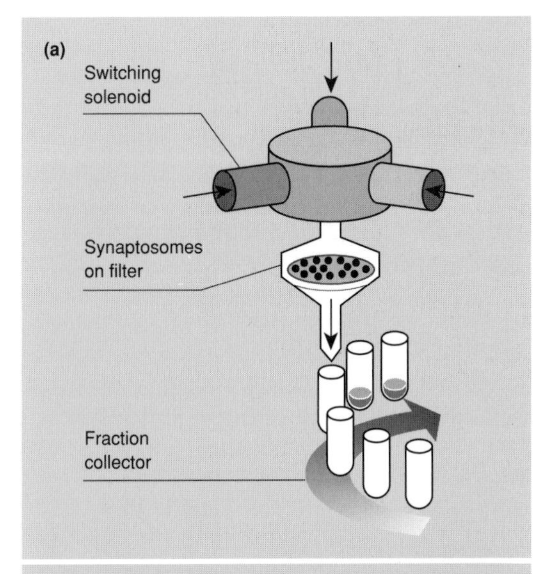

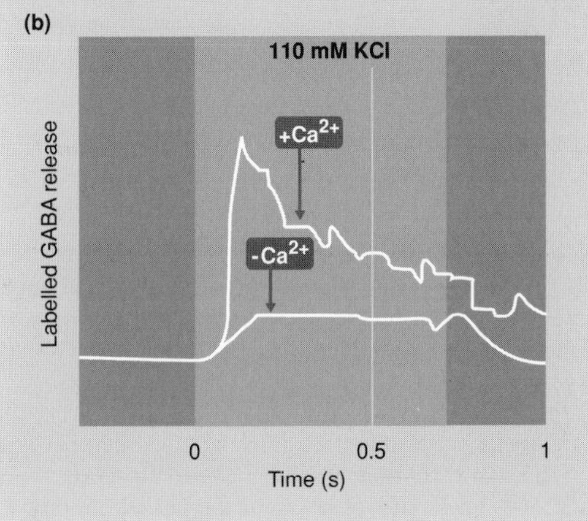

Fig. 9.9 High-speed superfusion of synaptosomes can monitor the kinetics of release of labelled amino acids. (a) High-speed superfusion through a bed of [^{3}H]-GABA loaded synaptosomes allows the kinetics of KCl-evoked release to be detected with a time resolution of about 30 ms. (b) Release of [^{3}H]-GABA evoked by 110 mM KCl in the presence and absence of Ca^{2+}. Adapted from Turner *et al.* (1989), by permission of Academic Press.

release and 'reserve' vesicles requiring prior transport to the membrane.

9.5.5 Metabolic glutamate and GABA release from the cytoplasm by a Ca^{2+}-independent pathway

Amino acid transmitters are present in the cytoplasm of their respective terminals at much higher concentrations than are Type II transmitters such as as ACh and catecholamines. Since the Na^+-coupled amino acid transporters discussed above operate close to thermodynamic equilibrium it follows that any factor which decreases the Na^+-electrochemical potential maintaining the glutamate gradient could result in a net efflux of the amino acid from the cytoplasm. It is most unlikely that significant amino acid leaves the cytoplasm by this route during the millisecond depolarization of an action potential, but in brain ischaemia the evidence is that a massive efflux of glutamate occurs by this route (Box 9.1). The efflux of glutamate is particularly sensitive since the stoichiometry of the transporter (*see* Fig. 9.6) is such than an increase in internal Na^+ and external K^+ and plasma membrane depolarization synergistically aid reversal of the carrier.

Two *in vitro* conditions under which this can be observed are when the membrane potential is chronically collapsed by elevated KCl (lowering the electrical component of the Na^+ electrochemical potential) and more spectacularly on addition of the Na^+-channel activating alkaloid veratridine (*see* Box 4.1) which collapses both the electrical and chemical components of the potential. Under these conditions a Ca^{2+}-independent efflux of metabolic glutamate or GABA is seen, which differs from the exocytotic release in being slow, continuous and ATP-independent. Aspartate, which can be regarded as a marker of cytoplasmic glutamate, is released in parallel with glutamate. This efflux must always be taken into consideration during *in vitro* experiments (Fig. 9.10).

Ca^{2+}-independent release of transmitter on addition of veratridine has sometimes been attributed to the Na^+ influx triggering the release of mitochondrial Ca^{2+} stores, since the mitochondrial Ca^{2+} efflux pathway is activated by Na^+. However, firstly veratridine continues to release labelled transmitter in synaptosomes which have been completely depleted of Ca^{2+},

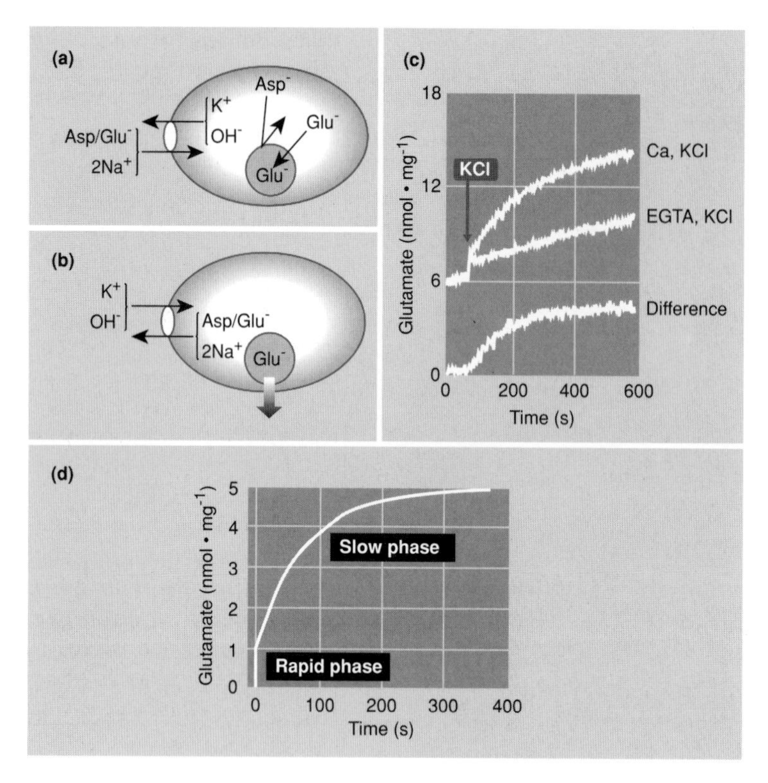

Fig. 9.10 Ca^{2+}-dependent and -independent release of amino acids from nerve terminals. (a) Pathways of glutamate (Glu) and aspartate (Asp) uptake. The acidic amino acid carrier can transport either amino acid while the vesicular carrier is specific for Glu; as a result vesicles only contain the latter amino acid. (b) Depolarization by KCl reverses the plasma membrane carrier by collapsing the membrane potential and lowering the K^+ gradient in/out. Asp and Glu are thus released from the cytoplasm by a Ca^{2+}-independent non-vesicular pathway superimposed on the Ca^{2+}-dependent exocytosis of Glu. (c) Fluorometric traces of KCl-evoked Glu efflux from synaptosomes: Ca, KCl — total efflux on depolarization (exocytotic plus cytoplasmic); EGTA, KCl — cytoplasmic release; difference — net Ca^{2+}-dependent exocytotic release. (d) Kinetic analysis of KCl-evoked Glu release showing biphasic kinetics obtained after correcting the net Ca^{2+}-dependent release in (b) for the kinetic limitations of the Glu assay. Data from McMahon & Nicholls (1991).

and secondly Ca^{2+} released from synaptosomal mitochondria is not competent to release transmitter since it is released into the bulk cytoplasm and not at the active zone.

In ischaemia, there is a rapid and massive increase in the extracellular concentrations of glutamate and aspartate (Box 9.1). This released amino acid is *excitotoxic*, due primarily to its ability to chronically activate postsynaptic NMDA receptors (*see* Section 9.6.3). The mechanism by which the amino acids are released is thus of some interest in the search for agents to limit the damage occurring to brain in stroke or global ischaemia. From first principles the amino acids released in ischaemia could have a vesicular origin, released perhaps in response to an increased presynaptic Ca^{2+}, or could originate from the cytoplasmic metabolic pool by reversal of the plasma membrane transporter.

The synaptosome can throw some light upon these alternative pathways. In common with other transmitters, the exocytosis of glutamate is energy-dependent, and any factor which significantly lowers the ATP level in the terminal decreases the extent of KCl-evoked glutamate release. The energetic demands of amino acid exocytosis from synaptosomes can only be met in full by aerobic glycolysis; inhibiting oxidative phosphorylation or mimicking hypoglycaemia by removing glucose from the medium results in

Box 9.1 Glutamate neurotoxicity

Hypoxia, hypoglycaemia, ischaemia, prolonged seizures and mechanical trauma can each lead to a pathological increase in extracellular glutamate. Cultured neurones can be destroyed by exposure to 100 μM glutamate for as little as 5 min. The cells may appear to recover following removal of the glutamate, but after a delay of perhaps several hours there is an irreversible increase in $[Ca^{2+}]_c$ and cell death. The neuronal injury occurs in two stages: the first is associated with acute neuronal swelling accompanying an influx of extracellular Na^+, while the second requires extracellular Ca^{2+} and intracellular $[Ca^{2+}]_c$ elevation predominantly via the NMDA receptor. Thus NMDA receptor antagonists can reduce the extent of neuronal injury both *in vivo* and *in vitro*. The mechanism by which elevated Ca^{2+} induces neuronal damage (if indeed Ca^{2+} is the direct causative agent) is currently under debate. Possibilities include free-radical generation, mitochondrial damage, pathological activation of Ca^{2+}-dependent enzymes such as PLA_2 and changes in gene expression.

a severe restriction of the extent of KCl-evoked glutamate exocytosis. Most extremely, when the bioenergetic effects of brain ischaemia are mimicked by parallel inhibition of mitochondrial oxidative phosphorylation and glycolysis, there is an immediate inhibition of KCl-evoked Ca^{2+}-dependent glutamate release. However, as the Na^+-electrochemical gradient slowly declines due to the lack of ATP to drive the $Na^+ + K^+$-ATPase, the plasma membrane glutamate carrier reverses and metabolic glutamate leaks out from the cytoplasm of the nerve terminal.

The nature of the ATP requirement for glutamate exocytosis is unclear. ATP is certainly required to accumulate glutamate within synaptic vesicles (*see* Section 9.4) although it is not clear whether efflux of glutamate into the cytoplasm occurs as soon as ATP is removed. Abolition of the pmf across the vesicle membrane with the protonophore FCCP does cause a time-dependent block of Ca^{2+}-dependent release although it is difficult to separate this effect from a general depletion of ATP caused by the protonophore. ATP is of course required for protein phosphorylation, although protein kinases do not appear to be involves in the exocytotic process itself (*see* Section 7.4.6). ATP is also required for the generation of GTP which is needed for the activation of small and trimeric G-proteins, and is required for the *in vitro* dissociation of membrane targetting complexes (*see* Fig. 7.14).

A pH gradient across the vesicle membrane does not appear to be essential for the maintenance of glutamate within vesicles, because high concentrations of permeant weak bases such as methylamine or ammonia which will abolish ΔpH across the vesicle membrane do not affect the Ca^{2+}-dependent release.

9.6 Glutamate receptors

Glutamate receptor research has followed a classic sequence in which the initial electrophysiological effects of applied glutamate were followed by detailed pharmacology, patch clamping and most recently molecular biology. The convulsive effects of glutamate and its ability to depolarize and excite single neurones were first reported in the 1950s. In the early 1980s it became clear that there were multiple ionotropic receptors for glutamate which could be distinguished pharmacologically by three selective agonists; *NMDA*, *quisqualate* (and subsequently *AMPA*) and *kainate* (KA). The distinction between AMPA and KA receptors has frequently been blurred and they are sometimes grouped as *AMPA/KA* or *non-NMDA* receptors. Non-NMDA glutamate receptors mediate transmission at a vast number of fast excitatory synapses. NMDA receptors have a more specialized role in memory and learning and may only be called upon during the learning process itself (*see* Section 13.5). In addition to the ionotropic glutamate receptors, a family of G-protein coupled *metabotropic receptors* exists which modulate transmission at glutamatergic and non-glutamatergic synapses.

9.6.1 The AMPA/KA receptor family (GluR)

Molecular cloning has identified four closely related members of a family of AMPA-selective ionotropic glutamate receptors, termed GluR1, GluR2, GluR3 and GluR4 (or GluRA to D) each containing 850–950 amino acids and sharing some 70–80% sequence homology (Fig. 9.11). While hydropathy plots indicate a large extracellular N-terminal domain, followed by four transmembrane segments and an extracellular cytoplasmic C-terminus, this is in conflict with immunocytochemistry using C-terminal-directed antibodies which are consistent with a cytoplasmic C-terminus. It is possible that TM4 is not a true transmembrane α-helix, but instead loops back into the cytoplasm. The subunits are much larger than those of the members of the MnAChR, GABA-A, glycine and 5-HT_3 receptor superfamily but share 30–40% homology with two other ionotropic glutamate receptors; the high-affinity KA receptor and the NMDA receptor. When expressed *in vitro* these receptors display a high affinity for AMPA, a binding site for the characteristic antagonist, CNQX, and a low-affinity binding site for KA, which can be antagonized by AMPA. By analogy to the nicotinic ACh receptor it is believed that the receptors function as pentameric complexes, either as homo-oligomers or following coexpression when they presumably form hetero-oligomers.

Each member of the receptor family exists in two forms, termed 'flip' and 'flop' which differ slightly in the sequence of a portion of the third cytoplasmic loop (Fig. 9.12). The two forms are generated by alternative splicing of the primary transcript. The GluR-flop mRNAs appear late in development and are generally less active than the flip variants; thus the GluR activity may be tuned down in the mature neurone.

Both the 'flip' and 'flop' forms of GluR2 display a further form of variation in which post-

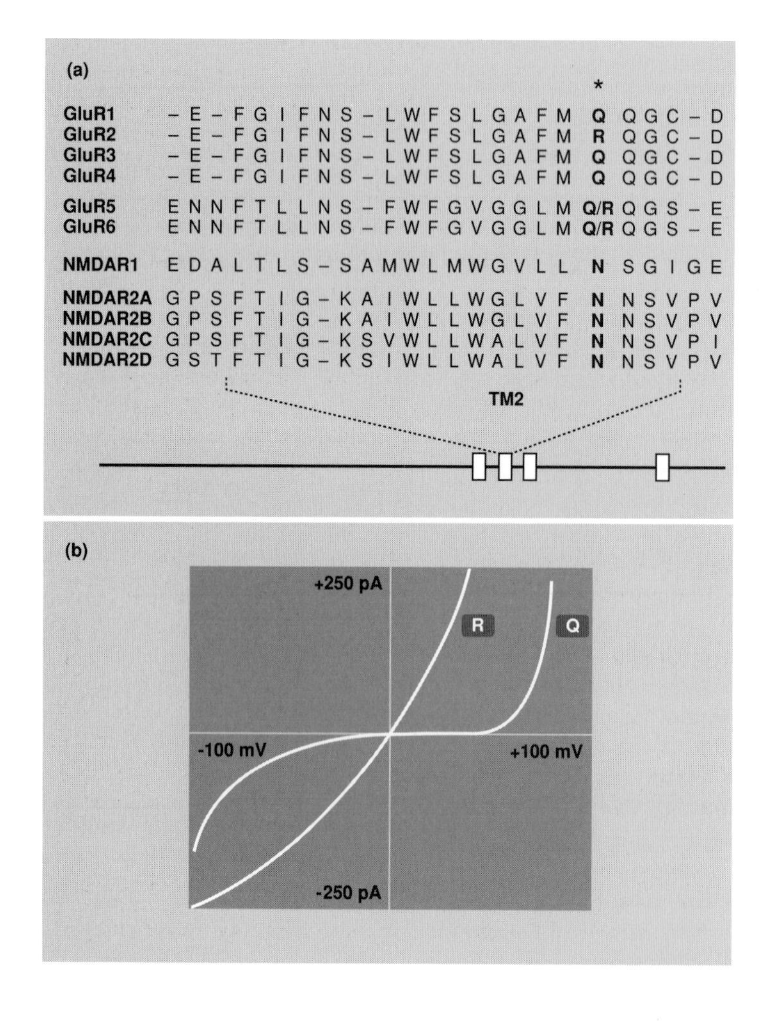

Fig. 9.11 The TM2 segment of ionotropic glutamate receptors and the control of ion conductance. (a) The putative TM2 segment of the ionotropic glutamate receptors. Sequences have been aligned for the overlap, leaving gaps where required. The bold residues (*) define the cation selectivities of the receptors: the DNA for GluR1–4 each codes for glutamine (Q) at this position; however, GluR2 posttranscriptional editing of the mRNA causes an arginine (R) to be substituted. The kainate receptors GluR5 and GluR6 can exist in both edited and non-edited forms. The corresponding residue in the NMDA receptor subunits is an asparagine (N). After Nakanishi (1992). (b) Schematic steady-state I/V relationship for whole cell Na^+ current for cells expressing homo-oligomers of recombinant GluR2(Q) and GluR2(R). GluR2(Q) shows a Ca^{2+} permeability relative to Cs^+ of 1.2 but with GluR2(R) this is reduced to 0.05.

translational editing of the mRNA removes an amino group from a specific adenine base, altering the codon for glutamine to arginine at a residue in the putative second transmembrane segment (TM2, *see* Fig. 9.11). The unedited GluR2(Q) form (which is present in low amounts at birth but decreases during development) gives a homo-oligomer with a high divalent cation conductance and little ability to conduct ions when depolarized beyond −50 mV. Homo- or hetero-oligomers containing the edited form, GluR2(R), have a low Ca^{2+} conductance and a more linear I/V relationship (*see* Fig. 9.11). These two functional modifications are not directly linked, however, since an *in vitro* mutation of the same residue from Gln to His alters the current/voltage relationship without suppressing the Ca^{2+} conductance. GluR2 is dominant in determining the ionic conductance of hetero-oligomers; without this subunit the receptors display a significant Ca^{2+} conductance; for example AMPA channels in cultured Purkinje neurones are Ca^{2+} conducting.

9.6.2 Kainate receptors

While KA interacts with GluR1−4 with low affinity, there is also a family of high-affinity KA receptor subunits (KA1, KA2, GluR5−7) which display only limited homology with the other cloned receptors and are 50 times less abundant than AMPA/KA receptors (*see* Fig. 9.11). KA1 and KA2 do not yield functional homo-oligomers but are functional in combination with GluR5 or GluR6. GluR6 possesses an arginine at position 621, which is homologous with the Q/R site in the putative TM2 of the AMPA receptor discussed above. This GluR6(R621) shows mild outward rectification which is converted to strong inward rectification when the arginine is changed to glutamine by site-directed mutagenesis. Thus this site plays a similar key role in both AMPA/KA and KA receptors.

The KA receptor is concentrated in a few specific areas of the CNS, generally complementary to the distribution of NMDA and AMPA receptors. There is a high concentration of KA receptors in the molecular layer of the chick cerebellum. These are located not on neurones but on Bergmann glia. During development these cells are essential to guide the parallel fibre projections from granule cells onto Purkinje dendrites, and in the adult they surround the parallel fibre/Purkinje cell spine synapses. Since glutamate is released at these synapses it will simultaneously depolarize the Bergmann glia, although the significance of this is not known.

Pharmacologically it is difficult to distinguish between the high-affinity KA receptor and the AMPA receptor because KA is not sufficiently selective. CNQX will compete with both AMPA and KA for binding to rat cortical membranes, with a five-fold ratio of binding affinity in favour of AMPA binding sites; NBQX, a tricyclic analogue of CNQX increases this selectivity to 30-fold.

The KA receptor is non-inactivating as long as glutamate is present. Perhaps because of this, KA is associated with excitotoxicity; injection of KA causes degeneration of neuronal cell bodies at the site of injection. The toxicity of KA is poten-

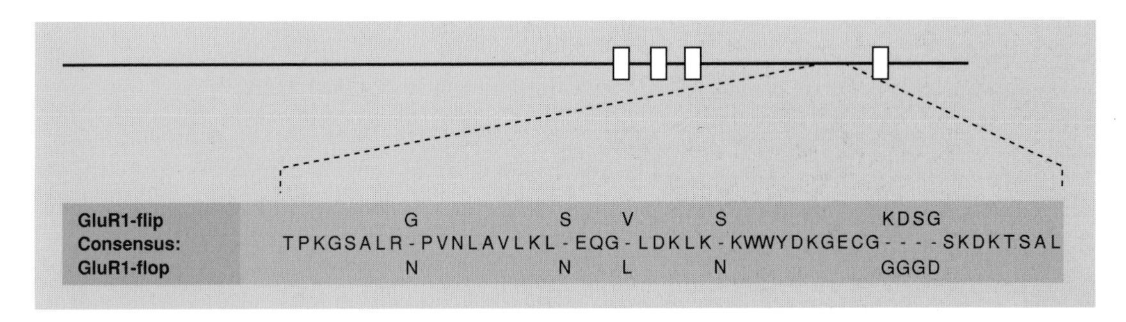

Fig. 9.12 Flip and flop: alternative splicing of GluRs. A short section of the third putative cytoplasmic loops of each of the GluR1−4 sequences exists in two alternatively spliced variants termed *flip* and *flop*. The amino acid substitutions between GluR1-*flip* and GluR1-*flop* are shown; related substitutions are also present in the other receptors. Data from Sommer *et al.* (1990).

tiated by the presence of an intact glutamatergic innervation into the area in question. KA can also cause damage at remote sites — perhaps due to the induction of epileptiform discharges in the cell's axon.

In addition to activating the KA subclass of receptors, KA also inhibits the acidic amino acid carrier at the presynaptic plasma membrane (see above), preventing re-uptake of glutamate (and aspartate). This may contribute to the neurotoxicity. The derivative dihydrokainate is also able to block the carrier, but lacks the ability to interact with the KA receptor.

9.6.3 The NMDA receptor

The NMDA receptor differs fundamentally from the AMPA/KA receptor in that application of agonist (NMDA or glutamate itself) under polarized conditions does not activate the integral ion channel. This is because the receptor is inhibited by physiological extracellular concentrations of Mg^{2+} which block the ion channel. Mg^{2+} behaves as though it were pulled into the channel and the membrane is independently depolarized, perhaps by a non-NMDA receptor acting in the close vicinity, then Mg^{2+} escapes from the ion channel and the receptor conducts ions. The NMDA receptor requires simultaneous presynaptic activity (the release of glutamate) and postsynaptic activity (the independent depolarization of the postsynaptic membrane). The NMDA receptor is therefore a novel combination of a ligand-activated and a voltage-operated channel, requiring both glutamate and depolarization for activity (Fig. 9.13). The NMDA receptor activates slowly, even if the membrane is depolarized, and displays a lengthy activated state of several hundred milliseconds in the continued presence of the agonist.

The NMDA-channel can conduct Ca^{2+} in addition to Na^+ and K^+. The ratio of permeabilities of $Ca^{2+} : Na^+$ is 10 : 1, although of course the much greater concentration of Na^+ in the extracellular medium will ensure a substantial Na^+ entry through the channel. In addition to glutamate, a second amino acid, glycine, has to be present to bind to an allosteric site on the receptor (Fig. 9.13). Glycine potentiates the binding of glutamate to the receptor, which is virtually inactive in the strict absence of glycine. Should glycine therefore be considered as a necessary cotransmitter at the NMDA complex, or is the 'regulatory' glycine site superfluous because it would always be occupied *in vivo*? Although the sensitivity of the site for glycine is very high (K_d about 0.2 μM) care must be taken before dismissing a regulatory role for glycine, since similar arguments were once used to argue against a neurotransmitter role for glutamate. Consistent with the high affinity of binding, glycine dissociates rather slowly from the binding site. The naturally occurring tryptophan metabolite *kynurenate* blocks the glycine site; however it has limited specificity, also interacting with KA/quisqualate responses. *HA966* is a competitive antagonist at the glycine site. MK-801 inhibits the receptor by blocking the channel (Fig. 9.13).

The first NMDA receptor, termed NMDAR1, was cloned in 1991 by combining an oocyte expression system with electrophysiological measurements (Fig. 9.14). The polypeptide shows some sequence homology with the AMPA/KA receptor subunits, with four putative transmembrane segments and a large extracellular N-terminal domain. Although a functional receptor could be assembled from homo-oligomers of the polypeptide, three further subunits have currently been identified, termed NMDAR2A, -2B and -2C which are only 20% homologous to NMDAR1. These three subunits possess distinctive patterns of expression in brain; 2A is present in forebrain and cerebellum, while the cerebellum contains the highest level of 2C but no 2B (Fig. 9.15). While homo-oligomers of NMDAR2 subunits are inactive, combinations of NMDAR1 with 2A or 2C gives channels with distinctive conductances and ion permeabilities; 1 + 2C shows substantially weaker Mg^{2+} inhibition under polarized conditions and inactivated more slowly after removal of glutamate or glycine when compared with 1 + 2A.

The NMDA receptor can be localized in the brain by receptor radiography using either the NMDA-displacable component of high-affinity [^{3}H]-glutamate binding, or by using high-affinity NMDA antagonists such as D-[^{3}H]-AP5 or [^{3}H]-CPP, while expression can be monitored by *in situ* hybridization (Fig. 9.15). The receptor is widely distributed in the CNS, particularly in the cerebral cortex and basal ganglia. Highest levels are found in the CA1 region of the hippocampus, where the Schaffer collateral pathway terminates.

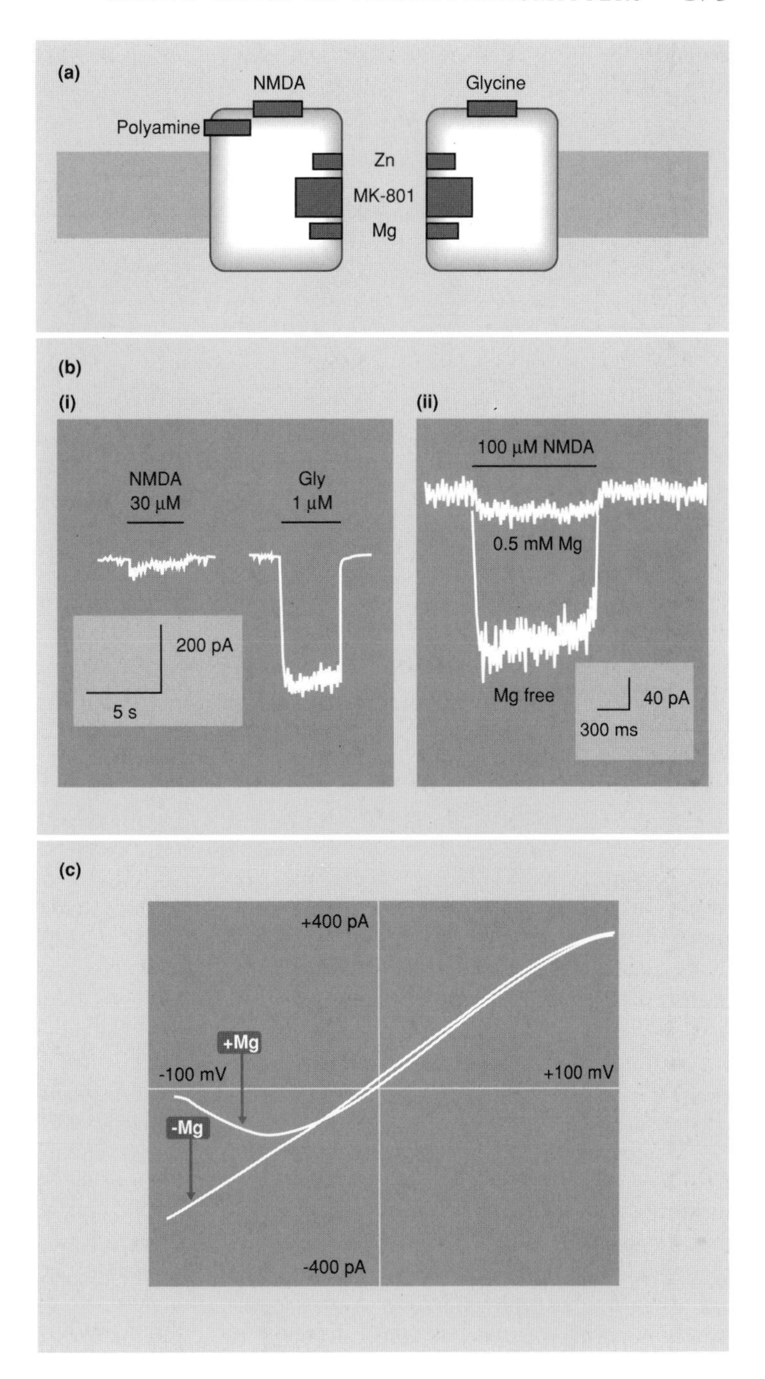

Fig. 9.13 The NMDA receptor. (a) Schematic structure of the NMDA receptor showing the regions of interaction with ligands. (b) (i) In the absence of Mg^{2+} (or under depolarized conditions) glycine (Gly) greatly potentiates the NMDA (glutamate)-evoked conductance; (ii) Mg^{2+} blocks NMDA-evoked conductance under polarized conditions. (c) The voltage-dependent Mg^{2+} block results in a anomalous current-voltage relationship for the receptor.

However, while the receptor distribution largely corresponds to that of AMPA binding sites, only select pathways appear to possess NMDA receptors, and even at NMDA receptor-using synapses it appears to be the AMPA/KA receptor which mediates the normal fast epsp.

A specific role for the NMDA receptor could be proposed following the finding that NMDA antagonists could be neuroprotective in models of ischaemia, and anticonvulsant in experimental epilepsy. At the same time it was found that NMDA receptors were involved in neuronal

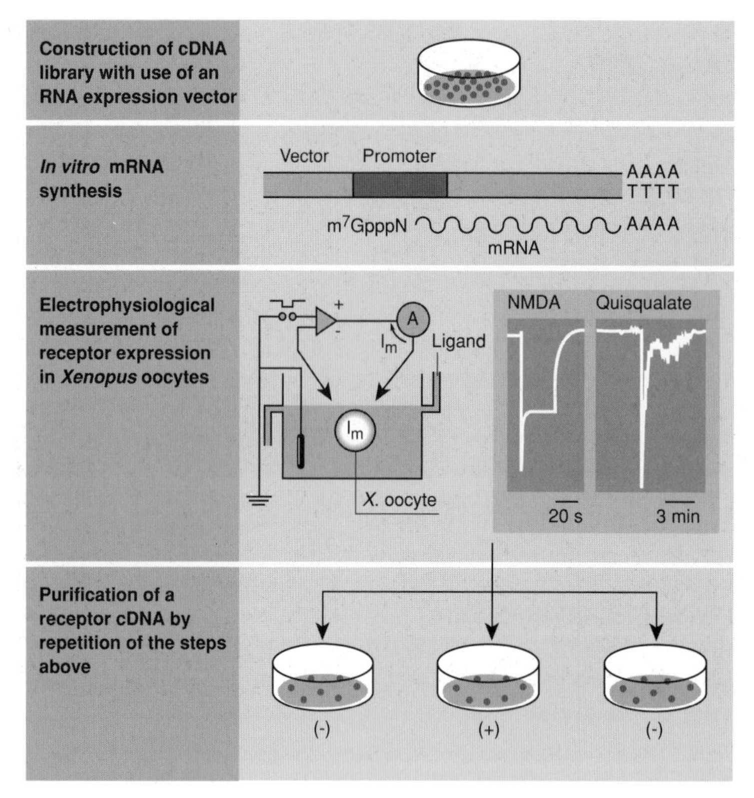

Fig. 9.14 Strategies for cloning the NMDA receptor. Poly (A^+) RNA isolated from rat forebrains and size fractionated on a sucrose gradient is assayed by injection into *Xenopus* oocytes. Electrophysiology is employed to detect NMDA-dependent ion currents (in Mg^{2+}-free media in the presence of glycine). The active mRNA fraction is inserted in a RNA expression vector in order to construct a cDNA library. The presence of a functional NMDA receptor cDNA clone is tested for by injecting oocytes with mRNAs synthesized *in vitro* from fractions of the cDNA library. Positive fractions are purified by serial subdivision until a single receptor clone is obtained. Adapted from Nakanishi (1992).

development as well as being necessary for the induction (but not the maintenance) of long-term potentiation (LTP), the most intensely investigated model of memory and learning (*see* Section 13.5) and full discussion of this role will be reserved until later.

When the blood supply to an area of brain is deprived of oxygen (hypoxia) or is cut off (ischaemia), cell damage occurs in two phases. Acute ischaemic cell damage appears to be due to osmotic swelling accompanying Na^+ entry into the cell. In addition, an uncontrollable increase in cytoplasmic free Ca^{2+}, $[Ca^{2+}]_c$, most likely underlies a delayed second phase of damage while free radical generation is also implicated. Thus hypoxia in the cerebral cortex is accompanied by a virtual disappearance of extracellular Ca^{2+} as it is accumulated into cells. While Ca^{2+} uptake would be expected, due firstly to a lack of ATP to drive the Ca^{2+}-ATPase expelling Ca^{2+} from the cell, secondly to influx via voltage-activated Ca^{2+}-channels following plasma membrane depolarization, and thirdly due to Na^+/Ca^{2+} exchange, the most potent damage appears to be due to Ca^{2+} entry through the NMDA receptor. NMDA antagonists are in many cases effective in protecting ischaemic neurones from the second phase of damage.

9.7 The metabotropic glutamate receptor (mGluR) family

In addition to the ionotropic glutamate receptors, a family of G-protein-coupled metabotropic glutamate receptors (currently with seven members) has been identified and cloned. The receptors may be either coupled positively to inositol phosphate turnover (mGluR1, mGluR5) or negatively to adenylyl cyclase activity and/or Ca^{2+}-channel activity (mGluR2, mGluR4, mGluR6, mGluR7). mGluR1 also exists in three alternatively spliced variants: α, β and c with increasingly truncated C-termini, while the other cloned receptors vary from 1171 (mGluR5) to 872 (mGluR2) residues.

The amino acid sequences for the receptors show surprisingly little homology with other G-protein-coupled receptors although retaining the 7-transmembrane sequence motif (Fig. 9.16). The predicted polypeptide for mGluR1α has 1199 amino acids and a relative

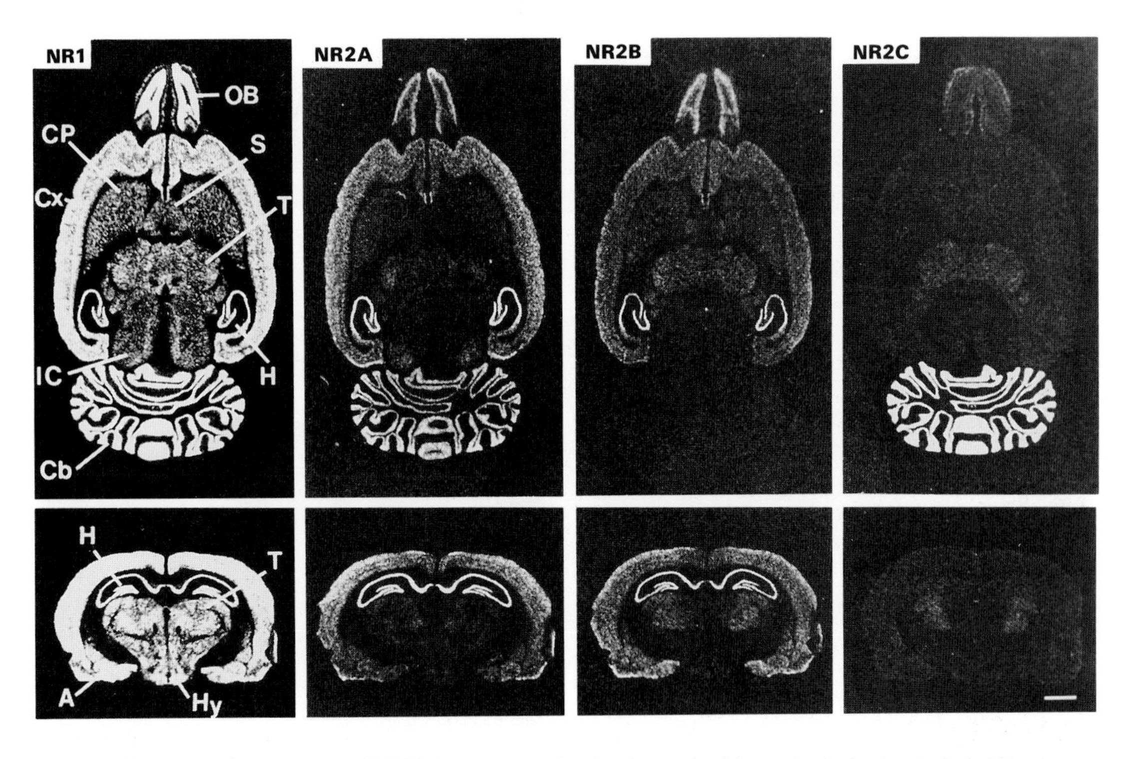

Fig. 9.15 Patterns of expression of NMDA receptor subunits determined in rat brain by *in situ* hybridization. There is a distinctive distribution of mRNAs for the Type 1, 2A, 2B and 2C NMDA receptor subunits. A, amygdala; Cb, cerebellum; CP, caudate putamen; Cx, cortex; H, hippocampus; Hy, hypothalamus; IC, inferior colliculi; OB, olfactory bulb; S, septal nuclei; T, thalamic nuclei. Bar = 1.6 mm. From Monyer *et al.* (1992), reproduced with permission from *Science*. © 1992 by the AAAS.

molecular mass of 133 kDa, and is thus considerably larger than other G-protein-coupled receptors (*see* Section 5.3). The predicted structure includes a large extracellular N-terminal sequence comprising half the total protein, more reminiscent of G-protein-coupled receptors for large glycoprotein hormones than those for small neurotransmitters.

The metabotropic receptors can be activated by quisqualate (which also is an agonist for the AMPA receptor), ibotenate and more specifically by 1S,3R-1-aminocyclopentane-1,3-dicarboxylate (1S,3R-ACPD) but not by L-aspartate, AMPA, NMDA or KA. While no entirely subtype-specific antagonists are currently available, the relative potencies of different agonists differ (Table 9.1): quisqualate is most effective at the PI-PLC-coupled mGluR1 and mGluR5 receptors, while 1S,3R-ACPD is more effective at the inhibitory mGluR2 and mGluR3 receptors L(+)-2-amino-4-phosphonobutyrate (L-AP4) is an agonist for an 'AP4' receptor which can most probably be identified with mGluR4 and is also active at the recently identified mGluR6 and mGluR7 receptors.

In situ hybridization reveals the expression of mGluR1 mRNA in different brain regions (Fig. 9.17). High levels of expression are found in the hippocampus, the cerebellum and olfactory bulb. In the cerebellum the enormous concentration of the receptor on Purkinje cells is particularly striking. In contrast, mGluR5 in the cerebellum is restricted to a small population of Golgi cells. mGluR6 expression is restricted to the retina.

mGluRs are present both presynaptically and postsynaptically; additionally the modulatory role in neurotransmission which these G-protein-coupled receptors play varies with brain area even for the same receptor subtype. It is therefore impossible to generalize about the action of these receptors. Thus in the hippocampus mGluR agonists inhibit release but also block a postsynaptic K^+-channel with a consequent facili-

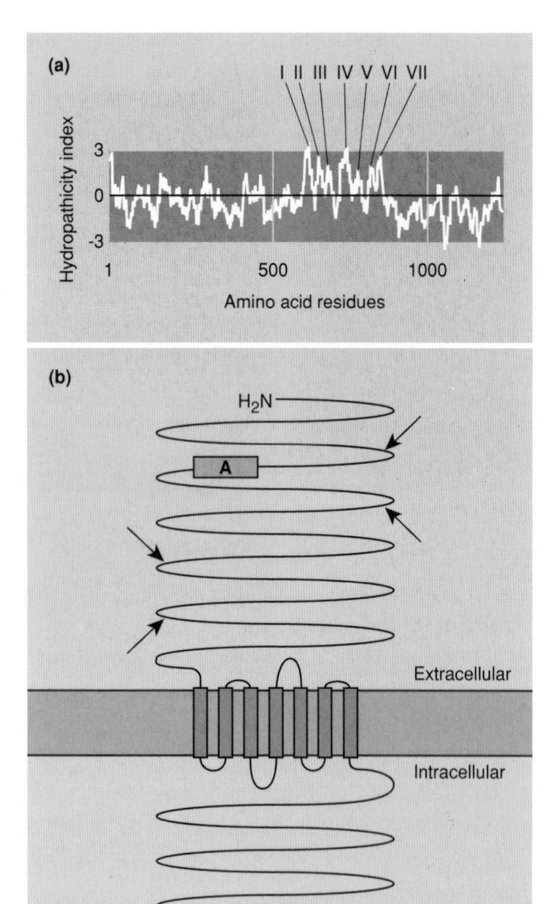

Fig. 9.16 The predicted transmembrane structure of the metabotropic glutamate receptor. (a) Hydrophobicity profile of mGluR1α showing the predicted positions of hydrophobic segments I–VII. (b) Predicted transmembrane structure. Note the enormous extracellular N-terminal domain. Segment A is a sequence of 24 consecutive uncharged residues. Note also that this receptor displays no sequence homology with the catecholamine family of metabotropic receptors. Adapted from Masu *et al.* (1991), reproduced with permission from *Nature*. © 1991 Macmillan Magazines Ltd.

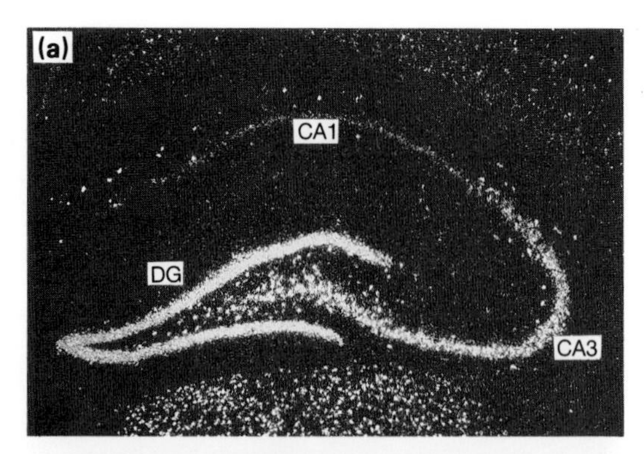

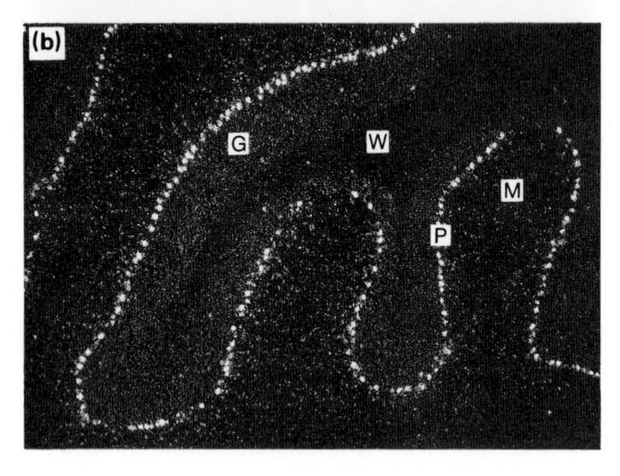

Fig. 9.17 Localization of mGluR1 expression in hippocampus and cerebellum by *in situ* hybridization with antisense mRNA. ^{35}S-labelled antisense RNA probes to the mGluR1α sequence were hybridized to sections of (a) hippocampus, and (b) cerebellum. In the hippocampus the receptor shows high levels of expression in the dentate gyrus (DG), CA3 pyramidal cells and CA1 extrapyramidal cells. In the cerebellum, Purkinje cell bodies (P) are intensely stained. ^{35}S-labelled RNA probe was prepared by *in vitro* transcription and hybridized to formaldehyde-fixed sagittal cryostat sections. Sections were dipped in emulsion, developed and counterstained with cresylviolet. G, granule layer; M, molecular layer; W, white matter. Adapted from Masu *et al.* (1991), reproduced with permission from *Nature*. © 1991 Macmillan Magazines Ltd.

tatory action and have been implicated in LTP (*see* Section 13.5). In the cerebral cortex the agonists can inhibit cAMP production probably at a postsynaptic site, but under particular circumstances they also facilitate glutamate release by blocking a presynaptic K^+-channel. In the cerebellum, the mGluR receptor paired with a concomitant activation of postsynaptic Ca^{2+}-channels can cause *long-term depression* of the synapses between parallel fibres and Purkinje dendritic spines (*see* Section 13.6).

Table 9.1 The mGluR family.

Subgroup	Receptor (amino acids)	Signal transduction	Agonist sensitivity	Characteristic mRNA expression sites
I	mGluR1 (1199 or 906)	PI-PLC	QA > Glu > Ibo > 1S,3R-ACPD	Purkinje cells, CA2–CA4 pyramidal cells, olfactory bulb
	mGluR5 (1171)			CA1–CA4 pyramidal cells, striatum, cortex
II	mGluR2 (872)	cAMP ↓	Glu > 1S,3R-ACPD > Ibo > QA	Golgi cells of cerebellum, cerebral cortex, olfactory bulb
	mGluR3			Cerebral cortex, dentate gyrus granule cells, glial cells
III	mGluR4 (912)	cAMP ↓	L-AP4 > Glu	Cerebellar granule cells, olfactory bulb

1S,3R-ACPD, 1S,3R-aminocyclopentane-1,3-dicarboxylate; Ibo, ibotenate; L-AP4, L(+)-2-amino-4-phosphonobutyrate; QA, quisqualate.

9.7.1 Presynaptic mGluR autoreceptors

PKC activators cause a large enhancement in the release of transmitter glutamate from guinea-pig cerebral cortical synaptosomes which have been stimulated to fire spontaneous action potentials by 4-aminopyridine (Fig. 9.18). The likely locus of PKC action is a K^+-channel equivalent to the delayed rectifier which controls action potential duration. Agonists for the metabotropic glutamate receptor (1S,3R-ACPD, quisqualate or ibotenate) mimic the action of phorbol ester indicating the presence of a presynaptic autoreceptor which increases release. PKC activity is not required for amino acid exocytosis itself, since KCl-evoked glutamate release (which does not involve transiently activated K^+-channels) is totally insensitive to PKC inhibitors which have a profound effect on the release evoked by 4AP.

A positive feedback is unusual, since most transmitters are subject to a negative feedback by autoreceptors which inhibit further release by activating K^+-channels and/or inhibiting Ca^{2+}-channels. However an inhibitory mechanism at the glutamatergic terminal might be counter-productive by preventing the terminal from responding to a high frequency stimulus — the condition required for the induction of LTP. On the other hand, a permanently engaged positive feedback could lead to the glutamatergic synapse running out of control. Perhaps to avoid this, presynaptic PKC activation requires synergistic activation both by receptor-generated diacylglycerol and by low concentrations of arachidonic acid. This fatty acid has been suggested to act as a 'retrograde messenger', generated postsynaptically by the activation of phospholipase A_2 and then diffusing across the synapse to increase glutamate release (*see* Fig. 13.5).

In many brain areas presynaptic glutamate autoreceptors *inhibit* the release of glutamate. For example hippocampal mossy fibre terminals which synapse onto CA3 neurones are a source of synaptosomes whose KCl-evoked glutamate release can be inhibited by L-AP4. The receptor on these terminals (probably mGluR4) may act via a 'membrane-delimited' G-protein interaction to inhibit Ca^{2+}-channels and/or activate K^+-channels.

9.8 GABA and glycine receptors

GABA is the main inhibitory transmitter in higher brain areas while glycine plays the analogous role in the brain stem and spinal cord. The transmitters activate the appropriate inhibitory ionotropic receptors — the GABA-A receptor and the glycine receptor. In addition GABA is active at the metabotropic GABA-B receptor, which may have both presynaptic and postsynaptic loci, while glycine is a coactivator of the NMDA receptor (*see* Section 9.6.3). The GABA-A and glycine receptors are closely homologous and will be considered together.

The GABA-A (Fig. 9.19) and glycine receptors

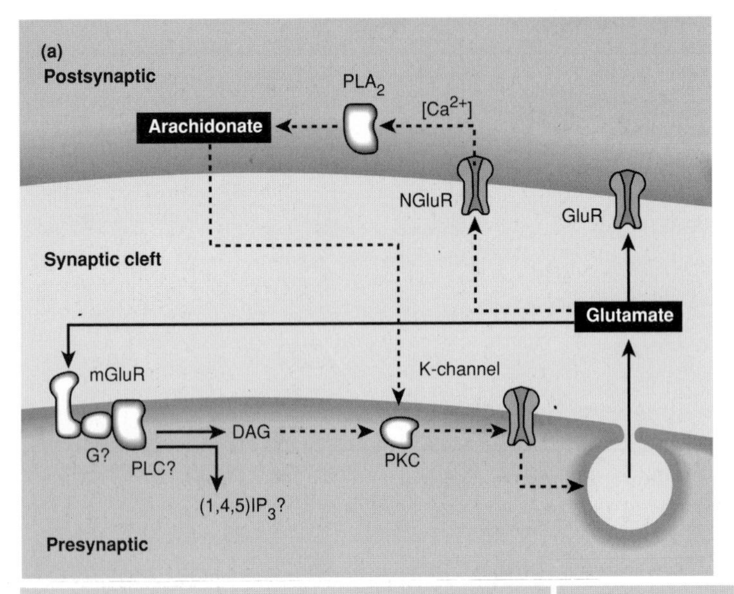

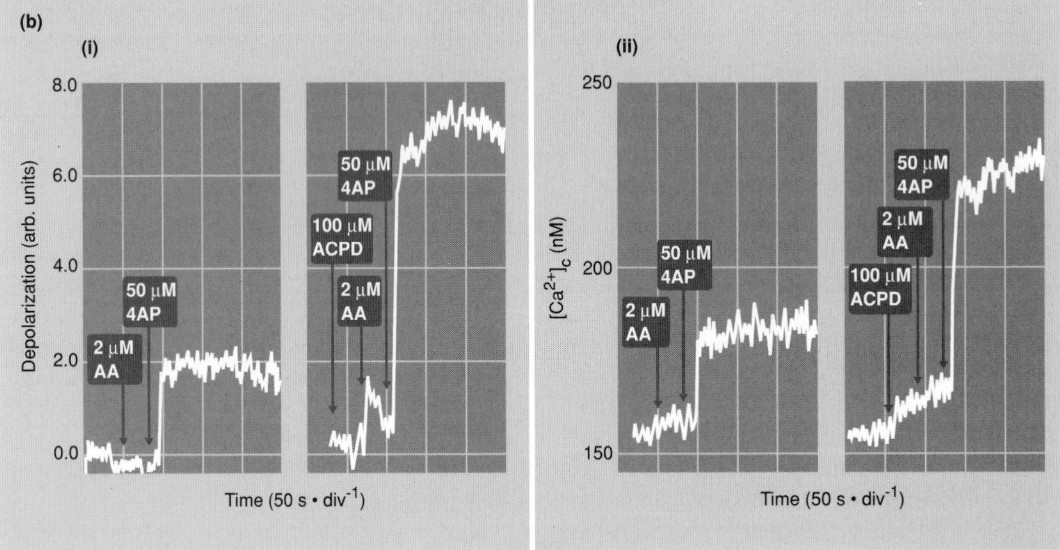

Fig. 9.18 Positive feedback by presynaptic metabotropic glutamate autoreceptors in cerebral cortex. (a) DAG generated by a presynaptic mGluR together with arachidonic acid (AA) synergistically activate PKC in cortical synaptosomes. PKC then appears to inhibit a presynaptic K^+-channel enhancing 4AP-evoked action potentials and increasing glutamate release. If the arachidonic acid originates postsynaptically in response to postsynaptic PLA_2 activation this could provide a possible mechanism for the fatty acid acting as a retrograde messenger in long-term potentiation. (b) Enhanced 4AP-evoked depolarization (i) and $[Ca^{2+}]_c$ elevation (ii). (c) (*Opposite*) Ca^{2+}-dependent glutamate release in the presence of arachidonic acid and the mGluR agonist 1S,3R-ACPD (ACPD). From Herrero *et al.* (1992), reproduced with permission from *Nature*. © 1992 Macmillan Magazines Ltd.

are predominantly postsynaptic, ionotropic and conduct Cl^- when the ligand is bound. The receptors are each composed of oligomers of subunits each possessing the four transmembrane domain motif characteristic of ionotropic receptors (*see* Fig. 4.6) and by analogy with the nicotinic ACh receptor are presumed to function as pentamers. Channel activation moves the membrane potential closer to the Cl^- equilibrium potential, which is usually quite negative (−60

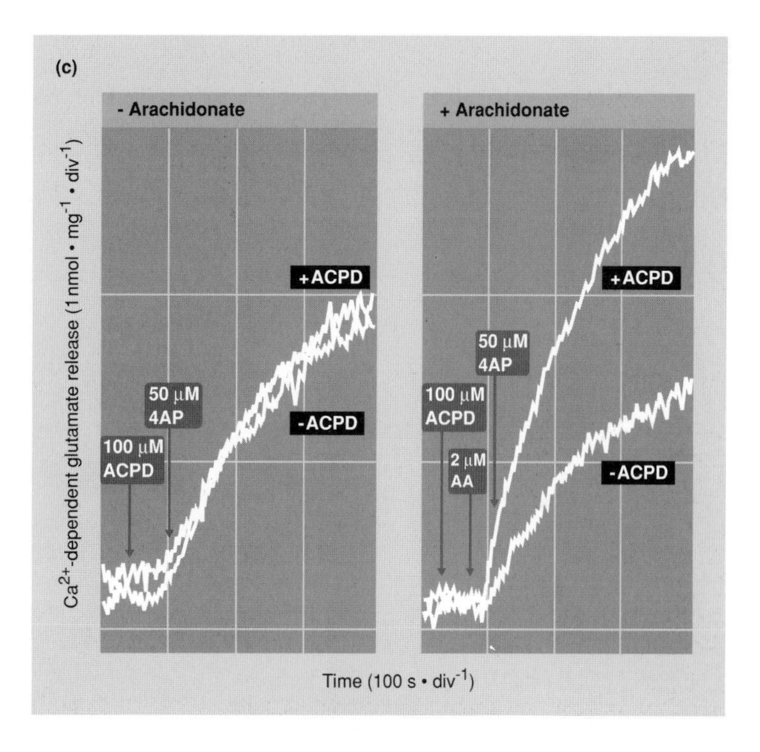

Fig. 9.18 Continued.

to −80 mV). This together with the concomitant decrease in the resistance of the plasma membrane antagonizes the tendency of excitatory receptors to depolarize the membrane sufficiently to reach the threshold at which action potentials fire.

The *GABA-A receptor* has a complex pharmacology, with binding sites for direct GABA agonists and antagonists together with multiple allosteric sites within the complex for the benzodiazepine (BDZ) tranquillizers, for the barbiturate CNS depressants and for both synthetic and endogenous steroids (Fig. 9.19).

The GABA-A receptor can be activated selectively by *muscimol* and 4,5,6,7-tetrahydroisoxazolo[5,4-c]pyridine-3-ol (*THIP*) and inhibited by *bicucculine.* The action of GABA is cooperative, suggesting the presence of two GABA binding sites, and consequently two β-subunits within the supposed pentameric structure. *Picrotoxin* inhibits by binding to the ion channel domain of the receptor.

A number of structurally diverse compounds enhance the action of GABA, these include *BDZs, barbiturates* and *pregnane steroids.* In animal models these compounds exhibit anxiolytic, anticonvulsant and sedative activity. BDZs increase the probability of channel opening, but have little effect on channel open time, whereas barbiturates and steroids prolong channel open time in the presence of GABA and may even activate channels in the absence of GABA, perhaps accounting for their general anaesthetic actions at high concentrations.

Molecular cloning has identified five classes of GABA-A subunits each of which exist in multiple isoforms. To date six α, four β, three γ, one δ and two ρ isoforms have been identified. By analogy to the most characterized ionotropic receptor, the muscle nicotinic receptor GABA-A receptors are generally assumed to function as pentamers. The enormous number of potential permutations of subunits (at least 1000 for a pentamer) appears to be somewhat restricted *in vivo* as, with some exceptions, most receptor isoforms appear to utilize a single type of α-subunit rather than a mixture. Even so there may be 50–100 combinations expressed in brain. The highest density of GABA-A receptors is found in the cerebral cortex and in the granule cell layer of the cerebellum. Figure 9.20 shows the brain distribution of six α-subunit isoforms — the

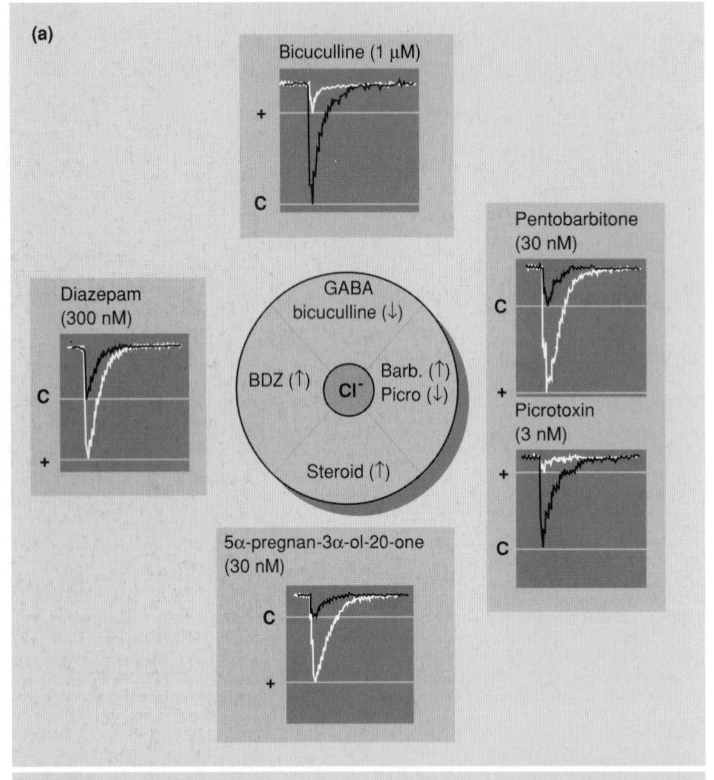

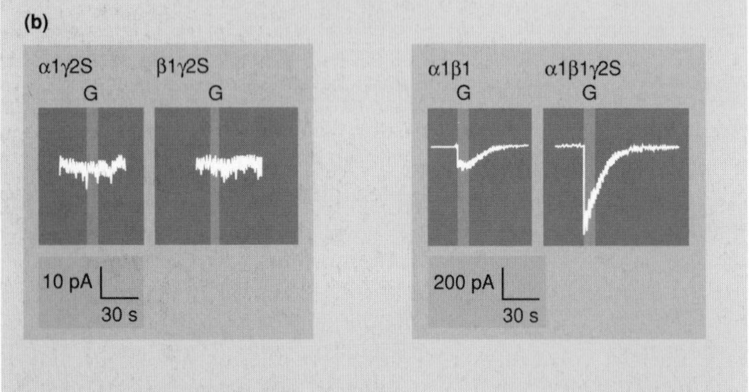

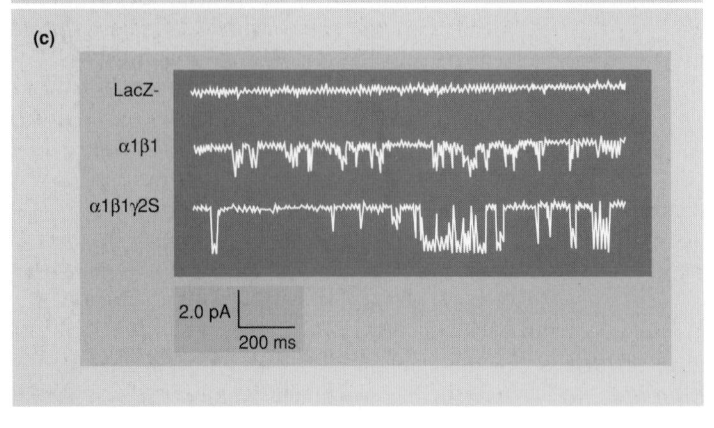

Fig. 9.19 GABA-A receptors. (a) Scheme of binding sites affecting GABA-A receptors (↓, inhibition of conductance; ↑, activation of conductance) with representative effects on whole-cell Cl^- currents in the presence of GABA (C, control; +, plus agent). Courtesy of J.J. Lambert. (b) Whole-cell currents in mammalian cells transiently expressing different combinations of GABA-A subunits; optimal conductance evoked by GABA (G) requires coexpression of α_1-, β_1- and γ_2-subunits. (c) Single-channel conductances for non-transfected cells (LacZ−) and cell transfected to express the indicated subunit combinations. The triple subunit combination shows greater single-channel conductance and prolonged opening. From Angelotti *et al.* (1993), by permission of the Society for Neuroscience.

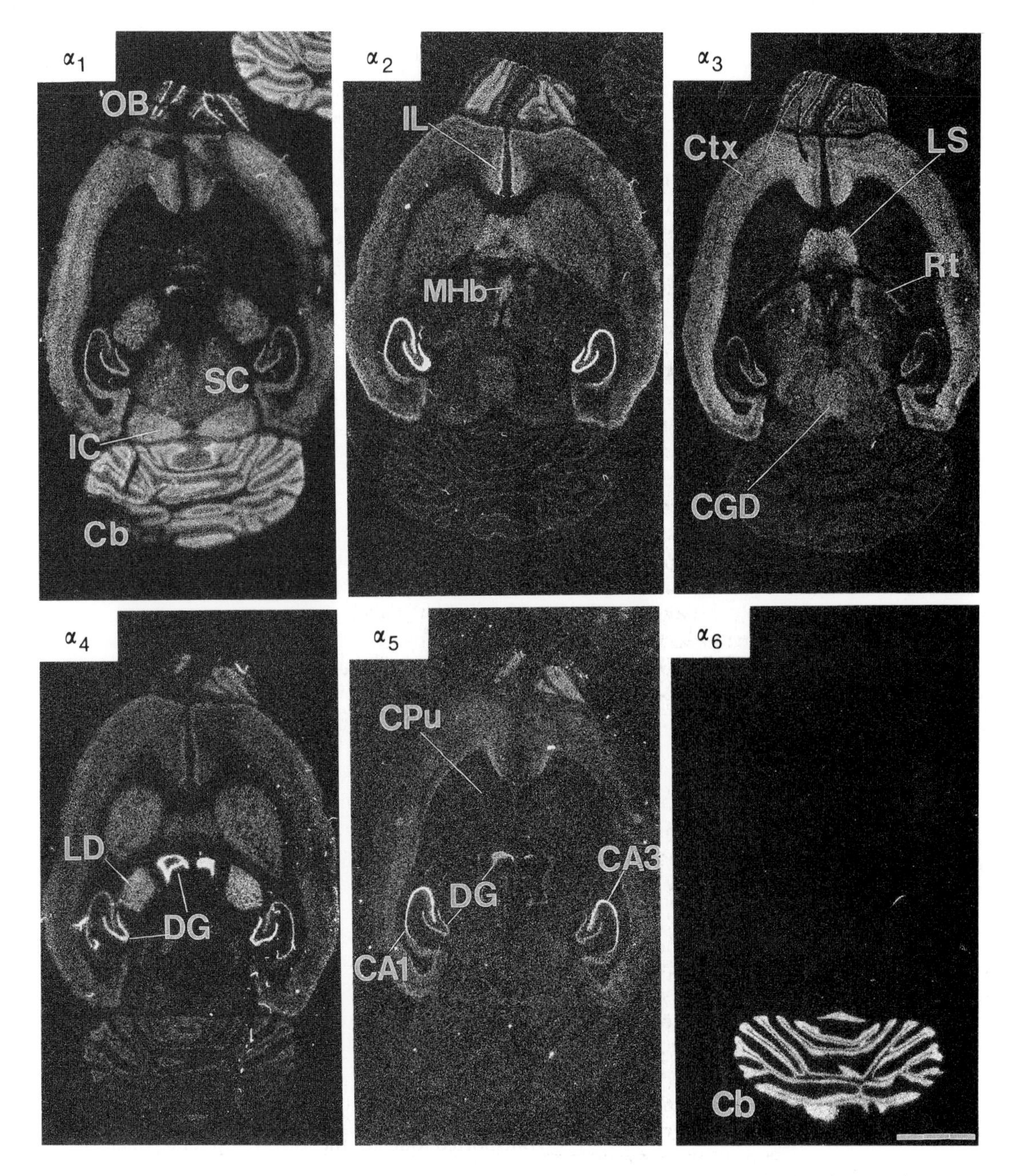

Fig. 9.20 Patterns of distribution of α-subunits of the GABA-A receptor in rat brain. The distribution of mRNAs for the GABA-A receptor subunits $\alpha_1-\alpha_6$ in horizontal sections of rat brain. Note that α_6 is uniquely expressed in the cerebellum. Bar = 4.4 mm. From Wisden *et al.* (1992), by permission of the Society for Neuroscience.

highly specific pattern of expression is evident, particularly the expression of α_6 which is limited to the cerebellum.

Recent developments in molecular biology have provided some insight into the relationship between subunit structure and pharmacological properties. A γ-subunit (usually γ_2) is required for BDZ sensitivity of expressed receptors, while the particular α-subunit isoform modulates the affinity of binding of different BDZ analogues. Based on the relative affinity of different BDZs, two pharmacological classes of BDZs binding to

CNS GABA-A receptor had been defined, termed Type I BDZ and Type II BDZ. Type I binding (characterized by selective binding of the agonist CL 218 872) was found in the cerebellum and other brain regions, while Type II binding was restricted to the hippocampus, striatum, spinal cord and neonatal cortex. When combinations of cloned subunits were coexpressed, it was found that $\alpha_1\beta_2\gamma_2$ gave a Type I pharmacology whereas $\alpha_2\beta_2\gamma_2$, $\alpha_3\beta_2\gamma_2$ and $\alpha_5\beta_2\gamma_2$ approximated to BDZ Type II pharmacology. In order to determine the structual determinants of this difference, chimeric α-subunits were constructed with elements derived from α_3 and α_1 — each coexpressed with β_2 and γ_2. It was found that a single residue (a Glu at position 225 of the α_3-subunit located on the N-terminal domain close to TM1) was an important determinant of BDZ binding, since its mutation to a Gly increased the binding affinity of the BDZ Type I specific ligand ten-fold in what otherwise was a BDZ Type II construct.

GABA-A receptors containing α_6-subunits are only found in the cerebellum where they are restricted to granule cells; the α_6-subunit has an Arg residue in the N-terminal region at a position occupied by a His in most other α-subunits, and this gives the granule cell receptor an unusual pharmacology with very low affinity for BDZs such as diazepam. A rat line has been described possessing a point mutation of this amino acid residue which gives diazepam sensitivity. In this animal diazepam causes anomalous motor impairment, suggesting that the granule cell GABA-A receptors may function in the cerebellar circuits which control motor function.

Progesterone metabolites can potentiate GABA responses in recombinant channels comprising homomeric β_1-, $\alpha_1\beta_1$- or $\alpha_1\beta_1\gamma_2$-subunits, while steroid antagonists also exist. No role has yet been assigned to the δ- and ρ-subunits; the former is expressed prominently by cerebellar granule cells while ρ is only found in the retina as a component of a receptor which is insensitive to bicucculine, BDZ, steroids and barbiturates and may represent a 'GABA-C receptor'.

The GABA-A receptor can be regulated by phosphorylation. The large intracellular loops present in each subunit class between TM3 and TM4 contains consensus sites for PKA, PKC or protein tyrosine kinases. Thus β-subunits contain potential PKA sites and γ-subunits contain a tyrosine kinase site. In addition an alternatively spliced form of the γ_2-subunit (γ_{2L}) possesses a PKC consensus site. This site appears to be required for the potent stimulation of GABA-A receptor activity caused by *ethanol* which may underlie the sedative effects of alcohol. PKA-dependent phosphorylation potently inhibits the GABA-A current.

In situ hybridization has revealed for the *glycine receptor* that the β-subunit in particular is widely expressed throughout the brain and that α_2- and α_3-subunits are expressed in higher brain areas as well as the brain stem and spinal cord (the areas known to contain most glycinergic synapses).

The conductance of the glycine receptor increases steeply with the concentration of the amino acid, suggesting that binding of three glycines may be required to activate the channel. The receptor is also voltage dependent; its probability of opening increases as the cell is depolarized.

The pharmacology of the receptor is hampered by the relative lack of specific agonists and antagonists. However the GlyR is inhibited by *strychnine*, an alkaloid from the seeds of the Indian tree, *Strychnos nux vomica*. Strychnine causes generalized excitation by blocking the inhibitory actions of glycine. The alkaloid binds with high affinity (*c.* 10 nM) to spinal cord synaptic membrane preparations. Although it displaces glycine, the binding sites are not identical and it is likely that strychnine and glycine bind to distinct sites on the GlyR. Since glycine also binds with high affinity to a strychnine-insensitive regulatory site on the NMDA-selective glutamate receptor, strychnine sensitivity is an essential criterion for analysing binding to the GlyR in unpurified preparations.

Currently, four 48 kDa α-subunits and one 58 kDa β-subunit of the glycine receptor have been identified. Strychnine is a natural photoaffinity ligand which becomes covalently attached to the 48 kDa polypeptide on UV illumination. The alkaloid can also be exploited in the purification of the GlyR by passing solubilized synaptic membrane proteins down a 2-aminostrychnine-agarose affinity column. The resulting complex contains a 48 kDa and the less abundant 58 kDa polypeptide, and additionally a 93 kDa polypeptide.

The 48 and 58 kDa components are highly

homologous glycoproteins believed to constitute the anion channel. Considerable structural homology was found between the primary sequence of the 48 kDa subunit and that of other ionotropic receptor subunits of the nicotinic ACh and GABA-A receptors consistent with a 'superfamily' of evolutionary related receptors. Thus hydropathy analysis indicates four transmembrane segments, whose positions are almost identical to those in the subunits of the MnAChR and the GABA-A receptor. The α-subunits are capable of reconstituting a functional homomeric channel.

The gene for the 58 kDa β-chain has been identified, and a total molecular weight of 250 000 estimated for the strychnine-binding core of the receptor, obtained by sedimentation analysis, would be consistent with three copies of the 48 kDa and two copies of the 58 kDa polypeptides. The function of the 93 kDa peptide is unclear, it is a peripheral membrane protein on the cytoplasmic face of the receptor and may thus be involved in the interaction of the receptor with the cytoskeleton.

9.8.1 The GABA-B receptor

The metabotropic GABA-B receptor, which has not to date been cloned, is less abundant than the GABA-A receptor. It can function both as a presynaptic inhibitory autoreceptor and a postsynaptic receptor. Baclofen is the characteristic agonist, while phaclofen is a weak GABA-B antagonist. The GABA-B receptor, in common with a group of metabotropic receptors may cause a coordinated inhibition of Ca^{2+}-channels and activation of K^{+}-channels by a direct linkage through a matrix of common G-protein complexes.

Further reading

Amino acid transmitters

Cuenod, M., Do, K.Q. & Streit, P. (1990) Homocysteic acid as an endogenous excitatory amino acid (Letter). *Trends Pharmacol. Sci.* **11**, 477–477.

Fonnum, F. (1984) Glutamate: a neurotransmitter in mammalian brain. *J Neurochem.* **42**, 1–11.

Meeker, R.B., Swanson, D.J., Greenwood, R.S. & Hayward, J.N. (1993) Quantitative mapping of glutamate presynaptic terminals in the supraoptic nucleus and surrounding hypothalamus.. *Brain Res.* **600**, 112–122.

Ottersen, O.P. (1989) Quantitative electron microscopic immunocytochemistry of neuroactive amino acids. *Anat. Embryol.* **180**, 1–15.

Uptake of amino acid neurotransmitters into terminals and glia

Amara, S.G. (1992) Neurotransmitter transporters: a tale of two families. *Nature* **360**, 420–421.

Bouvier, M., Szatkowski, M., Amato, A. & Attwell, D. (1992) The glial cell glutamate uptake carrier countertransports pH-changing anions. *Nature* **360**, 471–474.

Bowery, N.G. (1990) GABA transporter protein cloned from rat brain. *Trends Pharmacol. Sci.* **11**, 435–437.

Guastella, J., Nelson, N., Nelson, H., Czyzyk, L., Keynan, S., Miedel, M.C., Davidson, N., Lester, H.A. & Kanner, B.I. (1990) Cloning and expression of a rat brain GABA transporter. *Science* **249**, 1303–1306.

Kuhar, M.J. (1990) A GABA transporter cDNA has been cloned (Research news). *Trends Neurosci.* **13**, 473–473.

Nicholls, D.G. & Attwell, D.A. (1990) The release and uptake of excitatory amino acids. *Trends Pharmacol. Sci.* **11**, 462–468.

Pines, G., Danbolt, N.C., Bjorås, M., Zhang, Y., Bendahan, A., Eide, L., Koepsell, H., Storm-Mathisen, J., Seeberg, E. & Kanner, B.I. (1992) Cloning and expression of a rat brain L-glutamate transporter. *Nature* **360**, 464–467.

Radian, R., Ottersen, O.P., Storm-Mathisen, J., Castel, M. & Kanner, B.I. (1990) Immunocytochemical localization of the GABA transporter in rat brain. *J. Neurosci.* **10**, 1319–1319.

Uptake of amino acids into synaptic vesicles

Maycox, P.R., Deckwerth, T., Hell, J.W. & Jahn, R. (1988) Glutamate uptake by brain synaptic vesicles. *J. Biol. Chem.* **263**, 15 423–15 428.

Maycox, P.R., Hell, J.W. & Jahn, R. (1990) Amino acid neurotransmission: spotlight on synaptic vesicles. *Trends Neurosci.* **13**, 83–87.

Amino acid exocytosis

Clements, J.D., Lester, R.A.J., Tong, G., Jahr, C.E. & Westbrook, G.L. (1992) The time course of glutamate in the synaptic cleft. *Science* **258**, 1498–1501.

McMahon, H.T. & Nicholls, D.G. (1991) Transmitter glutamate release from isolated nerve terminals: evidence for biphasic release and triggering by localized Ca^{2+}. *J. Neurochem.* **56**, 86–94.

Nicholls, D.G. (1993) The glutamatergic nerve terminal. *Eur. J. Biochem.* **212**, 613–631.

Sihra, T.S. & Nicholls, D.G. (1987) 4-Aminobutyrate can be released exocytically from guinea-pig cerebral cortical synaptosomes. *J. Neurochem.* **49**, 261–267.

Turner, T.J., Pearce, L.B. & Goldin, S.M. (1989) A

superfusion system designed to measure release of radiolabeled neurotransmitter on a subsecond timescale. *Analytical Biochem.* **178**, 8–16.

Thomson, A.M. (1989) Glycine modulation of the NMDA receptor/channel complex. *Trends Neurosci.* **12**, 349–353.

Glutamate and neurotoxicity

Choi, D.W. & Rothman, S.M. (1990) The role of glutamate neurotoxicity in hypoxic-ischemic neuronal death. *Annu. Rev. Neurosci.* **13**, 171–182.

Mayer, M.L. & Westbrook, G.L. (1987) Cellular mechanisms underlying excitotoxicity. *Trends Neurosci.* **10**, 59–61.

Meldrum, B. & Garthwaite, J. (1990) EAA pharmacology: excitatory amino acid neurotoxicity and neurodegenerative disease. *Trends Pharmacol. Sci.* **11**, 379–387.

The AMPA/KA receptor family

Dingledine, R., Hume, R.I. & Heinemann, S.F. (1992) Structural determinants of barium permeation and rectification in non-NMDA glutamate receptor channels. *J. Neurosci.* **12**, 4080–4087.

Gasic, G.P & Heinemann, S.F. (1991) Receptors coupled to ionic channels: the glutamate receptor family. *Curr. Op. Neurobiol.* **1**, 20–26.

Nakanishi, S. (1992) Molecular diversity of glutamate receptors and implications for brain function. *Science* **258**, 597–603.

Sommer, B., Keinanen, K., Verdoorn, T.A., Wisden, W., Burnashev, N., Herb, A., Kohler, M., Takagi, T., Sakmann, B. & Seeburg, P.H. (1990) Flip and flop: A cell specific functional switch in glutamate-operated channels of the CNS. *Science* **249**, 1580–1585.

Kainate receptors

Miller, R.J. (1991) The revenge of the kainate receptor. *Trends Neurosci.* **14**, 477–479.

Sommer, B., Burnashev, N., Verdoorn, T.A., Keinänen, K., Sakmann, B. & Seeburg, P.H. (1992) A glutamate receptor channel with high affinity for domoate and kainate. *EMBO J.* **11**, 1651–1656.

The NMDA receptor

Barnard, E. (1992) Will the real NMDA receptor please stand up? *Trends Pharmacol. Sci.* **13**, 11–12.

Burnashev, N., Schoepfer, R., Monyer, H., Ruppersberg, J.P., Günther, W., Seeburg, P.H. & Sakmann, B. (1992) Control by asparagine residues of calcium permeability and magnesium blockade in the NMDA receptor. *Science* **257**, 1415–1419.

Monyer, H., Sprengel, R., Schoepfer, R., Herb, A., Higuchi, M., Lomeli, H., Burnashev, N., Sakmann, B. & Seeburg, P.H. (1992) Heteromeric NMDA receptors: molecular and functional distinction of subtypes. *Science* **256**, 1217–1221.

The metabotropic glutamate receptor family

Abe, T., Sugihara, H., Nawa, H., Shigemoto, R., Mizuno, N. & Nakanishi, S. (1992) Molecular characterization of a novel metabotropic glutamate receptor mGluR5 coupled to inositol phosphate/Ca^{2+} signal transduction. *J. Biol. Chem.* **267**, 13361–13368.

Anwyl, R. (1991) The role of the metabotropic receptors in synaptic plasticity. *Trends Pharmacol. Sci.* **12**, 324–326.

Baskys, A. (1992) Metabotropic receptors and slow excitatory actions of glutamate agonists in the hippocampus. *Trends Neurosci.* **15**, 92–96.

Herrero, I., Miras-Portugal, M.T. & Sanchez-Prieto, J. (1992) Positive feedback of glutamate exocytosis by metabotropic presynaptic receptor stimulation. *Nature* **460**, 163–166.

Martin, L.J., Blackstone, C.D., Huganir, R.L. & Price, D.L. (1993) Cellular localization of a metabotropic glutamate receptor in rat brain. *Neuron* **9**, 259–270.

Masu, M., Tanabe, Y., Tsuchida, K., Shigemoto, R. & Nakanishi, S. (1991) Sequence and expression of a metabotropic glutamate receptor. *Nature* **349**, 760–765.

Recasens, M., Guiramand, J. & Vignes, M. (1991) The putative molecular mechanism(s) responsible for the enhanced inositol phosphate synthesis by excitatory amino acids — an overview. *Neurochem. Res.* **16**, 659–659.

Schoepp, D.D., Bockaert, J. & Sladeczek, F. (1990) Pharmacological and functional characteristics of metabotropic excitatory amino acid receptors (Review). *Trends Pharmacol. Sci.* **11**, 508–508.

Schoepp, D.D. & Conn, P.J. (1993) Metabotropic glutamate receptors in brain function and pathology. *Trends Pharmacol. Sci.* **14**, 13–20.

Sladeczek, F., Recasens, M. & Bockaert, J. (1988) A new mechanism for glutamate receptor action: phosphoinositide hydrolysis (Review). *Trends Neurosci.* **11**, 545–549.

Tanabe, Y., Masu, M., Ishii, T., Shigemoto, R. & Nakanishi, S. (1992) A family of metabotropic glutamate receptors. *Neuron* **8**, 169–179.

Tanabe, Y., Nomura, A., Masu, M., Shigemoto, R., Mizuno, N. & Nakashini, S. (1993) Signal transduction, pharmacological properties, and expression patterns of two rat metabotropic glutamate receptors, mGluR3 and mGluR4. *J. Neurosci.* **13**, 1372–1378.

Thomsen, C., Kristensen P., Mulvihill, E., Haldeman, B. & Suzdak, P.D. (1992) L-AP4 is an agonist at the type IV metabotropic glutamate receptor which is negatively coupled to adenylate cyclase. *Eur. J. Pharmacol. Mol. Pharmacol.* **227**, 361–362.

GABA and glycine receptors

Angelotti, T.P., Uhler, M.D. & Macdonald, R.L. (1993) Assembly of $GABA_A$ receptor subunits: analysis of transient single-cell expression utilizing a fluorescent substrate/marker gene technique. *J. Neurosci.* **13**, 1418–1428.

Barnard, E.A., Darlison, M.G. & Seeburg, P. (1987) Molecular biology of the GABA-A receptor: the receptor/channel superfamily. *Trends Neurosci.* **10**, 502–508.

Baude, A., Sequier, J.-M., McKernan, R.M., Olivier, K.R. & Somogyi, P. (1992) Differential subcellular distribution of the α_6 subunit versus the α_1 and $\beta_{2/3}$ subunits of the GABA-A/benzodiazepine receptor complex in granule cells of the cerebellar cortex. *Neurosci.* **51**, 739–748.

Betz, H. (1991) Glycine receptors: heterogeneous and widespread in the mammalian brain. *Trends Neurosci.* **14**, 458–461.

Doble, A. & Martin, I.L. (1992) Multiple benzodiazepine receptors: no reason for anxiety. *Trends Pharmacol. Sci.* **13**, 76–81.

Farrant, M. & Cull-Candy, S.G. (1993) GABA-receptors, granule cells and genes. *Nature* **361**, 302–303.

Knapp, R.J., Malatynska, E. & Yamamura, H.I. (1990) From binding studies to the molecular biology of GABA receptors. *Neurochem. Res.* **15**, 105–112.

Langosch, D., Becker, C.M. & Betz, H. (1990) The inhibitory glycine receptor — a ligand-gated chloride channel of the central nervous system (Review). *Eur. J. Biochem.* **194**, 1–8.

Luddens, H. & Wisden, W. (1991) Function and pharmacology of multiple GABAA receptor subunits. *Trends Pharmacol. Sci.* **12**, 49–49.

Olsen, R.W. & Tobin, A.J. (1990) Molecular biology of GABAA receptors (Review). *FASEB J.* **4**, 1469–1469.

Stephenson, F.A. (1988) Understanding the GABA-A receptor: a chemically gated ion channel. *Biochem. J.* **249**, 21–32.

Storck, T., Schulte, S., Hofmann, K. & Stoffel, W. (1992) Structure, expression, and functional analysis of a Na^+-dependent glutamate/aspartate transporter from rat brain. *Proc. Natl. Acad. Sci. USA* **89**, 10955–10959.

Wieland, H.A., Luddens, H. & Seeburg, P. (1992) A single histidine in GABA-A receptors is essential for benzodiazepine agonist binding. *J. Biol. Chem.* **267**, 1426–1429.

Wisden, W., Laurie, D.J., Monyer, H. & Seeburg, P.H. (1992) The distribution of 13 GABA-A receptor subunit mRNAs in the rat brain. *J. Neurosci.* **12**, 1040–1062.

The GABA-B receptor

Bowery, N.G. (1989) GABA-B receptors and their significance in mammalian pharmacology (Review). *Trends Pharmacol. Sci.* **10**, 401–401.

10

Acetylcholine

10.1 Introduction

Acetylcholine (ACh) is synthesized in the nerve terminal cytoplasm from acetyl-CoA (AcCoA) and choline by the 68 kDa choline acetyltransferase, CAT (Fig. 10.1). Cytoplasmic AcCoA is in turn generated by citrate lyase from citrate which has been transported out of the mitochondrion. After dissociation from its receptor ACh is rapidly hydrolysed by acetylcholinesterase (AChE) to acetate and choline, the latter being reaccumulated into terminals by a Na^+-coupled transporter. This re-uptake may be the rate-limiting step in this metabolic cycle.

Only some 6% of the synapses in the rat cerebral cortex are cholinergic. The forebrain cholinergic system (Fig. 10.2) originates from a group of large cell bodies extending from the *medial septum* to the *nucleus basalis of Meynert*. These cells provide the cholinergic innervation for the hippocampus, the cerebral cortex and the amygdaloid complex.

ACh has a mainly facilitatory postsynaptic effect on cortical and hippocampal neurones, initially due to a direct depolarization, and more persistently to a prolonged decrease in a Ca^{2+}-activated K^+ current; a brief stimulation of cholinergic fibres can enhance the response of hippocampal neurones for tens of seconds.

10.2 Acetylcholine and the electric organ

Studies on preparations from the CNS (such as synaptosomes) are hampered by the heterogeneous nature of the transmitter content in the preparation. This is particularly true in the case of ACh which is the transmitter at a minority of synapses in most regions of the CNS. While the neuromuscular junction of vertebrates is cholinergic, this preparation is not suitable for most biochemical purposes. However the electric organ of *Torpedo marmorata* (Fig. 10.3), *Torpedo californica* or of the electric eel *Electrophorus* have proved enormously useful for the investigation of both pre- and postsynaptic aspects of cholinergic transmission.

The electric organ of these fish, which can weigh several hundred grams, is a stack of neuromuscular junctions, where the postsynaptic muscle cells have degenerated. In the case of *Torpedo marmorata*, two large electric organs are capable of generating a large potential difference in the surrounding seawater for navigation or stunning of prey. The electric organ contains stacks of individual cells, or *electrocytes*, one face of which contains high concentrations of nicotinic ACh receptor and is innervated by cholinergic terminals which cover about 80% of its surface (Fig. 10.4). The opposite face of the electrocyte is non-excitable. The tissue is controlled by paired *electric lobes* in the brain of the *Torpedo* which contain the cell bodies of the electromotor nerves innervating the electric organ. When the electric lobe discharges synchronously, the simultaneous stimulation of the cholinergic nerves innervating the stack causes each excitable face to depolarize. The non-excitable face of the cell retains its resting membrane potential, and since the extracellular

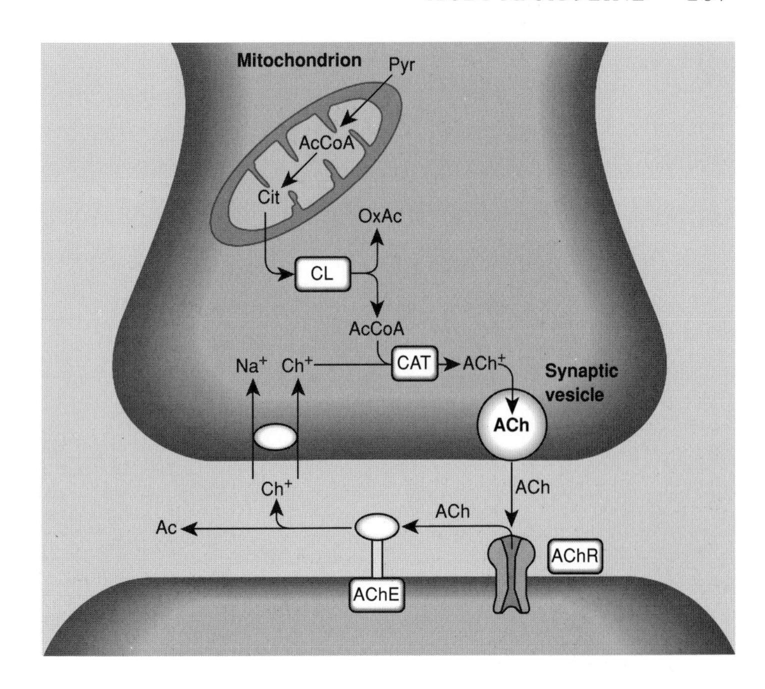

Fig. 10.1 Pathways of acetylcholine (ACh) metabolism. Pyruvate (Pyr) from glycolysis generates acetyl-CoA (AcCoA) within the mitochondrial matrix. This is converted to citrate (Cit) via the citric acid cycle. Citrate leaves the mitochondrion on the tricarboxylate carrier and is cleaved in the cytoplasm by citrate lyase (CL) to give AcCoA and oxaloacetate (OxAc). AcCoA and choline (Ch^+) generate ACh in the cytoplasm via choline acetyltransferase (CAT) which is accumulated into synaptic vesicles. After exocytosis and binding to ACh receptors (AChR), ACh is rapidly degraded by ACh esterase (AChE) and the choline group retrieved by a Na^+-coupled re-uptake pathway. ACh = $CH_3COO.CH_2CH_2\overset{+}{N}(CH_3)_3$.

medium between sequential electrocytes is electrically isolated, the potential across the stack will be the sum of the potentials across the multiple non-excitable faces — up to 50 V.

The tissue is thus a highly pure and concentrated source of many of the components of the cholinergic synapse. The ACh content of *Torpedo* electric organ is, at 1 μmol · g^{-1} tissue, 50 times greater than that of guinea-pig cerebral cortex. At least 80% of the tissue ACh is vesicular. The 'prism' (analogous to a brain slice), the cholinergic synaptosome and the cholinergic synaptic vesicle have all been exploited, while the tissue provided the source from which the first ionotropic receptor was purified.

Before considering these preparations in

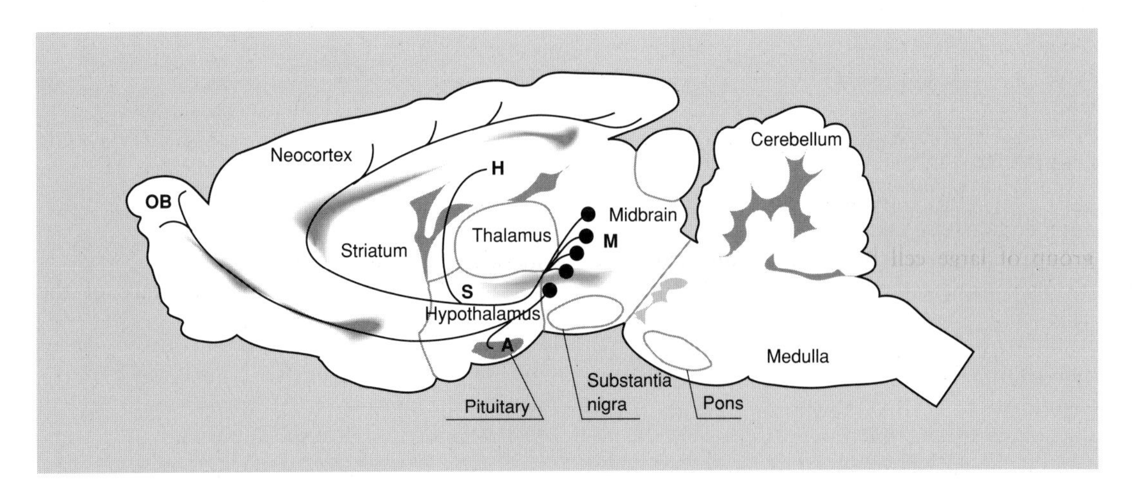

Fig. 10.2 Some major cholinergic pathways in the rat forebrain. A sheet of giant cholinergic cell bodies extends from the medial septal nucleus (S) to the nucleus basalis of Meynert (M). Their axons project to the cerebral cortex, olfactory bulb (OB), amygdala (A) and hippocampus (H). In addition the cerebral cortex contains a number of short cholinergic interneurones.

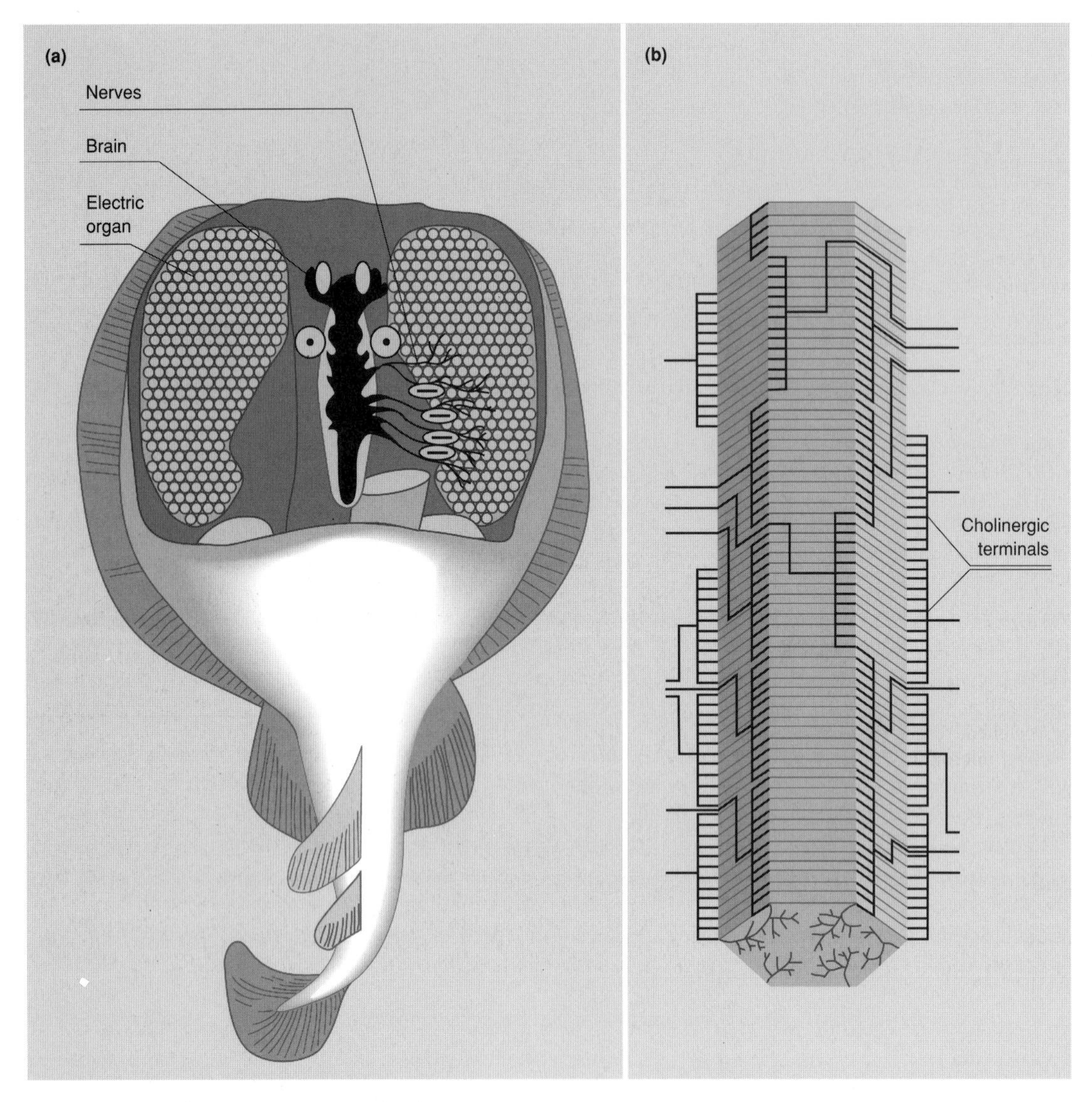

Fig. 10.3 *Torpedo* electric organ. (a) The electric organ together with its innervation from the brain. (b) A stack of electroplaque cells from the organ. One face of each cell possesses a high density of nicotinic acetylcholine receptors and is innervated by a cholinergic terminal (a source of pure cholinergic synaptic vesicles). When the nerves are stimulated synchronously the innervated faces depolarize. Since the extracellular fluid between electrolytes is electrically isolated, the potential across a stack of electrocytes is the sum of the *c*. 90 mV potentials across the non-excitable face of each cell. From Hucho (1986), by permission of VCH publishers.

detail however we shall consider one additional spin-off from this research, and that is the identification of a surface antigen on *Torpedo* synaptosomes which allows a pure preparation of mammalian CNS cholinergic synaptosomes to be prepared. The epitope for the antigen, termed *Chol-I*, is located on a family of minor gangliosides which are only expressed on the plasma membrane of cholinergic terminals from mammalian or non-mammalian sources. Antibodies to Chol-I can be raised by injecting purified synaptic membranes from *Torpedo* into sheep. As a demonstration of their specificity, the antibodies can activate the complement-induced lysis of the cholinergic subpopulation of rat cortical synaptosomes, liberating 85% of the specific cytoplasmic marker CAT, but only 6% of the general marker lactate dehydrogenase.

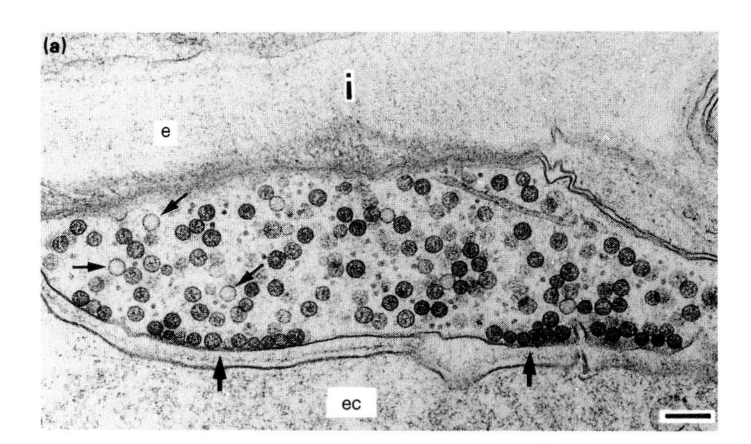

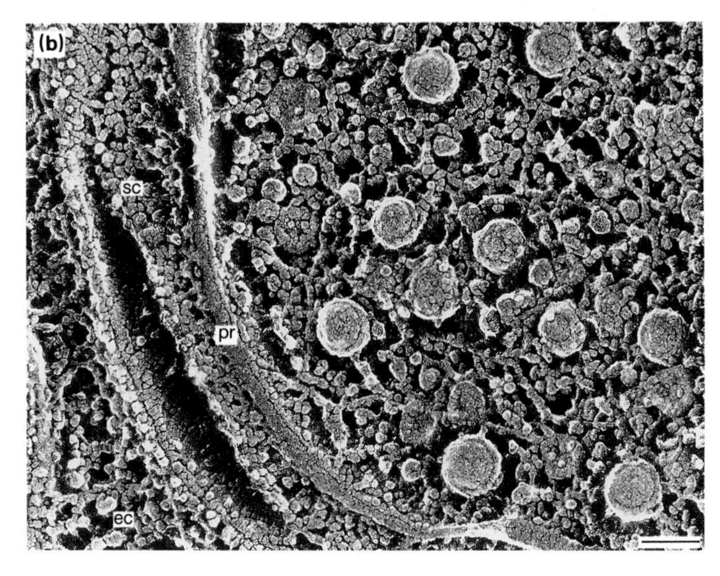

Fig. 10.4 Presynaptic terminals in the electric organ. (a) Thin section of the electromotor synapse from the electric ray after quick freezing and freeze-substitution with OsO_4. Most synaptic vesicles contain dark-staining material although some (small arrows) are electron lucid. Thick arrows show vesicles in apposition to the presynaptic membrane. e, extracellular space; ec, postsynaptic electroplaque cell; bar = 250 nm. (b) Quick-frozen, deep-etched, rotary shadowed electron micrograph showing synaptic vesicles apparently attached to cytoskeletal elements which are in turn attached to the presynaptic membrane (pr). ec, postsynaptic electroplaque cell; sc, synaptic cleft; bar = 100 nm. From Zimmermann *et al.* (1990), by permission of Academic Press.

Clearly it is more useful to select an intact cholinergic subpopulation than to destroy it, and this can be accomplished by starting with a crude mitochondrial P2 fraction from rat brain, pretreating with sheep polyclonal antibody, adding a mouse monoclonal anti-sheep IgG coupled to cellulose beads. The cholinergic synaptosomes can then be specifically sedimented by centrifugation of the cellulose beads. In this way an 18-fold purification of cholinergic synaptosomes from rat cerebral cortex can be obtained, leading to an estimate of 6% for the cholinergic synaptosomal subpopulation.

Despite being a homogeneous and concentrated source of cholinergic terminals, it is difficult to prepare synaptosomes from *Torpedo* electric organ, due in part to the toughness of the muscle-derived tissue. The yield can be improved by using young fish but even so the preparation lacks the metabolic and energetic integrity characteristic of the mammalian preparation and is much less enriched in ACh and CAT than would be expected for terminals which only account for 3% of the electric organ. However, even if the synaptosomes are imperfectly preserved, they can still be used as a source of presynaptic plasma membrane, which has, for example, been used as a source of the Na^+-coupled choline transporter.

Torpedo synaptic vesicles are somewhat larger than those from mammalian terminals (90 nm instead of 50 nm), and contain much more transmitter. Each newly synthesized *VP1* vesicle (*see* Section 7.4.7) contains some 200 000 molecules of ACh (equivalent to 6 nmol ACh per milligram of vesicle protein), together with 30 000 molecules of ATP. This should be compared with a content of some 2000 molecules of ACh in a mammalian CNS cholinergic synaptic vesicle. Nuclear magnetic resonance studies reveal that both components are free rather than complexed within the vesicle, and exert a high osmotic pressure. In the presence of external ATP the vesicles maintain an internal pH of 5.5. Most of the cholinergic synaptic vesicle proteins are closely homologous to corresponding proteins on mammalian small synaptic vesicles (*see* Fig. 7.11). Thus *Torpedo* synaptic vesicles are associated with synapsin I and also contain synaptophysin. With the exception of actin, no vesicle membrane proteins are found in plasma membrane preparations, indicating a rapid, selective and complete retrieval of the vesicle proteins from the plasma membrane following exocytosis.

Mammalian and *Torpedo* cholinergic synaptic vesicles contain ATP (Fig. 10.5). An ATP/ADP translocase distinct from that in the mitochondrial inner membrane has been proposed to allow ATP uptake into the vesicle, although as the mitochondrial carrier catalyses the strict 1 : 1 exchange of adenine nucleotides the function would have to be modified to allow net transport.

An ATP-dependent, uncoupler-sensitive, ACh uptake was first observed with cholinergic vesicles from PC12 cells. A similar uptake into *Torpedo* synaptic vesicles was subsequently demonstrated. Uptake driven by the V-type H^+-translocating ATPase could be inhibited by protonophores, by adding ammonia to inhibit the pH gradient or by adding valinomycin to abolish the pH gradient. In a K^+-containing medium, the uptake of [^{3}H]-ACh is inhibited by the K^+/H^+ antiport ionophore nigericin, which equates K^+ and pH gradients and thus in K^+-containing media essentially collapses ΔpH. This block of uptake suggests that the pH gradient is essential for the uptake of ACh, which in turn suggests that uptake occurs in exchange for protons. However, the uptake of ACh is also electrogenic which implies that the membrane potential component could also drive uptake. The stoichiometry of the transport has not been determined, but the electrogenicity of the transport makes the most probable mechanism an $ACh^+/2H^+$ exchange (Fig. 10.5).

ACh uptake is blocked specifically by the inhibitor *vesamicol* with an IC_{50} of about 40 nM. The inhibitor does not block the H^+-translocating ATPase, nor dissipate the proton gradient. However, vesamicol does not appear to inhibit the vesicular ACh transporter by competing with ACh, since [^{3}H]-vesamicol binding is not displaced by ACh. Rather it binds non-competitively to a regulatory site which may be on another protein (Fig. 10.5). The binding site density corresponds to about four sites per vesicle. Vesamicol inhibits quantal release from every cholinergic preparation which has been examined, presumably as a secondary consequence of

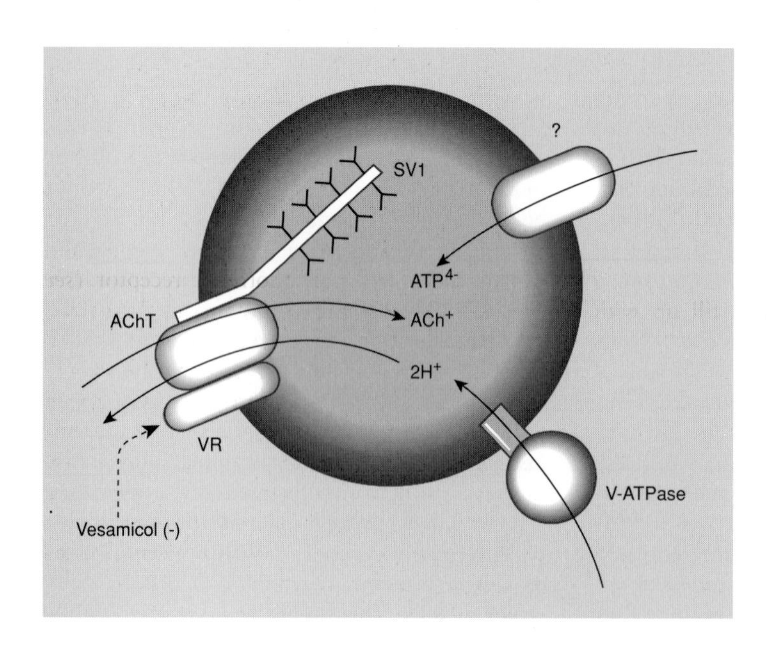

Fig. 10.5 Cholinergic synaptic vesicles. Schematic view showing characteristic features of a cholinergic synaptic vesicle: VR, vesamicol receptor, which is closely associated with the ACh transporter (AChT) and a proteoglycan which provides the SV1 epitope characteristic of synaptic vesicles. ACh^+ is probably accumulated in exchange for $2H^+$. ACh is co-stored with ATP but the mechanism of ATP transport is unclear. For clarity vesicular proteins common to SSVs in general have been omitted.

storage block. Vesamicol also inhibits non-quantal release from the motor-neurone terminal, which is believed to originate from the cytoplasm. It has been suggested that this could be due to the temporary insertion of synaptic vesicle membrane into the plasma membrane between exo- and endocytosis. The vesicular ACh transporter/vesamicol receptor complex which has been partially purified copurifies with the cholinergic synaptic vesicle proteoglycan which contains the SV1 antigenic epitope.

10.2.1 Cytoplasmic vs vesicular release of ACh

ACh is synthesized from AcCoA and choline in the terminal cytoplasm by the soluble cytoplasmic enzyme choline acetylase. Because there is a relatively high concentration of ACh in the cytoplasm there is some 'non-quantal' release of the transmitter by direct leakage from the cytoplasm; while this exceeds the 'resting' exocytotic release of the transmitter it does not increase significantly on stimulation.

Since it is difficult to prepare fully functional synaptosomes from *Torpedo*, most compartmental analysis of the terminals has been carried out with the perfused electric organ or with tissue prisms. The perfused organ can be labelled with ACh precursors to follow the kinetics of turnover of the cytoplasmic and vesicular pools, isolated by subsequent tissue fractionation. In resting tissue the cytoplasmic pool only accounts for 20% of the total tissue ACh and even less in stimulated tissue. Cytoplasmic ACh turns over rapidly within the cytoplasm.

As discussed in Section 7.4.7, the kinetics of synaptic vesicle recycling in *Torpedo* prisms has been subjected to a detailed analysis. Two synaptic vesicle populations termed VP1 (reserve) and VP2 (recycling), have been separated. When electrical stimulation generates a pool of recycling VP2 vesicles (*see* Fig. 7.20) these fill up with ACh with a specific activity characteristic of this cytoplasmic pool. In contrast, the fully loaded VP1 reserve vesicles fail to exchange with the cytoplasm in the time course of the experiments. The ACh released on stimulation has the same specific activity as the VP2 pool and hence the cytoplasm; thus these experiments cannot be used to distinguish between a cytoplasmic and vesicular origin of the released transmitter. However false transmitters formed from choline analogues differing in chain length do not equilibrate between the cytoplasm and recycling vesicles. The ratio of released natural ACh to 'false transmitter' is thus diagnostic of the origin of the pool. The results show that resting release, or leakage, takes place from the cytoplasm, whereas stimulated release occurs from the VP2 pool.

ACh release from terminals can be monitored continuously by a sensitive technique in which choline generated as the product of AChE is assayed by a bioluminescent reaction (Fig. 10.6). The technique has been reported to be sufficiently sensitive to register the release of ACh from single active zones with the giant terminal comprising the chick ciliary ganglion calyx (*see* Fig. 7.6).

10.3 Nicotinic ACh receptors

ACh receptors are either nicotinic (specifically activated by the tobacco alkaloid nicotine) or muscarinic (activated by muscarine from the mushroom *Amanita muscaria*). Isoforms of the ionotropic nicotinic receptors are found at peripheral neurones (notably the vertebrate neuromuscular junction) and centrally. Muscarinic receptors, which account for at least 95% of the ACh receptors in the CNS, are metabotropic, and were originally classified as M_1 or M_2 on the basis of pharmacology, but molecular genetics have currently shown the presence of five genes labelled m1–m5, which fall into 'even' (m2, m4) and 'odd' (m1, m3, m5) subgroupings on the basis of their G-protein/effector coupling (Table 10.1).

10.3.1 The muscle nicotinic ACh receptor

The nicotinic receptor was introduced in Section 4.4 as the archetypal ionotropic receptor (*see* Fig. 4.7). The receptor expressed at the neuromuscular junction, and also in *Torpedo*, binds, and is inhibited by, *α-bungarotoxin* from the Malaysian banded krait (not to be confused with the β-toxin from the same venom which has PLA_2 activity and inhibits K^+-channels, *see* Box 4.2). In the brain and sympathetic ganglia the dominant isoform of the receptor binds nicotine but not the toxin although there is also an α-bungarotoxin-sensitive receptor which has only recently been characterized (*see* Section 10.3.2).

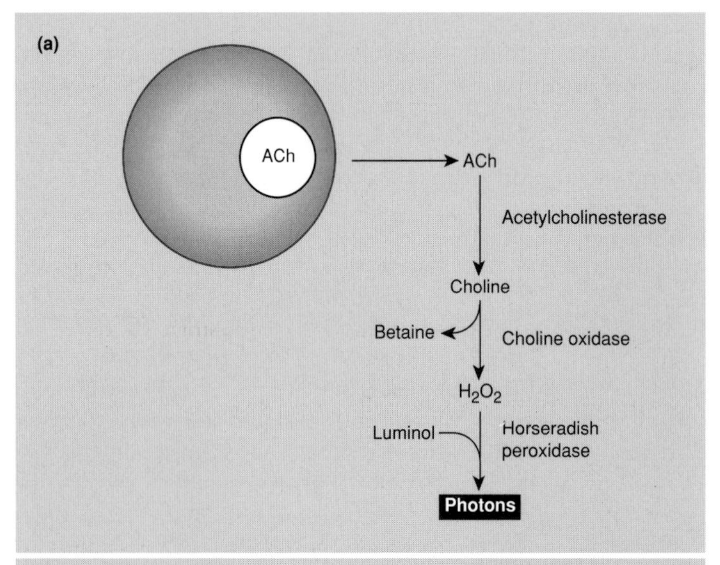

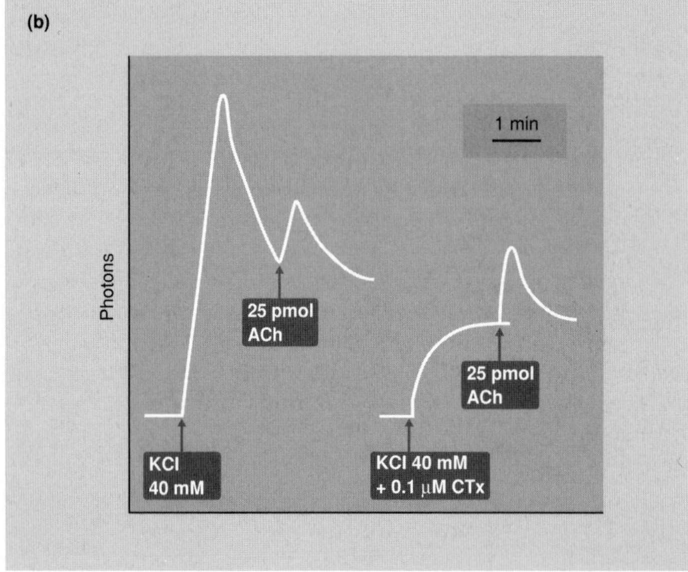

Fig. 10.6 Bioluminescent assay for the release of endogenous ACh. (a) The assay for released ACh relies on its hydrolysis to choline which is then oxidized to betaine yielding H_2O_2 which reacts with luminol to give photons in the presence of horseradish peroxidase. (b) A representative assay showing the KCl-evoked release of ACh from chick brain slices and its potent inhibition by ω-conotoxin-GVIA. Adapted from Lundy *et al.* (1989).

Toxin affinity binding was exploited to identify and purify the receptor from *Torpedo*. The affinity of the muscle receptor for α-bungarotoxin is too high to allow reversible binding, but an acetylated toxin from the venom of the cobra *Naja naja* covalently attached to an affinity column specifically retained the receptor when solubilized membrane proteins were passed through the column. cDNA to each of the four constituent polypeptides has been cloned and sequenced.

The muscle nicotinic receptor (MnAChR) is a 295 kDa pentamer composed of four different polypeptide chains. A different gene encodes each subunit, which have molecular masses of 53–60 kDa and show a high degree of mutual homology. During development and in denervated muscle a 2α.β.γ.δ-subunit composition is expressed, which responds to ACh with a slow-activating, low-conductance channel, while in adult muscle a 2α.β.δ.ε composition gives a fast, high-conductance channel (Fig. 10.7).

Image reconstruction of electron micrographs of the reconstituted MnAChR (*see* Fig. 4.7) are consistent with five subunits arranged cylindrically around a central hydrophilic channel

Table 10.1 Characteristics of cloned muscarinic receptors.

	M_1	M_2	M_3	M_4	M_5
Molecular weight of human receptor*	51 387	51 681	66 085	53 014	60 120
Amino acids (rat)	460	466	589	478	531
Major effector systems	PI-PLC ↑	cAMP ↓	PI-PLC ↑	cAMP ↓	PI- PLC ↑
Characteristic neuronal locations	Cerebral cortex, striatum, hippocampus	Very rare	Forebrain and thalamic and brain stem nuclei	Cortex, striatum, hippocampus	Rare

* Excluding glycosylation.

7–9 nm in diameter. Each subunit possesses four membrane-spanning α-helical regions termed M1–M4, and structural models predict that each subunit contributes one transmembrane α-helix to form the lining of the pore. The receptor is a large structure, extending some 6.5 nm out of the membrane and ions must pass through a long cylindrical vestibule before reaching the 'neck' of the channel.

Most evidence is consistent with the five uncharged M2 helices from the subunits forming the 'neck', while the negatively charged residues bracketing these regions confer the cationic specificity on the channel. The cation sensitivity may be determined by the net negative charge adjacent to each end of this region; thus the closely homologous anion-selective members of the ionotropic receptor superfamily, the GABA-A

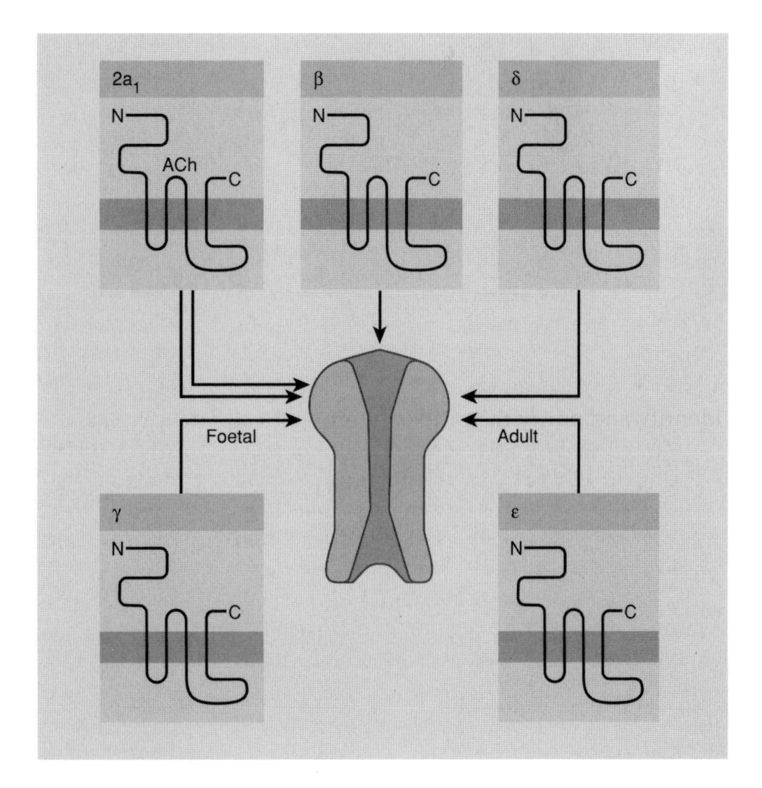

Fig. 10.7 Assembly of the muscle nicotinic ACh receptor. Five genes encode α_1-, β-, γ-, δ- and ε-subunits. The composition of the foetal and adult receptors are respectively 2α.β.γ.δ and 2α.β.δ.ε. ACh binds to α-subunits.

receptor and the glycine receptor (*see* Section 9.8.1) have positive charges bracketing their equivalent M2 helices.

Photoaffinity analogues of non-competitive antagonists which bind to the open channel are found to label the M2 helices; while the point mutation of an anionic or neutral residue adjacent to the cytoplasmic end of the M2 helix into a positively charged residue produces a decreased conductance and an inwardly rectifying channel. Conversely a positive charge introduced at the external face of the M2 helix resulted in an outwardly rectifying channel. The lower conductance of the foetal receptor (in which a γ-subunit substitutes for the ϵ-subunit of the adult) is consistent with this model since the γ-subunit has a greater density of inhibitory positive charge at the cytoplasmic end of the M2 helix. The narrowest region of the channel may be very short, suggesting that the M2 helices are not parallel, but spread apart towards the cytoplasmic face.

The M1 helix is also highly conserved between the different subunits and is found in the GABA and glycine receptors. Two ACh molecules bind to the α-subunits in order to open the channel, which can then conduct up to ten million ions per second. The non-competitive antagonist quinacrine can photoaffinity label the M1 helix of the open but not closed MnAChR-channel, suggesting that the conformational change which opens the channel exposes the M1 helices to the lumen of the channel.

The MnAChR is an *in vitro* and *in situ* substrate for a number of protein kinases. PKA phosphorylates the γ- and δ-subunits, PKC phosphorylates the δ-subunit, while a protein tyrosine kinase phosphorylates β-, γ- and δ-subunits. Exposure of the receptor to ACh for seconds rather than milliseconds results in desensitization. The PKA-phosphorylated, reconstituted receptor desensitizes much more rapidly than the non-phosphorylated channel and consistent with this cAMP elevation by forskolin also increases the rate of desensitization in intact muscle. At the neuromuscular junction the signal responsible for activating PKA may be calcitonin gene-related peptide (CGRP) which is located in vesicles within the terminals, while ACh itself, via muscarinic receptors, may autoregulate the MnAChR via the PKC-mediated phosphorylation.

MnAChRs are packed at a density of about $10^4 \cdot \mu m^{-2}$ opposite the ACh release sites. *Agrin* is a 200 kDa extracellular matrix protein which is stably associated with the synaptic basal lamina of muscle cells. Antibodies to agrin disrupt the clustering of receptors, while recombinant agrin can induce such clustering. The receptor is firmly immobilized in the membrane as shown by epr (electron paramagnetic resonance) studies of *Torpedo* electrocyte membranes. A *43 kDa protein* also appears to play an integral role in anchoring the receptor, and after removal from receptor-rich membranes by alkaline treatment the mobility of the MnAChR is increased. The 43 kDa protein may be a major component of the postsynaptic density, which is no longer visible after this extraction. The protein can be cross-linked to the β-subunit of the receptor and may serve to link the receptor to the cytoskeleton, since direct connections can be observed between the MnAChR and bundles of cytoskeletal intermediate filaments.

10.3.2 The neuronal nicotinic ACh receptor

The CNS effects of nicotine are predominantly excitatory. Cloned nucleic acid sequences have revealed a family of structurally related neuronal nicotinic ACh receptors (NnAChR) on neurones. Sequences are recognized as candidate NnAChRs on the basis of sequence homology with subunits of MnAChR. NnAChR-subunits are quite homologous with each other except that the ACh-binding α-subunit has a pair of adjacent cysteine residues involved in binding the transmitter. The presence of a similarly located cysteine pair in neuronal sequences is taken as evidence for an 'α-like' subunit. If the muscle α-subunit is called α_1, then α_2, α_3, α_4 and α_5 homologues have to date been characterized in neuronal tissue. Two additional subunits, α_7 and α_8, will be discussed separately below. Several neuronal 'non-α' subunits, i.e. with homology to the α-subunits but lacking the cysteine pair, have to date been identified in rat or chick brain. These can functionally substitute for the muscle β-subunit and are termed the β_2–β_5-subunits.

Cloned mRNAs for any of the neuronal α-subunits together with the non-α-subunit can express functional NnAChRs in *Xenopus* oocytes, while a high-affinity nicotine-binding component from brain contains α_4 and non-α

polypeptides. Although these data do not exclude the possibility of more subunits in the NnAChR complex, they do suggest that the neuronal receptor may have a simpler subunit composition than the muscle receptor, requiring just α- and non-α-subunits, although the number of copies of each polypeptide in the functional receptor is not known. An apparent molecular mass of >250 kDa for the nicotine-binding component mentioned above would suggest the presence of multiple copies of the subunits of molecular mass 80 kDa (α-subunits) and 50 kDa (non-α-subunit). *In situ* hybridization indicates that α_4 is the most widespread of the neuronal α-subunits (Fig. 10.8). RNA which hybridizes to 'non-α' probes is generally found wherever α-subunit RNA exists, suggesting that the different α-subunits may in many cases form NnAChRs with a common non-α-subunit. The reason for the heterogeneity of NnAChRs is not known. No major functional differences have been reported for the different forms, and their electrophysiology is similar to the MnAChR.

NnAChRs can be regulated by second messengers. Increased cAMP in cultured chick ciliary ganglion neurones as a result of PKA activation by vasoactive intestinal peptide (VIP) increases the peak conductance of cell-attached patchs without affecting the single channel behaviour. Since the neurones appear to possess ten times as many receptors assayable by antibody than can be detected electrophysiologically, this suggests that there may be 'silent' receptors on the surface of the cell which can be reactivated by cAMP.

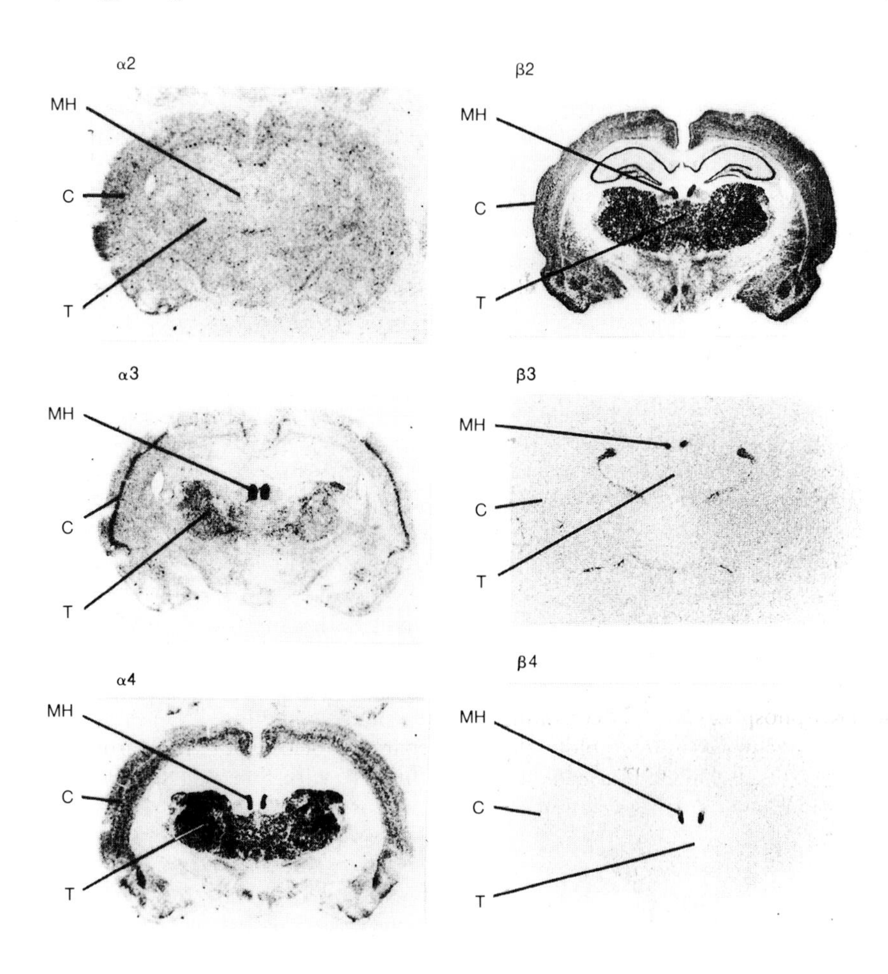

Fig. 10.8 Neuronal nicotinic receptors. Pattern of expression of α- and β-subunits of neuronal nicotinic ACh receptors (NnAChR) shown in coronal sections of rat brain at the level of the thalamus (T). C, cortex; MH, medial habenula. From Deneris *et al.* (1991).

This regulation is not found for muscle and *Torpedo* receptors, where the main effect of elevated cAMP is phosphorylation of the γ- and δ-subunits with an increase in the rate of agonist-induced desensitization, i.e. the rate at which the receptor becomes inhibited in the prolonged presence of the agonist. PKC activation, probably by substance P, enhances the rate of desensitization of the NnAChR in sympathetic neurones.

NnAChR localization in brain has been investigated by autoradiography of bound nicotine, or by using a monoclonal antibody, mAb 270, to the non-α-subunit. Surprisingly, in view of the potency with which α-bungarotoxin binds to the MnAChR, the distribution of toxin binding in brain does not coincide to that of the NnAChR. Furthermore, in contrast to nicotine, the toxin does not competitively inhibit the binding of radiolabelled ACh to neuronal membranes. However a minor component of the venom from *Bungarus multicinctus*, termed *neuronal bungarotoxin* recognizes and inhibits a distinct $\alpha_3\beta_2$ combination when expressed in oocytes. Oocyte expression can also be used to investigate the single-channel conductances of αβ-subunit combinations; for each combination investigated, two conductance states were found which probably correspond to different stoichiometric assemblies of α- and β-subunits. The problematic nature of the α-bungarotoxin binding protein has recently been resolved by the cloning of two proteins α_7 and α_8 which bind the toxin and show moderate sequence homology with the established neuronal nicotinic α-subunits. *In situ* hybridization of mRNA for α_7-subunits resembled the pattern of α-bungarotoxin binding in the brain. α_8-subunits are less widespread. Unlike the 'classical' neuronal AChRs which require coexpressed α- and β-subunits for function, α_7 homo-oligomers gave functional, fast-inactivating, currents in response to ACh. It is possible that the receptors have a significant conductance to Ca^{2+} as well as the more usual monovalent conductances.

10.4 Muscarinic receptors

Metabotropic muscarinic receptors are the dominant ACh receptors in the mammalian CNS. The classic agonist *muscarine* from the fly-agaric mushroom *Amanita muscaria* first excites and then blocks muscarinic receptors. The effects of muscarine are reversed by *atropine* and *scopolamine*. *Carbachol* is a non-selective agonist with some nicotinic action. Pharmacologically, it has been possible to divide muscarinic receptors into two subclasses, M_1 and M_2, based upon agonist specificity and effect. M_1 receptors were characterized as having a high affinity for *pirenzipine* and causing turnover of phosphoinositides, whereas M_2 receptors had a low affinity for pirenzipine and appeared to act by inhibiting adenylyl cyclase.

Subsequent genetic analysis has allowed five genes (m1–m5) to be detected, each coding for distinct muscarinic receptors (*see* Table 10.1). m1 and m2 were cloned as cDNAs by purifying enough receptor to obtain a partial amino acid sequence, synthesizing an oligonucleotide probe and using this to screen a cDNA library by DNA hybridization. cDNAs for the m3 and m4 receptors were isolated from a rat cerebral cortex library by using an oligonucleotide probe derived from a likely conserved sequence of the previously cloned M_1 receptor. This region was chosen by comparison with the corresponding region of another sequenced metabotropic receptor, the β-adrenergic receptor, belonging to the same 'superfamily', looking for homologies. No introns were found in the coding sequence and thus a cDNA was not required to define the sequences of the receptors. The nature of the receptors was confirmed by expressing RNA derived from the cDNA into *Xenopus oocytes* or by transfecting mammalian cell lines lacking muscarinic receptors with mammalian expression vectors containing the cloned cDNAs or genes. Function was detected electrophysiologically in the case of the oocytes (using the indirect assay of a Ca^{2+} activation of an endogenous Cl^- channel), or by agonist binding in the case of the mammalian transfected cells.

The amino acid sequences derived from the clones show a hydropathy profile consistent with the seven hydrophobic transmembrane segments which is a feature common to all G-protein-coupled receptors which have been sequenced (*see* Section 5.3). The muscarinic receptor subtypes have 90–98% sequence identity with each other, most of the substitutions occurring in the N-terminus or in the central portion of the large cytoplasmic loop.

The tissue distribution of the muscarinic subtypes can be determined either by Northern blot analysis of mRNA or by *in situ* hybridization to rat brain. M_2 is present at high concentration in heart but has a restricted distribution in brain, while M_1, M_3 and M_4 are abundant in brain but absent from heart. Less is known about the distribution of M_5; M_1, M_3 and M_5 all stimulate phosphoinositide hydrolysis by PI-PLC via a pertussis-toxin insensitive mechanism (indicating that G_i is not involved), while M_2 and M_4 inhibit adenylyl cyclase by a pertussis toxin-sensitive mechanism, indicating that a G_1-type G-protein mediates the coupling.

Carbachol-evoked inositol phosphate production and Ca^{2+} liberation has been studied in many neuronal and non-neuronal cells. In cerebellar granule cells (*see* Section 2.2) this response is M_3-mediated and results in the production of $(1,4,5)IP_3$, $(1,3,4,5)IP_4$ and lower phosphates. Associated with this is a biphasic increase in $[Ca^{2+}]_c$, with an initial spike caused by the release of internal stores and a second sustained elevation dependent on the presence of external Ca^{2+}. These responses are discussed in more detail in Section 5.8. There is still a very incomplete picture as to the function of the Ca^{2+} liberated in this manner. In suitable transfected mammalian cells the $(1,4,5)IP_3$ liberated Ca^{2+} leads to the activation of Ca^{2+}-dependent K^+-channels. The odd-numbered receptors also stimulate the production of arachidonic acid. Some caution is needed with the interpretation of such *in vitro* transfections, since it is possible that an overexpressed receptor can, by virtue of its high concentration, talk to G-proteins which do not normally associate with the receptor.

10.4.1 Presynaptic regulation of ACh release

At the neuromuscular junction and in the striatum, feedback inhibition of ACh release is mediated by adenosine (derived from co-released ATP) acting on presynaptic adenosine A1 receptors. However, in the cerebral cortex the predominant mechanism is via activation of inhibitory presynaptic muscarinic auto-receptors. Both mechanisms may coexist in the hippocampus, although it is not known whether this is due to heterogeneous terminal populations, or whether both mechanisms coexist on the same terminals.

Muscarinic agonists, including ACh itself, inhibit the release of ACh evoked by electrical stimulation and high KCl. There is no agreement as to which muscarinic receptor subtype(s) are involved. Information has had to be gathered by classical agonist/antagonist specificities, rather than gene classification. Generally, presynaptic autoreceptors fit into the pharmacological M_2 rather than M_1 classification, but there are exceptions; in contrast, the postsynaptic soma/dendritic muscarinic receptors are frequently M_1. As with virtually all studies on presynaptic receptors, there is no consensus on mechanism. The muscarinic autoreceptor could functionally decrease release of ACh by either blocking the transmission of the nerve impulse at some point on the varicose axon or by leaving propagation unimpaired but decreasing the coupling between depolarization and transmitter release. The major ionic hypotheses discussed are an increase in K^+ conductance, hyperpolarizing the varicosity (which could either diminish release at that varicosity alone or block further propagation) or an inhibition of Ca^+ entry which would be more likely to act via the first mechanism. Evidence for K^+-channel activation comes from the finding that the K^+-channel inhibitors 3,4-diaminopyridine and tetraethylammonium (*see* Box 4.2) not only increase the release of ACh from hippocampal slices but also abolish the inhibitory effect of M_2 agonists.

The sea hare *Aplysia*, which is discussed in more detail in Chapter 13, has simple ganglia with large cells allowing individual neurones to be identified and facilitating electrophysiology of terminal currents. A defined buccal ganglion presynaptic neurone, known to be cholinergic, possesses both nifedipine-sensitive and ω-conotoxin-GVIA-sensitive Ca^{2+}-channels; however only the latter is coupled to transmitter release. Three presynaptic receptor-mediated mechanisms modulating ACh release have been characterized; histamine shifts the voltage-dependency of the Ca^{2+}-channel to more positive values, and is hence inhibitory, while the peptide FLRFamide facilitates transmission by an opposite effect on the voltage dependency. Both mechanisms are G-protein coupled and can be inhibited by injected GDP-γ-S. An endogenous peptide *bucculin* causes a G-protein-independent inhibition of the nifedipine-insensitive Ca^{2+} current and hence ACh release. Thus multiple regulatory mechanisms can converge on a single class of channel.

Nicotine will directly induce the release of ACh from synaptosomes without the need for independent depolarization, indicating an action of presynaptic nicotinic receptors. The presence of an ionotropic receptor on a nerve terminal is somewhat unexpected, since its activation might make the release of ACh independent of an incoming action potential. The physiological role of this receptor and its interaction inhibitory muscarinic responses remains to be fully clarified.

10.5 Acetylcholinesterase

ACh is degraded by AChE to choline and acetate (Fig. 10.9). More than one molecular form of the enzyme has been described, in addition to pseudocholinesterases whose physiological functions are unclear. The AChE isoforms are generated from a single gene by alternative splicing, show similar catalytic activity, but differ in their C-termini and mode of assembly. The homologous class of AChEs consists of monomers, dimers or tetramers of 65 kDa glycoprotein subunits, which may be soluble or equipped with a C-terminal lipid anchor to locate the enzyme on the outer face of the plasma membrane. Heterologous forms of the enzyme are linked via disulphide bonds to structural subunits which provide a collagen-like tail. This form is localized to synapses and the collagen tail may further locate the enzyme at the basal lamina rather than the plasma membrane.

ACh is irreversibly inhibited by organophosphorus nerve gases such as diisopropylfluorophosphate (DFP) which phosphorylates the serine-OH of the esteratic site. Reversible inhibitors such as physostigmine form carbamylate bonds at the serine hydroxyl in the active site at which DFP binds.

10.5.1 Choline re-uptake

After hydrolysis of ACh by AChE the liberated choline can be retrieved into the terminal. A Na^+-coupled cotransporter is responsible, which

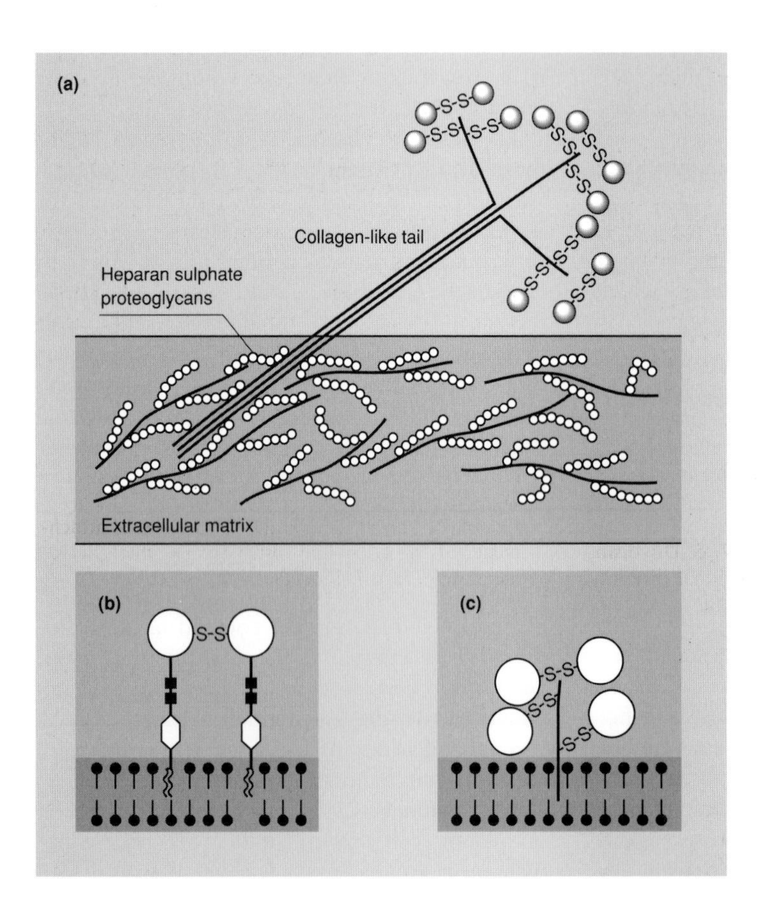

Fig. 10.9 Acetylcholinesterase (AChE). AChEs have a variety of linkages. (a) Asymmetric AChE has a long collagen-like tail which interacts with glycosaminoglycan chains of extracellular matrix proteoglycans. (b) Dimeric G_2 AChE is coupled to the plasma membrane via a phosphatidylinositol glycolipid. (c) Tetrameric G_4 AChE is anchored through disulphide bonds to a membrane-associated 20 kDa peptide. Adapted from Inestrosa & Perelman (1989).

can be inhibited by *hemicholinium-3*. This and related inhibitors have been used to identify and purify the carrier. The transporter from insect brain appears to be a single polypeptide of M_r 86 000 kDa and that from *Torpedo* a complex of 43 000 and 57 000 peptides.

Further reading

ACh release

Hucho, F. (1986) *Neurochemistry, Fundamentals and Concepts*. Weinheim: VCH.

Israel, M. & Lesbats, B. (1981) Chemiluminescent determination of AcCh and continuous detection of release from *Torpedo* synaptosomes. *Neurochem. Int.* **3**, 81–90.

Lundy, P.M., Stauderman, K.A., Goulet, J.C. & Frew, R. (1989) Effect of omega-conotoxin GVIA on Ca^{2+} influx and endogenous acetylcholine release from chicken brain preparations. *Neurochem. Int.* **14**, 49–54.

Trudeau, L.-E., Baux, G., Fossier, P. & Tauc, L. (1993) Transmitter release and calcium currents at an *Aplysia* buccal ganglion synapse. Characterization. *Neurosci.* **53**, 571–580.

Whittaker, V.P. (1984) The structure and function of cholinergic vesicles. *Biochem. Soc. Trans.* **12**, 561–576.

Whittaker, V.P. & Roed, I.S. (1981) New insights into vesicle recycling in a model cholinergic system. *Methods Neurosci.* **1**, 151–173.

Zimmerman, H., Volknandt, W., Henkel, A., Bonzelius, F., Janetzko, A. & Kanaseki, T. (1990) The synaptic vesicle membrane: origin, axonal distribution, protein components, exocytosis and recycling. In *Neurotransmitter Release*, F. Clementi & J. Meldolesi, eds. London: Academic Press, pp. 13–26.

Nicotinic ACh receptors

Changeux, J.P. (1990) The TIPS lecture — the nicotinic acetylcholine receptor — an allosteric protein prototype of ligand-gated ion channels. *Trends Pharmacol. Sci.* **11**, 485–485.

Deneris, E.S., Connolly, J., Rogers, S.W. & Duvoisin, R. (1991) Pharmacological and functional diversity of neuronal nicotinic acetylcholine receptors. *Trends Pharmacol. Sci.* **12**, 34–40.

Ferns, M.J. & Hall, Z.W. (1992) How many agrins does it take to make a synapse? *Cell* **70**, 1–3.

Mitra, A.K., McCarthy, M.P. & Stroud, R.M. (1989) Three-dimensional structure of the nicotinic acetylcholine receptor, and location of the major associated 43-kD cytoskeletal protein, determined at 22 A by low dose electron microscopy and X-ray differaction to 12.5. *J. Cell Biol.* **109**, 755–774.

Nastuk, M.A. & Fallon, J.R. (1993) Agrin and the molecular choreography of synapse formation. *Trends Neurosci.* **16**, 72–76.

Steinbach, J.H. & Ifune, C. (1989) How many kinds of nicotinic acetylcholine receptor are there? *Trends. Neurosci.* **12**, 3–6.

Stroud, R.M., McCarthy, M.P. & Shuster, M. (1990) Nicotinic acetylcholine receptor superfamily of ligand-gated ion channels. *Biochemistry* **29**, 11009–11023.

Muscarinic receptors

Allgaier, C., Hertting, G. & Gottstein, P. (1992) Muscarinic receptor-mediated regulation of electrically evoked ACh release in hippocampus: effects of K-channel blockers. *Pharm. Pharmacol. Lett.* **2**, 191–194.

Bonner, T.I. (1989) The molecular basis of muscarinic receptor diversity (Review). *Trends Neurosci.* **12**, 148–151.

Bonner, T.I. (1992) Domains of muscarinic acetylcholine receptors that confer specificity of G protein coupling. *Trends Pharmacol. Sci.* **13**, 48–50.

Hulme, E.C., Birdsall, N.J.M. & Buckley, N.J. (1990) Muscarinic receptor subtypes (Review). *Annu. Rev. Pharmac. Toxicol.* **30**, 633–673.

Lechleiter, J., Peralta, E. & Clapham, D. (1989) Diverse functions of muscarinic acetylcholine receptor subtypes. *Trends. Pharmacol. Sci. Supp*, 34–38.

Acetylcholinesterase and choline re-uptake

Inestrosa, N.C. & Perelman, A. (1989) Distribution and anchoring of molecular forms of acetylcholinesterase. *Trends Pharmacol. Sci.* **10**, 325–329.

Silman, I. & Futerman, A.H. (1987) Modes of attachment of acetylcholine esterase to the surface membrane. *Eur. J. Biochem.* **170**, 11–22.

11

Monoamine and purine neurotransmitters

11.1 Introduction

The catecholamines which function as neurotransmitters in the mammalian CNS are dopamine (DA), noradrenaline (NA) and adrenaline (Adr), each derived from tyrosine (Fig. 11.1). The monoamine transmitters additionally include the indoleamine 5-hydroxytryptamine (5-HT, serotonin), synthesized from tryptophan, and histamine synthesized from histidine (Fig. 11.2). Adenosine triphosphate (ATP) and its dephosphorylated form, adenosine, are purinergic transmitters.

Catecholamine and serotonin-secreting neurones represent a minute fraction of the total neurones in brain; perhaps 50 000 in the rat and two million (out of more than 10 000 million) in man. Dopamine predominates in rat brain, with perhaps 40 000 DA neurones, while there may be only 10 000 NA neurones and very few adrenaline-secreting neurones (identified histochemically by the presence of the final enzyme in the catecholamine sequence, phenylethanolamine N-methyltransferase, PNMT). Despite the small number of cell bodies, the remarkable branching of the axons of these cells allows the catecholamines to exert a major influence in brain. Thus a dopaminergic neurone with its cell body in the rat *substantia nigra* may have 500 000 release site (boutons or varicosities) along its highly branched axon in the neostriatum, and thus be able to act as a local neuromodulator, synchronously regulating a vast number of neurones. The numbers of 5-HT cell bodies in the CNS are considerably fewer than for the catecholamines; despite this, every area of the brain receives a 5-HT input.

Monoaminergic axons carry impulses relatively slowly ($<1\ m \cdot s^{-1}$) and their postjunctional effects are generally slow acting via metabotropic rather than ionotropic receptors.

11.1.1 Monoaminergic pathways

The cell bodies of neurones secreting catecholamines and serotonin are largely located in the brain stem from which their axons project forwards (Fig. 11.3). The *locus coeruleus* is the principle source of NA innervation to most brain areas. In the rat there are three types of monoamine input into the cortex: DA fibres originate largely from the *substantia nigra*, NA fibres from the *locus coeruleus* while 5-HT cell bodies are found throughout the *raphe nuclei*. The *pineal*

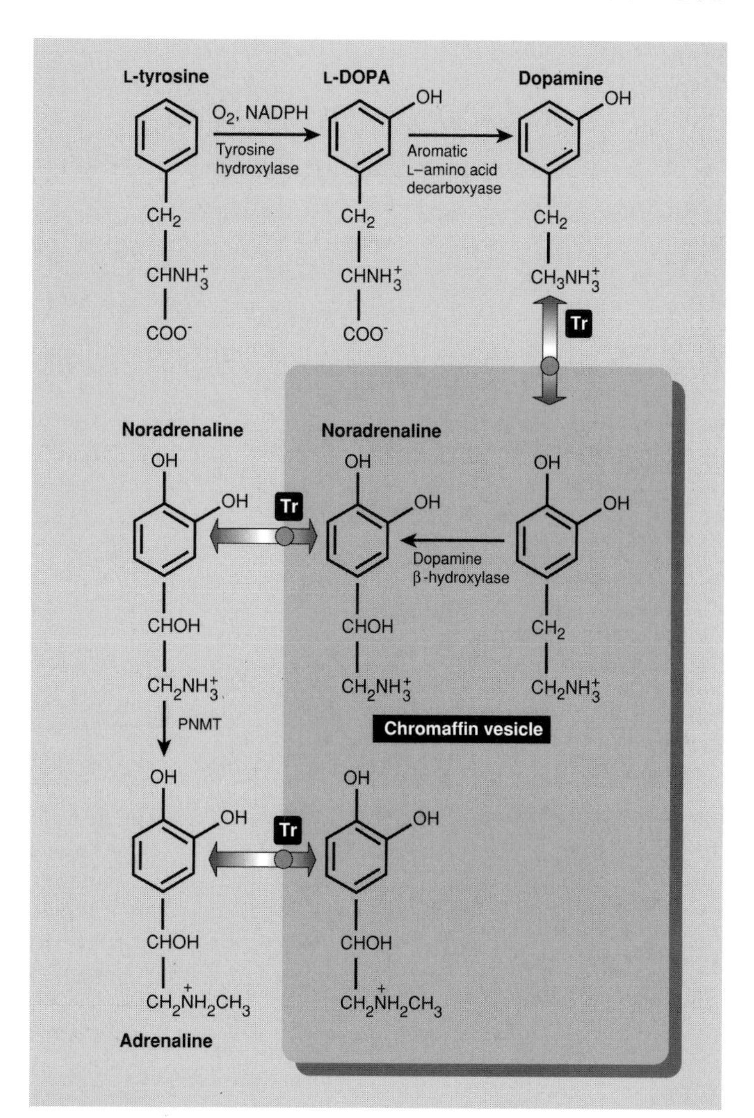

Fig. 11.1 Catecholamine synthesis. The catecholamines dopamine, noradrenaline and adrenaline are derived by a common pathway from tyrosine. PNMT, phenylethanolamine *N*-methyltransferase; Tr, common H^+-linked, reserpine-sensitive catecholamine transporter.

gland has a high concentration of 5-HT, and has been used to elucidate the 5-HT pathway of metabolism.

Catecholamine and 5-HT pathways were originally detected by the Falck and Hillarp histofluorescence technique and subsequently by the glyoxylic acid fluorescence method. In the original studies, few direct synaptic contacts were seen associated with monoamine-containing varicosities, and this lead to the view that monoamines predominantly act as *neuromodulators* sometimes at some distance from their site of release, performing a role analogous to that of hormones. However, there is debate as to the proportion of varicosities (or boutons) which make synaptic contacts. Original studies used electron microscopic autoradiography to determine the proportion of those varicosities capable of accumulating labelled 5-HT or NA which formed conventional synapses onto other neurones. Only 5–20% of varicosities were reported to be associated with synapses. Recently however, serial sectioning of catecholamine and 5-HT varicosities identified by immunocytochemistry has shown that some 90% of vesicle containing varicosities form conventional synapses with postsynaptic targets. This suggests that release at conventional synapses is likely to be of major significance.

Each monoamine appears to innervate specific

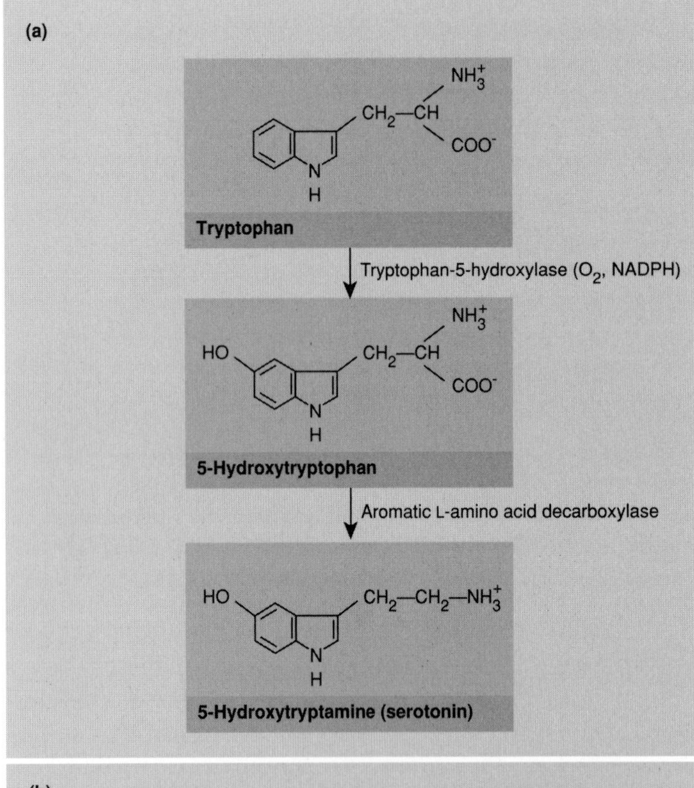

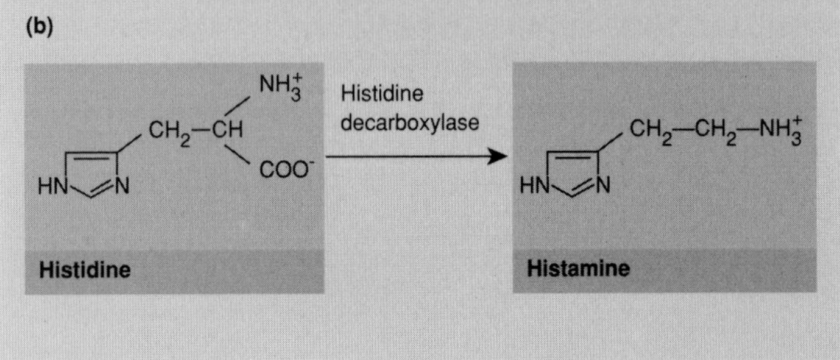

Fig. 11.2 5-Hydroxytryptamine and histamine synthesis. Serotonin (a) and histamine (b) are derived from tryptophan and histidine respectively. Note the parallel to corresponding reactions in catecholamine biosynthesis shown in Fig. 11.1.

layers of the cerebral cortex; thus 5-HT and NA are largely complementary in their distribution in the visual cortex, while both show a high degree of spatial organization within given layers.

11.1.2 Transmitter synthesis

The enzymes of the complete sequence from tyrosine to adrenaline (*see* Fig. 11.1) are coded for by a single gene or a closely coordinated set of genes. The first enzyme on the pathway for catecholamine synthesis from tyrosine is *tyrosine hydroxylase* (TH). Alternative splicing of a single primary transcript can lead to mRNAs encoding four isoenzymes. The enzyme is rate limiting, consistent with the very low concentrations of its product L-3,4 dihydroxyphenylalanine (L-DOPA). The adrenal enzyme has a molecular weight of 280 kDa, and is composed of four identical subunits. The enzyme uses molecular oxygen, the enzyme being oxidized to an inactive form. Regeneration of the reduced TH-H_2 occurs by transfer of electrons from tetrahydrobiopterin, which is in turn re-reduced by NADH.

TH is activated during nerve terminal stimulation; this may be a consequence of an elevated $[Ca^{2+}]_c$ activating CaMKII and phosphorylating the enzyme at Ser^{19}. PKA, predominantly phos-

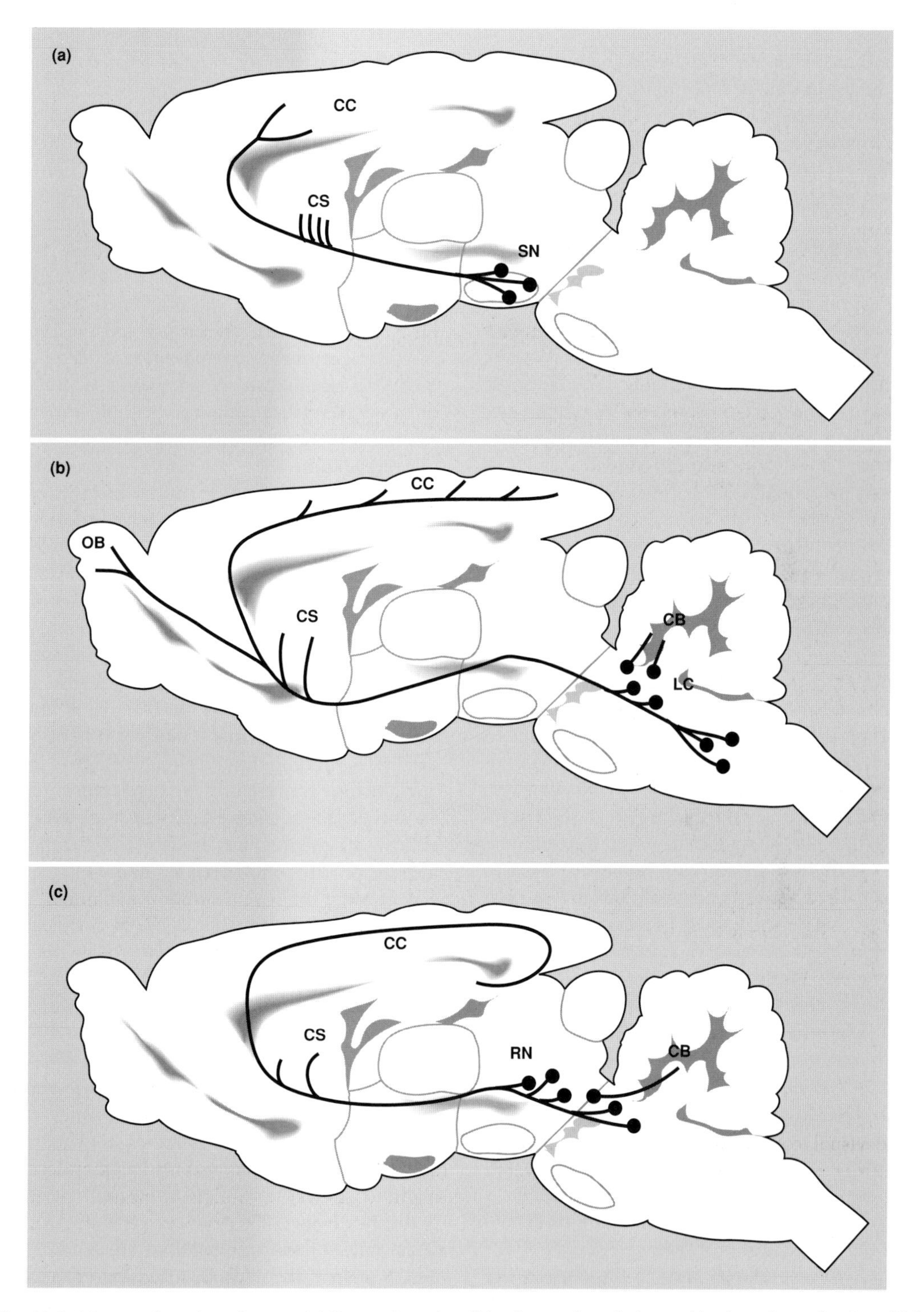

Fig. 11.3 Monoaminergic pathways. (a) Dopaminergic cell bodies are largely located in the substantia nigra (SN) and project forwards to the corpus striatum (CS) and cerebral cortex (CC). (b) Most noradrenergic cell bodies are in the locus coeruleus (LC) and project to many brain areas, including the cerebral cortex, cerebellum (CB), corpus striatum and olfactory bulb (OB). (c) Serotonergic cell bodies in the Raphe nuclei (RN) predominantly innervative the cerebral cortex, corpus striatum and cerebellum. 5-HT concentrations are highest in the amygdala, hypothalamus and septal area. 5-HT concentrations in the cortex and cerebellum are ten times lower.

phorylates TH at Ser^{40}, activating the enzyme by decreasing feedback inhibition from synthesized catecholamines.

The next enzyme, *dopamine decarboxylase (DDC)* is the terminal synthetic enzyme in dopaminergic neurones. It is more correctly termed L-aromatic amino acid decarboxylase, since it has broad specificity, also decarboxylating the precurser of 5-HT. The 'suicide' substrate monofluoromethyldopa covalently labels a 50 kDa monomer; the native enzyme may be a dimer.

The third enzyme, *dopamine β-hydroxylase* (DBH) contains 2 Cu atoms per catalytic subunit which undergo valency changes; ascorbate reduces cupric to cuprous copper, which is re-oxidized when NA is formed from DA; molecular oxygen is also a substrate. The enzyme from rat consists of four 88 kDa subunits and is associated with noradrenergic synaptic vesicles, although the extent to which the enzyme is a soluble constituent of the vesicle lumen, in which case it would be co-released with the transmitter, is unclear. Thus apart from the signal sequence there is no obvious membrane spanning region, and the nature of its binding to the membrane (if any) is unknown. The loss of DBH during exocytosis would limit the ability of noradrenergic SSVs to undergo local recycling, since there is no means of replenishing the enzyme at the terminal, whereas this would not be a hindrance to large dense-core vesicle exocytosis which do not undergo local recycling (*see* Section 8.1).

Phenylethanolamine-N-methyltransferase is present only in the very few adrenaline-secreting neurones, and also in the adrenal chromaffin cells. The enzyme is soluble in the cytoplasm, which means that the synthesis of adrenaline from tyrosine involves three changes of compartment; DA is transported into vesicles, NA is transported out to the cytoplasm and finally adrenaline is reaccumulated into the storage vesicles (*see* Fig. 11.1). The enzyme transfers a methyl group from *S*-adenosylmethionine (SAM) onto the amine group of NA.

The sequence of reactions in the pathway of *serotonin* synthesis is closely analogous to that for the catecholamines, the starting amino acid being tryptophan instead of tyrosine (*see* Fig. 11.2). *Tryptophan hydroxylase* produces 5-hydroxytryptophan (5-HTP) by a reaction analogous to tyrosine hydroxylase.

A pteridin cofactor is again involved. This is the only enzyme unique to the 5-HT pathway. The enzyme is less well characterized than TH, but appears to be a tetramer of 59 kDa subunits. Tryptophan hydroxylase can be regulated by CaMKII.

5-HTP is decarboxylated to 5-HT by the same *L-aromatic amino acid decarboxylase* as for L-DOPA. In the pineal gland, and also in the retina and some hypothalamic and midbrain regions, 5-HT is further metabolized to *N*-acetylserotonin and *melatonin* (*N*-acetyl-5-*O*-methylserotonin). The *N*-acetylating enzyme has quite a wide distribution, for example it is found within both granule cells and Purkinje cells of the cerebellum.

Tryptophan is also the source of *kynurenine*, by tryptophan pyrrolase (also called indoleamine 2,3-dioxygenase), which catalyses oxidative cleavage of the amino acid. Kynurenine is on the pathway to nicotinic acid, and a further intermediate is *quinolinate*. Both quinolinate and kynurenate interact with glutamatergic transmission (*see* Section 9.6); quinolinate can activate NMDA receptors and is excitotoxic in excess, while kynurenate (or particularly synthetic haloderivatives) inhibit the glycine site of the receptor.

Histamine is present at high concentrations in *mast* cells (*see* Section 8.2.2) which are found in almost all mammalian tissues and which are mediators of inflammation and allergy. In addition the brain also contains histamine, with the highest concentration in the hypothalamus, although care must be taken to distinguish neuronal from mast cell histamine. The cell bodies of histaminergic neurones (which may be identified by *in situ* hybridization against mRNA for histidine decarboxylase) are confined to specific regions within the posterior ventral hypothalamus but send out a diffuse innervation to the whole forebrain.

Histamine is formed by the decarboxylation of histidine (*see* Fig. 11.2). *Histidine decarboxylase* is distinct from 5-HTP- and DOPA-decarboxylases and is thus only synthesized in neurones which possess the dedicated decarboxylase. The 'suicide' inhibitor α-fluoromethylhistidine can inhibit the enzyme specifically and deplete neuronal histamine stores allowing the behavioural aspects of the transmitter to be determined. Histamine depletion most notably decreases wakefulness. No re-uptake transporter has been reported to date for histidine, which may indicate that the transmitter is degraded after release from the receptor.

ATP is present in the lumen of catecholamine, serotonin and ACh-containing small and large-dense core synaptic vesicles in both the CNS and the sympathetic nervous system. ATP can either act at purinergic (P) receptors, or be hydrolysed to adenosine. *Adenosine* is present in the CSF at about 0.1 μM, and this concentration is increased dramatically during periods of hypoxia and ischaemia. There are two main sources of extracellular adenosine (Fig. 11.4), firstly from ATP co-stored and co-released with ACh and catecholamines and subsequently degraded by ecto-nucleotidases, and secondly from cytoplasmic adenosine released under low energy conditions by thermodynamic reversal of a plasma-membrane Na^+ coupled adenosine transporter. The latter may be quantitatively the more important.

11.2 Vesicular storage of catecholamines

Neuronal catecholamines and 5-HT are stored within both small (50 nm) and large (>75 nm) synaptic vesicles. Both classes of vesicle show electron-dense contents but for different reasons. Small vesicles contain catecholamine and ATP but no neuropeptide, the electron density is due to the catecholamine contents, thus depletion of catecholamine by the vesicular catecholamine transport inhibitor *reserpine in vivo* results in the vesicles becoming electron lucid. In contrast the neuropeptides which are exclusively located within the large vesicles give an electron density which is retained even after reserpine depletion of the catecholamines in these large dense-core vesicles (LDCVs). It is unlikely that the extremely small amounts of neuropeptides can be responsible for the retained electron density. It is more probable that it is due to the presence of chromogranins

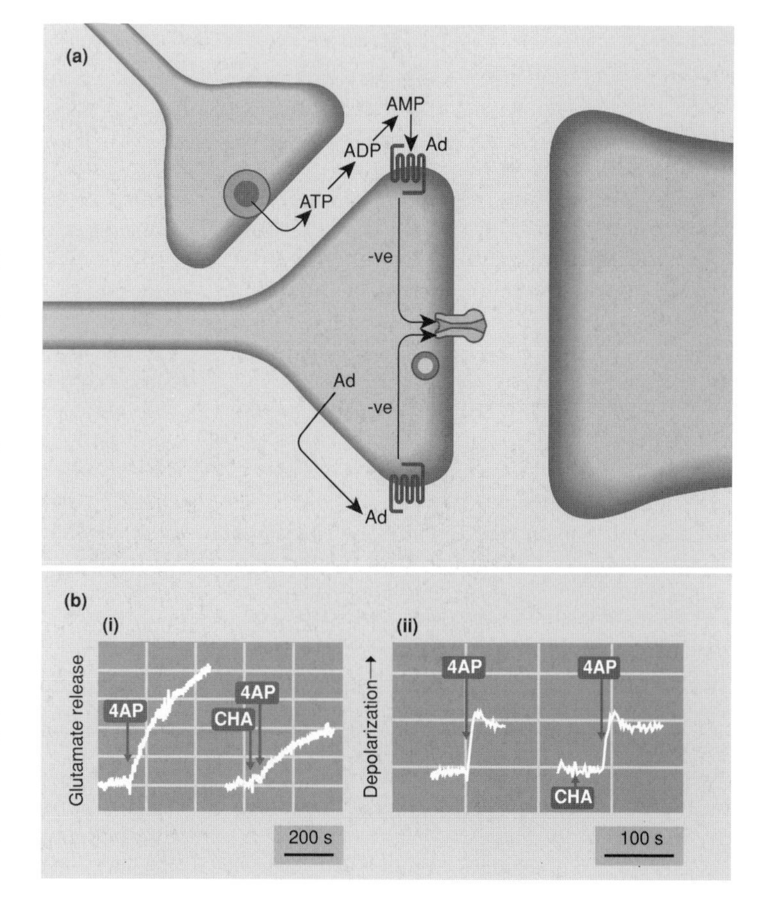

Fig. 11.4 Adenosine as a presynaptic modulator. (a) ATP co-stored with Type II neurotransmitters, can be released exocytotically and act either directly on P_2 purinergic receptors or, as shown here, be hydrolysed to adenosine (Ad) which can act on presynaptic A_1 receptors. Adenosine can also be released directly from the cytoplasm under low energy conditions when cytoplasmic adenosine concentrations increase. (b) The adenosine A_1 agonist cyclohexyladenosine (CHA) inhibits the exocytosis of glutamate from cerebral cortical synaptosomes evoked by 4-aminopyridine (4AP) (i); no effect on depolarization (measured with a cyanine dye) is seen (ii). The decreased release is due to inhibition of the Ca^{2+}-channel rather than activation of K^+-channels. Data from Barrie & Nicholls (1993).

which are the major soluble proteins within LDCVs, but which are absent from the small vesicles of sympathetic nerves.

11.2.1 Large dense-core synaptic vesicles

A pure preparation of peripheral LDCVs can be made from bovine splenic nerve. The proportion of LDCVs in noradrenergic varicosities increases with the size of the animal, presumably because the greater distances over which LDCVs must be transported from the perikaryon demands a greater standing population of the vesicles; in man almost half of all noradrenergic terminal vesicles, or 80–90% of stored NA, are accounted for by LDCVs. These closely resemble the more easily studied chromaffin vesicle (*see* Sections 11.2.3 and 11.2.4) in function and biochemical composition (Fig. 11.5). In particular, both preparations contain chromogranin A and chromogranin B, secretogranin II, enkephalins and neuropeptide Y. These proteins appear to be absent from small adrenergic vesicles. A variety of roles have been proposed for chromogranins and the related secretogranins, including an action as precursors for biologically active peptides such a pancreastatin which inhibits stimulated insulin secretion. Additionally, chromogranins, which carry a high number of negative charges, may assist the storage of positively charged catecholamines or Ca^{2+} within the vesicles. In the trans-Golgi network chromogranin aggregation evoked by low pH and increased Ca^{2+} may be a stage in the biogenesis of LDCVs.

11.2.2 Small catecholaminergic synaptic vesicles

Relatively little is known about these vesicles in contrast to cholinergic SSVs, since a convenient preparation comparable to that from *Torpedo* is lacking and only partially purified preparations are available. Since they can be depleted by reserpine in a similar manner to LDCVs one can assume that their transport and metabolic properties are similar. The most fundamental difference is the lack of the soluble chromogranins and neuropeptides characteristic of LDCVs, and also the specific localization and synaptophysin (*see* Section 7.4.4) on small but not large dense-core vesicles.

A specific 5-HT-binding protein (SBP) has been described which may play a role in vesicular storage. It is co-released with 5-HT. In synaptosomes, SBP has a molecular weight of 45 kDa, and is derived from a 56 kDa precursor. In the presence of Fe^{2+} and K^+ SBP binds 5-HT with nanomolar affinity, however in the presence of extracellular concentrations of Na^+ and Ca^{2+} it greatly reduces its affinity, allowing the 5-HT to dissociate immediately after exocytosis. It is not clear whether the chromogranins found in catecholaminergic vesicles play a similar role.

The origin of small and large adrenergic vesicles is debated (*see* Section 8.1). LDCVs (which contains peptides and proteins) must be synthesized in the cell body Golgi apparatus, while the small vesicles may be separately synthesized by the constitutive pathway feeding early endosomes within varicosities. One problem is that NA synthesis requires vesicular dopamine β-hydroxylase; if the soluble contents of small vesicles, including this enzyme, are co-secreted from small vesicles, it is not immediately apparent how they can replenish their catecholamine for a second cycle of local exocytosis without returning to the cell body.

11.2.3 Chromaffin vesicles

Chromaffin cells from the adrenal medulla have made two contributions to the understanding of transmitter release. Intact or permeabilized cells provide insight into LDCV exocytotic mechanisms, as discussed in Section 8.2.1, while the isolated vesicles provide the most detailed model for the transport and metabolic properties of catecholamine-containing LDCVs.

It is easier to investigate the biochemistry of chromaffin cell storage vesicles ('chromaffin vesicles') than the aminergic LDCVs which they closely resemble since they may readily be obtained in bulk in a pure form not contaminated with other transmitters. Chromaffin vesicles (for historical reasons usually called chromaffin granules) appear to share the same transport mechanisms as adrenergic synaptic vesicles.

Chromaffin vesicles (Fig. 11.5) have a diameter of 280 nm (much larger than even neuronal LDCVs) and maintain an internal pH of about 5.7. Their lumen contains catecholamine, ATP (approximately one molecule per four catecholamines), Ca^{2+}, Mg^{2+}, neuropeptide and chromogranins. All these components, together

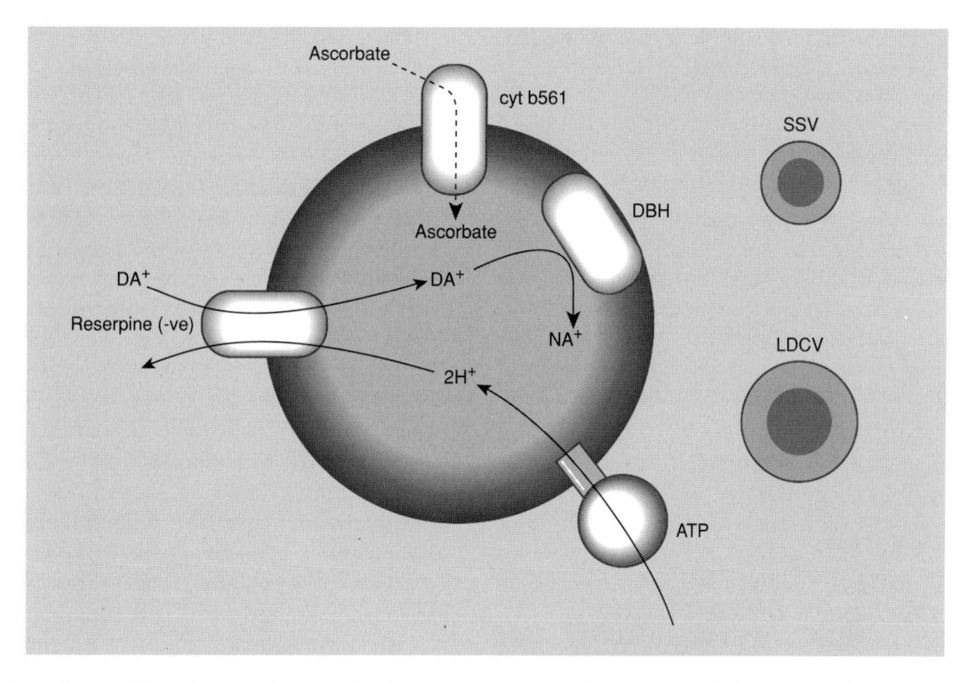

Fig. 11.5 Chromaffin vesicles and catecholaminergic synaptic vesicles. The catecholamine (DA^+) is accumulated in exchange for $2H^+$ by a reserpine-sensitive carrier. Dopamine β-hydroxylase (DBH) attached to the lumenal face of the vesicle membrane hydroxylates DA to NA; this requires ascorbate which is reduced by electrons entering the vesicle through the transmembrane cytochrome b561. The 260 nm diameter vesicle is shown to scale beside a small synaptic vesicle (SSV) and a large dense-core vesicle (LDCV).

with DBH are co-released during vesicle exocytosis.

11.2.4 Vesicular catecholamine transport

The chromaffin vesicle was the first non-bioenergetic organelle for which a specific H^+-translocating ATPase was found. Chromaffin vesicles contain a V-type ATP-driven proton pump (*see* Section 4.3) generating a proton-motive force (pmf), comprising a ΔpH (acid in the lumen of the vesicle) and a positive-inside membrane potential. This gradient then drives the uptake of catecholamine through a transporter (Fig. 11.5). The uptake mechanisms have provided a model for subsequent investigations of amino acid and ACh accumulating synaptic vesicles.

To date, two H^+-dependent monoamine transporter clones have been identified encoding cDNAs for the chromaffin vesicle and brain vesicular monoamine transporters. The predicted structure has 12 transmembrane segments, cytoplasmic N- and C-termini and a large lumenal 1–2 glycosylated loop (*see* Fig. 4.2).

Isolated chromaffin vesicles accumulate adrenaline in an ATP-dependent manner. The total internal concentration of some 0.5 M is much higher than in the cytoplasm (where it is 50–500 μM), although much of the vesicular catecholamine is complexed to ATP and chromogranins. Uptake can be inhibited by reserpine or by collapsing the pmf with a protonophore. ATP-dependent uptake of protons and acidification of the interior can be detected by following the fluorescence quenching of 9-aminoacridine (*see* Section 3.4.1). Rather than using intact granules, which already contain high concentrations of catecholamines, a simpler preparation is the chromaffin granule 'ghost', which is derived from the intact granule by osmotic shock, losing its soluble contents before re-sealing. The transport properties are retained in this preparation.

The ΔpH component of the pmf is essential for the uptake of adrenaline (or 5-HT) since the

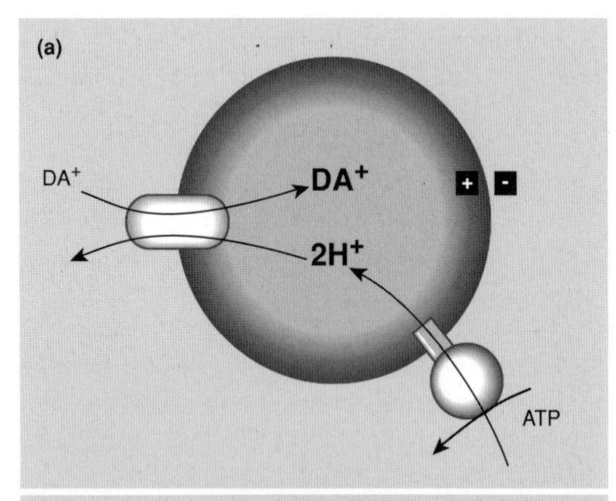

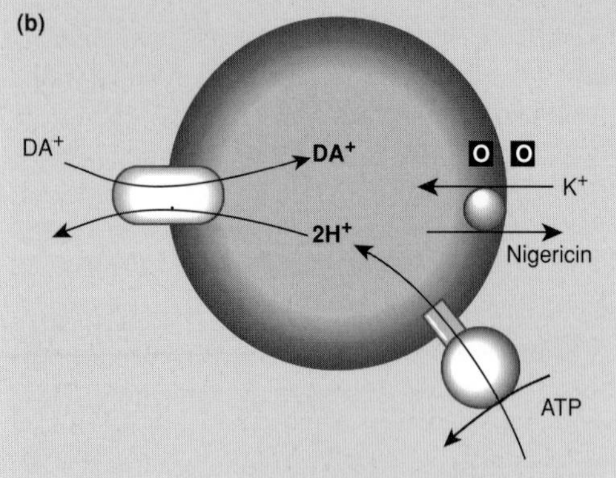

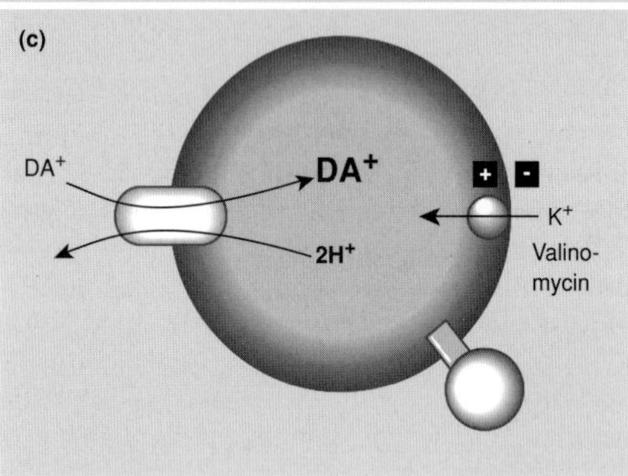

Fig. 11.6 The stoichiometry of the vesicular catecholamine transporter. (a) Catecholamine uptake driven by membrane potential and pH gradient; (b) the K^+/H^+ antiport ionophore nigericin decreases uptake by abolishing ΔpH when vesicles are suspended in a high K^+ medium; (c) a membrane potential generated by valinomycin and an inward K^+ gradient can drive uptake in the absence of ATP hydrolysis.

reserpine-sensitive carrier appears to have low specificity (Fig. 11.6). Thus addition of nigericin (the K^+/H^+ antiport ionophore which equalizes K^+ and H^+ concentration gradients across membranes) to membrane vesicles suspended in K^+ medium to discharge any pH gradient inhibits the ATP-driven uptake of adrenaline. Under these conditions the vesicles display a high membrane potential but low pH gradient. A direct role of ΔpH can be shown in the absence of ATP by generating an artificial pH gradient which can drive adrenaline uptake; however transport is also electrogenic since a membrane potential, positive inside, generated by suspending K^+-free vesicles in KCl and adding valinomycin, also drives uptake. Taking these data together the probable mode of uptake is of the protonated catecholamine in exchange for two protons (*see* Fig. 11.5). This should be contrasted with the electrophoretic uptake of the glutamate anion deduced from applying similar techniques to glutamatergic vesicles (*see* Fig. 9.7).

The thermodynamic consequences of this stoichiometry can now be discussed. Weak bases which permeate across bilayer membranes in the neutral, unprotonated form accumulate within acidic compartments in response to the pH gradient:

$$\frac{[RNH_3^+]_{lumen}}{[RNH_3^+]_{cyto}} = \frac{[H^+]_{lumen}}{[H^+]_{cyto}} \tag{11.1}$$

At first it was thought that such a mechanism could account for catecholamine accumulation. However, since uptake depends on both the pH gradient and the transvesicular membrane potential, $\Delta\psi$, this indicates that a net transfer of charge occurs during the operation of the transporter. Furthermore, the magnitude of the amine gradient achieved in vesicle ghosts is too high to be accounted for by equation 11.1. A stoichiometry allowing for the inward transport of the cationic form of the catecholamine (the predominant form at neutral pH) in exchange for two protons gives the following thermodynamics (although it should be emphasized that it is not possible on thermodynamic grounds to distinguish this from exchange of neutral catecholamine for one proton):

$$\frac{[RNH_3^+]_{lumen}}{[RNH_3^+]_{cyto}} = \left(\frac{[H^+]_{lumen}}{[H^+]_{cyto}}\right)^2 \cdot 10^{\frac{\Delta\psi}{60}} \tag{11.2}$$

Given a pH gradient of 1.5 units (inside acid) and a membrane potential of 50 mV (inside positive) this can maintain an equilibrium concentration gradient of free catecholamine approaching 10 000. To maintain this high gradient across the vesicle membrane it is essential that the catecholamine, in either its neutral or protonated form, is not able to leak out across the membrane. The importance of hydroxylation in the biosynthesis of catecholamines and 5-HT may lie in increasing the hydrophilicity and decreasing the non-specific permeability of these compounds. *Tyramine*, which differs from dopamine in lacking one hydroxyl group, can be accumulated into vesicles on the reserpine-sensitive carrier, but is sufficiently hydrophobic to leak out continuously, leading to a decreased accumulation and a futile cycling across the vesicle membrane. Tyramine can accumulate in patients treated with monoamine inhibitors (*see* Section 11.5.1), particularly if tyramine-rich food such as blue cheese is ingested. Under these conditions there is a massive catecholamine release leading to a hypertensive crisis. Tyramine probably competes with catecholamine for uptake into vesicles, but because it never accumulates to the same extent as the catecholamine it would not equally inhibit efflux of catecholamine on the carrier. Vesicles would thus deplete themselves of catecholamine.

Reserpine inhibits the transport of adrenaline, NA, DA, 5-HT and histamine into synaptic and chromaffin vesicles. This broad spectrum of activity suggests that the respective vesicles possess identical or closely related monoamine transporters which facilitate accumulation of all the monoamines. However reserpine also appears to promote the direct exocytosis of NA into the synaptic cleft, rather than leakage into the terminal cytoplasm, since the latter pathway would allow monoamine oxidase (*see* Section 11.5.1) to produce oxidation products and this is not observed.

ATP-dependent catecholamine transport into chromaffin vesicles is inhibited by reserpine with a K_i of about 0.1 nM. Reserpine inhibits transport, but not the ATPase or the generation of the pmf. By blocking the catecholamine carrier it can distinguish between carrier-mediated and unspecific permeation and can also distinguish vesicular from plasma membrane transport. Although reserpine is very hydrophobic and tends to bind non-specifically, specific binding of [^{3}H]-reserpine can be demonstrated. Binding is dependent on the pmf across the vesicle membrane: thus it is ATP-dependent and inhibited by protonophores such as FCCP. The carrier may exist in two conformations with different affinities for reserpine. Tetrabenazine also inhibits the carrier, but binds to only one of the conformations.

The mechanism of ATP transport across the membrane is incompletely resolved. It has been suggested that vesicles have an atractyloside-sensitive ATP transporter similar to the ATP/ADP exchanger of the mitochondrial inner membrane but able also to operate in a uniport mode. A similar transport mechanism may exist on *Torpedo* vesicles (*see* Section 10.2). Mitochondrial contamination can be ruled out since the protein binds antibody against a vesicle-specific protein.

11.2.5 Electron transfer into chromaffin vesicles

DBH requires electrons for the hydroxylation of dopamine (*see* Fig. 11.5), and ascorbate is the most effective *in vitro* electron donor. Although the vesicles contain some 20 mM ascorbate, there is no import, and thus some mechanism must exist for its regeneration. *Cytochrome b561* is an integral component of the chromaffin vesicle membrane, with six membrane-spanning domains which transfers electrons from external ascorbate to ascorbate within the vesicle. Cytochromes are one electron donors and ascorbate in both the cytoplasm and the vesicles appears to cycle between the fully reduced ascorbate and the free radical semi-dehydroascorbate. Finally, cytoplasmic ascorbate is regenerated by a mitochondrial NADH-coupled reductase. Cytochrome b561 may also play a role in the processing of neuropeptide precursors in LDCVs, by supplying electrons to a mono-oxygenase responsible for amidating glycine residues.

11.3 Exocytosis of biogenic amines and ATP

The protein chemistry of LDCV exocytosis is discussed in Chapter 8, here we are concerned with the regulation of that release in intact terminals and neurones.

Catecholamine release has been extensively investigated in the peripheral nervous system.

Sympathetic stem cells can express up to a dozen peptide neurotransmitters or classical transmitter metabolic pathways; in tissue culture the cocktail which is expressed can depend on culture conditions, secretory activity and external signals. Thus single cultured neurones can even be persuaded to switch from being mainly cholinergic to mainly adrenergic. In mature brain it is the rule rather than the exception that individual sympathetic neurones synthesize, store and release not only NA but also ATP (about 2% of NA on a molar basis) and a neuropeptide such as neuropeptide Y or met-enkephalin (*see* Section 12.3).

Individual sympathetic varicosities contain from 50 to 1000 SSVs and between 5 and 30% of this number of LDCVs. Since only LDCVs contain lumenal proteins synthesized in the cell body, such as dopamine β-hydroxylase, only LDCVs can synthesize NA. Small adrenergic synaptic vesicles must obtain their NA by uptake from the cytoplasm of NA leaking from LDCVs or retrieved into the terminal.

A slow rate of electrical stimulation, particularly under the influence of α_2 autoinhibition (*see* Section 11.4.2) causes NA and ATP release from SSVs. With high frequency stimulation, or if α_2 autoreceptors are inhibited, the contents of LDCVs are preferentially released; thus the same neurone has changed from being 'noradrenergic' to partly peptidergic (a mechanism for this frequency-dependent switch is discussed in Section 7.3). There is some evidence that SSVs are released from 'preferred sites' while LDCVs are released randomly; however the distinction between directed and ectopic release is less clear-cut than for non-sympathetic central neurones, since sympathetic boutons (varicosities) lack well defined prejunctional grids, or active zones.

It is not known if LDCVs release their entire content at each exocytotic event. Since their proteins and polypeptides are synthesized in the cell body they may be too 'expensive' to be lost by all-or-none exocytosis. As the postsynaptic effects of the catecholamines are overwhelmingly metabotropic, it is not possible to show quantal release electrophysiologically with the same ease as for ACh. However at those peripheral synapses where the co-released ATP has an ionotropic effect the results are consistent with quantal release of the nucleotide.

Even though a sympathetic nerve might possess 20 000 varicosities and each varicosity might contain 1000 SSVs, the amount of detectable NA released by a single nerve impulse might be only 10^{-5} of the total, corresponding to one or two quanta from the entire terminal arborization. Electrophysiological analysis of the depolarizations evoked by the co-released ATP suggests that the average sympathetic nerve varicosity only responds (and then with the release of the contents of a single SSV) to about 1% of low frequency action potentials passing through it. In the presence of the K^+-channel inhibitors 4-aminopyridine (4AP) and TEA to enhance exocytosis, the maximal release which could be evoked corresponded to about four quanta per varicosity per minute. This is vastly slower than for cholinergic or amino acid transmitter release, and may reflect the existence of a single 'preferred release site' at which vesicles can only be released in single file (monoquantal release) with a 15 s latency for reoccupation.

Noradrenaline release at sympathetic and CNS varicosities can be inhibited by prejunctional α_2 *autoreceptors*. This is by far the most intensely investigated example of presynaptic inhibition by an inhibitory autoreceptor. *Clonidine* (an α_2 agonist) inhibits the release of NA evoked by nerve stimulation (i.e. enhances feedback inhibition due to prejunctional autoreceptors), while *yohimbine* (an α_2 antagonist) enhances such release by abolishing this autoregulation and increasing the probability of monoquantal release. *In vivo*, any local build-up of NA will activate the receptors and decrease the probability of release. As with other inhibitory autoreceptors there are two main mechanisms by which this regulation might occur; decreased Ca^{2+}-channel activity or enhanced K^+-channel activity (*see* Fig. 5.11) although it has also been suggested that the receptors may decrease the coupling between Ca^{2+} entry and NA release. Protein kinase C does not appear to be involved in this autoreceptor mechanism.

Dopamine release can be inhibited by presynaptic *D_2 autoreceptors*, whose modulation by antipyschotic drugs plays a role in the therapeutic treatment of schizophrenia (Box 11.1). The receptor is distinct from the NA-selective α_2 receptor. D_2 agonists can also inhibit the synthesis of dopamine, even under conditions where there is no release, suggesting a direct effect on D_2 receptors regulating tyrosine hydroxylase. However by far the majority of D_2 receptors are

Box 11.1 Schizophrenia

About 1% of the population will develop schizophrenia. Diagnosis of this psychosis is not straight-forward, but symptoms include delusions (for example, that the subject is being controlled by external agents), hallucinations and disordered thoughts and mood. While there is little information on the neurochemical changes in schizophrenia, amphetamine, which releases dopamine from varicosities, can induce a state resembling schizophrenia, while dopamine D_2 antagonists are effective in the treatment of symptoms with a rank order which correlates well with their affinity for brain D_2 receptors. This 'dopamine' hypothesis is however complicated by the low antipsychotic action of D_2 antagonists relative to the speed of receptor blockade. The action of D_2 antagonists is complex; as well as the expected postsynaptic block, the inhibition of presynaptic inhibitory autoreceptors will *increase* release of the transmitter in the short term. Eventually those terminals which possess presynaptic autoreceptors may become inactive due to adaptation and a state called inactivation block. Since D_2 antagonists inhibit dopaminergic synapses, a side-effect is the development of a Parkinsonian syndrome (*see* Box 11.3). Antipsychotic drugs (also called major tranquillizers or neuroleptics) can manage the symptoms of the disorder but the degree of recovery of the patient is very variable.

located postsynaptically on intrinsic neurones in the corpus striatum where it is established that they activate K^+-channels and hence hyperpolarize the membrane.

D_2 agonists are still effective in the presence of K^+-channel inhibitors such as 4AP. This does not disprove an action of presynaptic α_2 receptors in activating K^+ conductance (see, for example, the regulation of glutamate release, Fig. 9.18), but rather indicates that the activatable channels are 4AP resistant. Despite this the K^+-channel/Ca^{2+}-channel discussion is not resolved.

It is difficult to study *adrenaline release* in the CNS since there are so few adrenergic neurones relative to noradrenergic neurones, and since the properties of the two classes are very similar. Thus [^{3}H]-NA will label adrenergic in addition to noradrenergic cells. The most detailed information on adrenaline release, and indeed catecholamine release from LDCVs in general has come from studies of the chromaffin cell.

The inhibitory effects of adenosine on the release of a number of transmitters, notably glutamate, suggest that it is a powerful endogenous neuroprotective agent. There are two sources of extracellular *adenosine*. Firstly adenosine is formed by the hydrolysis by ectonucleotides of 5′-AMP, which is itself a hydrolysis product of ATP co-released with a variety of transmitters including catecholamines and ACh (*see* Fig. 11.4). The small amount of ATP released by synaptosomes (picomoles per milligram of protein) make it unlikely that the massive amounts of amino acid released in brain are accompanied by significant amounts of ATP. Also a preparation of synaptosomes from prawn muscle, containing a mixture of GABAergic and glutamatergic terminals released glutamate and GABA but not ATP in a Ca^{2+}-dependent manner, suggesting that amino acids are not co-released with ATP.

ATP is stored in cholinergic synaptic vesicles of the peripheral and central nervous systems, as well as *Torpedo*, in adrenergic vesicles, chromaffin granules and 5-HT-containing granules of platelets. Co-release of ATP with NA and other non-amino transmitters is a general phenomenon. Affinity-purified rat striatal cholinergic synaptosomes (*see* Section 10.2), show a molar ratio of ACh : ATP release of 9 : 1, comparing rather well with a ratio of contents for synaptic vesicles from the same preparation of 7 : 1.

Once ATP is released, it is rapidly hydrolysed by *ectonucleotidases* to adenosine. The final stage of hydrolysis, from AMP to adenosine, is catalysed by 5′-nucleotidase, which at least in the electric organ appears to be on glial membranes rather than neurones. In contrast the first enzyme, ecto-ATPase, appears to be associated with the presynaptic plasma membrane. It is not known whether ADP is hydrolysed by the triphosphatase or whether an additional enzyme is required. From the activity of the ecto-ATPase, ATP may persist in the synaptic cleft for much longer than ACh which is hydrolysed very rapidly.

The second source of adenosine is cytoplasmic. Adenosine is retained in the cytoplasm by a plasma membrane transporter which may be

Na^+-coupled. Thus, as in the case of amino acids, conditions such as ischaemia which lead to prolonged collapse of Na^+ gradients would be expected to result in a Ca^{2+}-independent release of adenosine. In view of the potent ability of adenosine to inhibit the release of glutamate and other transmitters (*see* Fig. 11.4), this could be an important protective mechanism against glutamate-induced excitotoxicity and is discussed in more detail below (*see* Section 11.4.2).

11.4 Receptors

Receptors for these 'Class II' transmitters (Table 11.1) are metabotropic with rare exceptions, such as the ionotropic 5-HT_3 receptor. While the original classifications (and consequent nomenclature) were based on pharmacological criteria, cloning of virtually all the pharmacologically classified receptors has now been accomplished and inevitably a much greater degree of heterogeneity is being discovered than was previously suspected. All the metabotropic receptors belong to a G-protein-coupled superfamily with the characteristic hydropathy plot consistent with the presence of 7-transmembrane domains (*see* Section 5.2). The receptors will now be discussed in relation to their dominant second messenger coupling.

11.4.1 Receptors coupled to adenylyl cyclase activation

Three *β-adrenoceptors* $β_1$, $β_2$ and $β_3$ can be defined at the pharmacological level. All act via G_s to increase cAMP. Destruction of noradrenergic neurones by 6OH-DOPA (*see* Section 11.6) decreases $β_1$ but not $β_2$ receptor density, suggesting that the former are involved in neuronal function, although $β_2$ receptors could still be on postjunctional neurones. The $β_2$ receptor has been the most intensively investigated of the

Table 11.1 Neuronal receptors for monoamines and purines.

Receptor	Second messenger	Amino acids	V–VI	C-terminus
$α_{1A}$	(1,4,5)IP_3/DAG ↑	560	73	161
$α_{1B}$	(1,4,5)IP_3/DAG ↑	515	71	164
$α_{1C}$	(1,4,5)IP_3/DAG ↑	466	67	137
$α_{2A}$	cAMP ↓ G→K^+ ↑	450	156	21
$α_{2B}$	cAMP ↓	450	179	21
$β_1$	cAMP ↑	477	80	97
$β_2$	cAMP ↓	413	53	91
D_1	cAMP ↑	446		
D_{2S}				
D_{2L}				
D_3 D_4				
D_5				
5-HT_{1A}	G-K^+ ↑			
5-HT_{1B}	cAMP ↓			
5-HT_{1C}	(1,4,5)IP_3/DAG ↑			
Adenosine A_1	cAMP ↓ G→K^+ ↑			
Adenosine A_2	cAMP ↑			
Hist H_2	cAMP ↑	359		
Hist H_3				

Information primarily from *TIPS Receptor Nomenclature Supplement*, Jan. 1991. Dopaminergic receptors are classified as D_1, D_2 or D_3 while 5-HT receptors are classified as 5-HT_1, 5-HT_2, 5-HT_3 (ionotropic) or 5-HT_4 and histamine receptors as H_1, H_2 and H_3. Purinergic (ATP) receptors are classified as P_1, P_2 and P_{2x} (ionotropic) (for purinergics see *Bean TIPS* **13**, 87–90 (1992)).

7-transmembrane loop superfamily and the generic information obtained from these studies is discussed in Section 5.3.

Desensitization is the process whereby the response evoked by a transmitter declines in the continued presence of the agonist. *Homologous desensitization* is specific to the desensitizing agent itself, while *heterologous desensitization* will decrease the response to multiple classes of activator. Desensitization can occur at the level of the receptor, G-protein or effector: receptor desensitization can involve slow downregulation of the amount of receptor (by enhanced degradation or diminished synthesis), more rapid sequestration (temporary internalization of the receptor) or receptor uncoupling. This last process, which has been investigated in particular detail in the case of the β_2 receptor, may be of most physiological relevance.

Two distinct mechanisms of phosphorylation-dependent desensitization appear to occur. Firstly the coupling of the β_2-adrenoceptor to G_s can be regulated by PKA-dependent phosphorylation on the third cytoplasmic loop (where G-protein interaction is believed to occur) and the proximal region of the C-terminal cytoplasmic tail (Fig. 11.7). Since PKA will be activated in the cell by the G_s-mediated production of cAMP, this will provide an inherent negative feedback which is believed to be of particular relevance when low agonist concentrations are present, for example, in response to circulating adrenergic drugs.

In contrast at synapses, where high local concentrations of catecholamines will occur, desensitization may be due to activation of a specific *β-adrenergic receptor kinase* (βARK). cDNAs for two isoforms, βARK1 and βARK2 have currently been identified. βARK can phosphorylate a series of four serine residues close to the C-terminus of the receptor (Fig. 11.7), but only when the receptor is activated by binding of agonist. Phosphorylation itself has little effect on receptor G-protein coupling, but facilities the binding of a 48 kDa protein termed β-arrestin, analogous to the arrestin of visual transduction (*see* Section 6.2), which effects uncoupling of the receptor from G_s.

Dopaminergic D_1 and D_5 receptors are positively coupled via G_s to adenylyl cyclase, while the second subfamily (D_2, D_3 and D_4, *see* Section 11.4.2) inhibit cyclase and/or interact directly with ion channels. D_1 receptors are more common than D_2 receptors and have a relatively low (micromolar) affinity for dopamine. The 446 residue D_1 receptor is closely homologous to the β-adrenoceptor discussed above and shows the characteristic long cytoplasmic C-terminal region of a G_s-linked receptor and a small third cytoplasmic loop. Most of the neurones expressing the D_1 receptor belong to the striatonigral projection system (*see* Fig. 11.3) and also express substance P. The D_1 and D_5 receptors are highly homologous, although levels of expression of the latter are generally much lower than for the D_1 receptor.

The *adenosine A_2 receptor* elevates cAMP and requires micromolar adenosine. The *histaminergic H_2 receptor* is shorter (at 359 amino acids) than most G-protein-linked receptors, although it retains the seven hydrophobic domain structure. Interestingly it lacks the characteristic pair of serine residues in the fifth transmembrane domain which are postulated to bind the catechol hydroxyls of the catecholamines. The receptor is positively coupled to adenylyl cyclase via G_s. In addition a presynaptic H_3 receptor has been described.

11.4.2 Receptors coupled to adenylyl cyclase inhibition and/or G-protein-coupled ion channel modulation

As a generalization, the receptors which will now be discussed can be located either presynaptically or postsynaptically and display closely related responses. Their presynaptic action appears to be mediated via channel modulation (typically an inhibition of Ca^{2+}-channels and/or activation of K^+-channels, both of which are inhibitory), whereas receptors located on the cell soma or dendrites appear to act either by channel modulation or by lowering cAMP levels. As discussed in Section 5.4.1, the mechanism by which cAMP is lowered is not entirely clear. It may involve competition for βγ G-protein subunits with the stimulatory G_s. In any case there is little evidence that this mechanism functions presynaptically, where a direct 'membrane-delimited' interaction between G-protein and ion channel appears to be the preferred mechanism.

An *α_2-adrenoceptor* was originally defined as that mediating the presynaptic inhibition of NA release. Molecular cloning has to date identified three genes, denoted α_{2A}, α_{2B} and α_{2C}. The

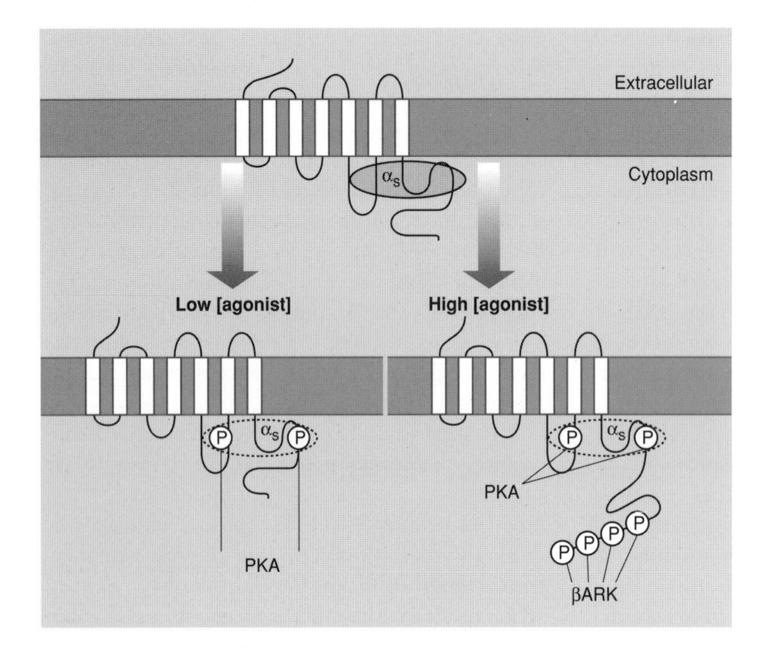

Fig. 11.7 β-receptor desensitization. The β-adrenoceptor is shown schematically with the region of interaction with $G\alpha_s$. At low agonist concentration the activation of G_s, adenylyl cyclase and hence protein kinase A (PKA) can result in phosphorylation of the receptor (P) disrupting G-protein interaction and causing desensitization. Other receptors may also be phosphorylated (heterologous desensitization). At high agonist concentration βARK is also activated, causing multiple phosphorylation of the C-terminus of the receptor. Adapted from Lefkowitz *et al.* (1990).

sequences are highly homologous in their membrane-spanning regions, but more divergent in their loops, although each possesses a very large third cytoplasmic loop and a relatively short cytoplasmic C-terminal domain.

D_2-like dopaminergic receptors (currently classified as D_2, D_3 and D_4) inhibit cAMP production and/or act directly on Ca^{2+}- and K^+-channels, inhibiting and activating respectively. They have a high (nanomolar) affinity for dopamine in equilibrium binding studies. The D_2 receptor exists in two isoforms, D_{2L} and D_{2S} which differ in length by 29 amino acids within the large third cytoplasmic loop without detectable functional effect. The isoforms are derived from the same gene by differential RNA splicing. Although this third intracellular loop is implicated in G-protein interaction (*see* Section 5.3) no consistent difference in G-protein coupling has so far been reported.

D_2 receptors may act both postsynaptically and as prejunctional autoreceptors; thus mRNA for the receptor is found both in the regions to which the major dopaminergic terminals project (particularly on enkephalin and ACh-releasing neurones) and also in dopaminergic cell bodies, for example, in the substantia nigra. D_2 receptors are implicated in the therapeutic treatment of schizophrenia (*see* Box 11.1).

Recently, a new dopamine receptor has been cloned which has been designated D_3, exhibiting 50% overall homology and 80% transmembrane domain homology with the established D_2 receptor. Unlike the short and long forms of the D_2 receptor a distinct gene is involved. The second messenger system has not been identified, but the homology with D_2 is sufficient to make an inhibitory coupling to adenylyl cyclase likely, in which case it may be considered as a D_2 subtype.

Serotonergic 5-HT_1 receptors are most abundant in the cortex. The 5-HT_{1A} receptor occurs in high concentrations in limbic regions and in the dorsal raphe nucleus. *8OH-DPAT* is a selective agonist, while *spiperone* is an antagonist. The receptor hyperpolarizes cells by increasing K^+ conductance. The hyperpolarization induced by the agonist is slow and shows similar kinetics to that produced by GABA-B or α_2-adrenoceptor activation in appropriate cells. The receptor also inhibits adenylyl cyclase, although this does not mediate the K^+ conductance increase.

The 5-HT_{1B} receptor (in rat, mouse or hamster) and the closely related 5-HT_{1D} receptor in guinea-pig and human are involved in presynaptic autoregulation of 5-HT release from the terminals of raphe neurones. In addition the 5-HT_{1B} receptor can act presynaptically on heteroreceptors to inhibit the release of glutamate, ACh and GABA.

Since no analogous receptor has been detected on cell bodies, where it would be accessible to electrophysiological analysis, the ionic mechanisms are still unestablished, although they probably act via the inhibition of Ca^{2+}-channels, in contrast to the soma-dendritic 5-HT_{1A} subclass discussed above which increases membrane K^+ conductance. The *histaminergic H_3 receptor* was first identified as an inhibitory autoreceptor regulating release of the transmitter from terminals in the cerebral cortex. The same receptor may mediate the inhibitory effect of histamine on 5-HT release in the cortex.

The *adenosine A_1 receptor* has a nanomolar affinity for adenosine, and has effects common to this group of receptors (depression of cAMP or direct coupling via G-proteins to ion channels). Adenosine has a strong inhibitory action on neurotransmission, both pre- and postsynaptically. For reasons of accessibility, most work on the coupling of receptors to ion channels has been carried out on postsynaptic preparations where the channels can be monitored directly. It is however important to realize that the actions of the same pharmacologically identical receptor at pre- and postsynaptic locations may differ considerably, even though the adenosine A_1 receptor responds to the same agonists both pre- and postsynaptically. The A_1 receptor is present at high concentration in the hippocampus, closely associated with dendritic zones which are particularly susceptible to ischaemic damage.

The presynaptic receptor will inhibit the evoked release of a number of transmitters, including ACh from the neuromuscular junction and striatum, but not the cerebral cortex where muscarinic feedback inhibition predominates. Both mechanisms may operate in the hippocampus; the reason for this heterogeneity is not known. The A_1 receptor is also able to inhibit glutamate release from cerebellar granule cells and synaptosomes. In the latter preparation there is good evidence that the receptor decreases Ca^{2+}-channel activity rather than activating presynaptic K^+-channels and hyperpolarizing the membrane, since A_1 agonists are effective in modulating release from KCl-stimulated synaptosomes (*see* Fig. 11.4), conditions under which K^+-channel activation would fail to inhibit release. GABA exocytosis is not sensitive to adenosine.

11.4.3 Receptors coupled to phospholipase C

The *α_1-adrenoceptors*, *serotonergic 5-HT_{1C}* and *5-HT_2* receptors, and the *histaminergic H_1* receptor are each coupled, generally via G_i, to phospholipase C and hence elevate (1,4,5)IP_3 and diacylglycerol. Three α_1-adrenoceptor receptor genes have been cloned, their sequences display relatively short i3 intracellular loops and long C-termini, in contrast to α_2 receptors. Agonists for the 5-HT_2 receptor are excitatory and are frequently hallucinogenic (e.g. LSD, (+)-lysergic acid diethylamide). In neurones the 5-HT_2 receptor selectively inactivates the inward rectifying K^+-channel.

11.4.4 Ionotropic receptors for 5-HT and ATP

Although the transmitters discussed in this chapter are overwhelmingly metabotropic, there are two exceptions. The *serotoninergic 5-HT_3* receptor is ionotropic, and mediates rapid synaptic transmission. The receptor possesses cationic nonselective ligand-gated channel conductances similar to the nicotinic ACh receptor, although with a significant Ca^{2+} conductance. The cloned receptor (5-HT_3RA) is homologous to the nicotinic receptor with four transmembrane segments and presumably functions as a pentamer. The channel can be partially inhibited by (+)-tubocurarine. In the CNS, the receptor may be present on GABAergic inhibitory inter-neurones and other locations.

The *purinergic P_{2x} receptor*, although to date not cloned, may provide a second example of an ionotropic receptor. The purinergic P_2 terminology receptors originated in the need to distinguish ATP receptors (P_2) from adenosine receptors (P_1). Although this terminology is superfluous it is still used. ATP mediates excitatory synaptic currents in the CNS which can be blocked by suramin or by the desensitizing purinergic agonist $\alpha\beta$-methylene ATP. These ionotropic responses show a distribution similar to those for ACh, suggesting that it may act as a cotransmitter at some cholinergic synapses where it is known to be co-stored, and has been ascribed to a purinergic P_{2x} receptor. In the adrenergic sympathetic nervous system, ATP has an ionotropic excitatory action, which in smooth muscle might be mediated by a Ca^{2+}-permeable ATP-activated channel. ATP may be the only truly ionotropic transmitter in sympathetic nerves triggering fast depolarization of target cells. In most sympathetic systems, presynaptic effects of ATP are lacking. ATP co-released from cholinergic terminals does not appear to have a direct ionotropic action.

11.5 Termination of biogenic amine neurotransmission

The predominantly modulatory role played by catecholamines and 5-HT means that factors which control the duration of elevated transmitter levels in the synaptic cleft can have powerful controlling effects on brain activity. This, in turn, has made the re-uptake and degradation pathways for these transmitters (Fig. 11.8) a central focus for pharmacological intervention (Box 11.2). We shall first consider the two enzymes for catecholamine catabolism: monoamine oxidase (MAO) and catechol-O-methyltransferase (COMT).

11.5.1 Monoamine oxidases

MAO oxidizes primary aromatic amines (and other amines more slowly) to the corresponding aldehydes (Fig. 11.8). Aldehyde dehydrogenase further oxidizes the product to the corresponding acid (the predominant fate for dopamine), while alcohol dehydrogenase reduces the aldehyde to alcohols (mainly for NA). MAO exists as two isoenzymes: MAO-A and MAO-B. MAO-A has a higher affinity for biogenic amines and is selectively inhibited by *clorgyline*; MAO-B has a higher affinity for dietary amines and is selectively inhibited by *deprenyl*. Both enzymes are flavoproteins associated with the outer mitochondrial membrane and are located both pre- and postjunctionally, as well as in glia. They thus do not directly degrade the amines in the synaptic cleft, but rather after re-uptake into surrounding cells. The enzymes are not brain specific, indeed the highest levels of both forms are usually found in liver. In adult brain, MAO-A is located mostly in catecholamine neurones, while MAO-B is concentrated in serotonergic neurones, astrocytes and radial glia. Monoamine oxidase inhibitors have been used as antidepressant drugs (Box 11.2).

Box 11.2 Affective disorders and monoamine metabolism

Depression has been treated by *monoamine oxidase inhibitors*, particularly those acting on MAO-A. It is not clear which monoamine is relevant. A dangerous side-effect which has limited their use is the 'blue cheese reaction': foods such as blue cheese and red wine contain a high concentration of tyramine which is normally degraded by peripheral MAO. In the presence of the inhibitor this is no longer possible and accumulated tyramine can enter peripheral adrenergic varicosities, displacing noradrenaline and precipitating dangerous hypertension. *Tricyclic antidepressants* are potent inhibitors of the Na^+-coupled re-uptake transporters for noradrenaline, dopamine or 5-HT. While both classes of inhibitors conform to the 'monoamine hypothesis' of depression (that depression may be due to lowered monoamine transmitter concentrations in the synapse) this is almost certainly oversimplistic. The use of Li^+ salts in the control of affective disorders is discussed in Box 5.1.

11.5.2 Catechol-*O*-methyl transferase

COMT is a cytoplasmic enzyme which transfers a methyl group from *S*-adenosyl-methionine to the *m*-OH of the catechol (Fig. 11.8). It has a broad localization, in glia and dendrites. This is the major pathway for the catabolism of catecholamines.

11.5.3 Re-uptake

Re-uptake is the main means of inactivation of catecholamines. Na^+-coupled transporters for dopamine must preceed any degradation via MAO. Distinct re-uptake carriers for DA, NA and 5-HT have been cloned belonging to a common superfamily of plasma membrane neurotransmitter transporters. It is not clear whether there is a distinct adrenaline transporter or whether the catecholamine is reaccumulated via the NA transporter.

It is possible to inhibit either pathway. *Cocaine* is a rather non-specific inhibitor of dopamine re-uptake with a K_i of 1 μM, while *tricyclic anti-*

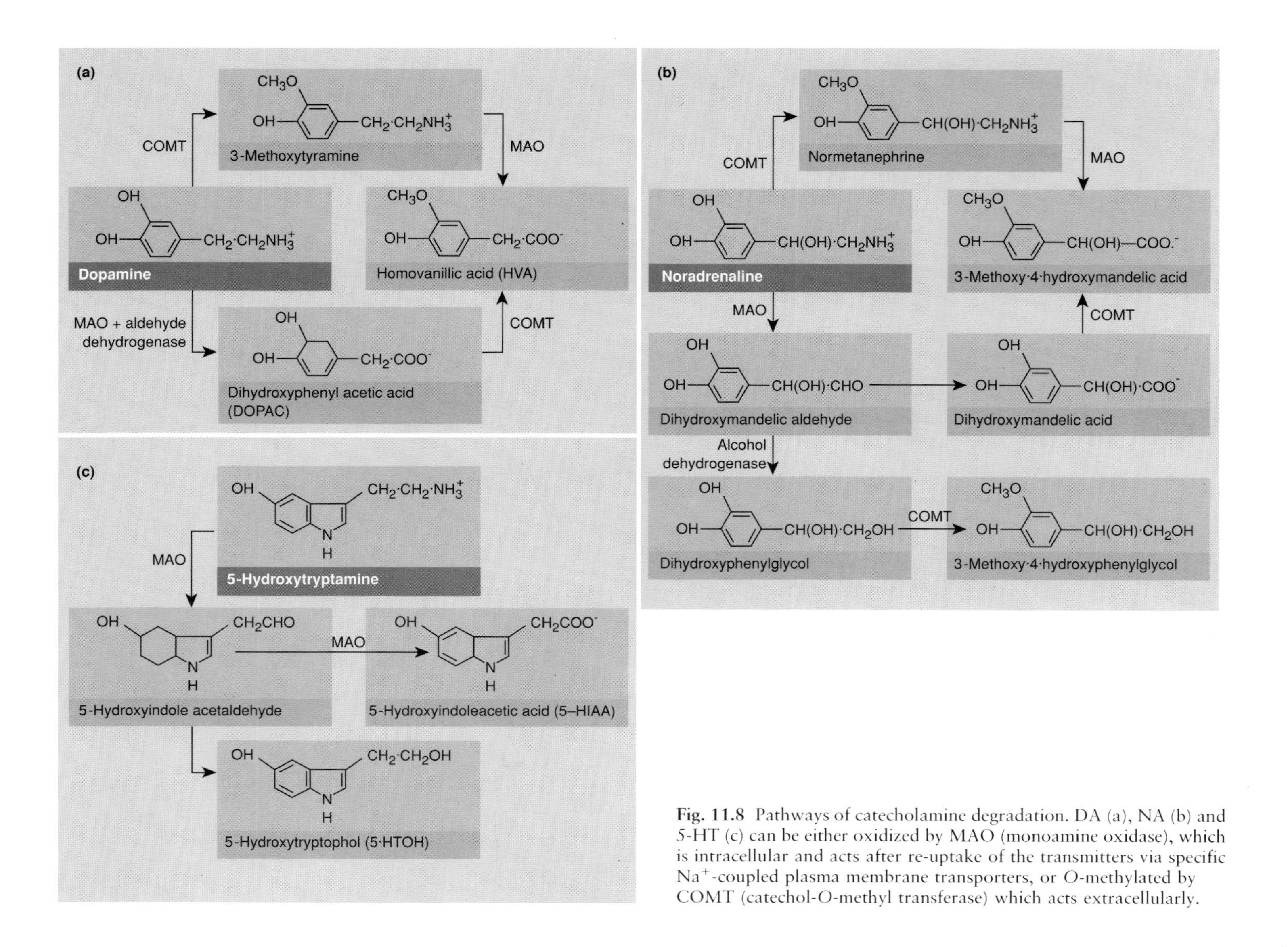

Fig. 11.8 Pathways of catecholamine degradation. DA (a), NA (b) and 5-HT (c) can be either oxidized by MAO (monoamine oxidase), which is intracellular and acts after re-uptake of the transmitters via specific Na^+-coupled plasma membrane transporters, or *O*-methylated by COMT (catechol-*O*-methyl transferase) which acts extracellularly.

depressants are more effective as NA and 5-HT uptake inhibitors; *desipramine* is the most potent NA uptake inhibitor. Tricyclics are similar to the phenothiazines which block postjunctional receptors. *Buproprion*, related to amphetamine is a relatively selective dopamine uptake inhibitor. The spectrum of tricyclic antidepressants inhibiting 5-HT uptake differs from that for either of the catecholamines, consistent with the existence of a specific Na-coupled 5-HT uptake transporter. Since the synaptic vesicle catecholamine transporter appears also to transport 5-HT, this plasma membrane specificity is important to maintain the specificity of the terminals. *Imipramine* and *panuramine* are relatively specific for 5-HT, whereas desipramine is a much weaker inhibitor of 5-HT transport than of NA.

No specific re-uptake system has been found for histamine. Thus the precursor, histidine must be employed in order to label histamine stores in terminals. Although there is a transport system for histamine into rat brain synaptosomes, it has the puzzling property of being Na^+-independent and is therefore unlikely to be of relevance as a concentrative re-uptake mechanism. Enzymatic breakdown is therefore likely to be solely responsible for terminating the action of the transmitter. Histamine can be degraded by either oxidative deamination to imidazoleacetate by diamine oxidase or by methylation by histamine-*N*-methyltransferase followed by oxidative deamination by MAO-B.

11.6 Neurotoxins acting on catecholamine transmission

6-Hydroxydopamine (6OH-DA) is selectively accumulated into dopaminergic and noradrenergic boutons. The accumulation of high concentrations of the agent, followed by its oxidation in the bouton leads to the destruction of the neurone, either as a consequence of a superoxide radical formed during the oxidation, or because of the toxic nature of the *p*-quinone formed. Direct injection into specific sites such as the substantia nigra causes selective degeneration. The combination of 6OH-DA with a selective inhibitor of NA re-uptake can limit damage to the latter neurones.

One dopaminergic-specific neurotoxin which has provided a considerable impetus to investigations into the nature of the neurodegenerative Parkinson's disease (Box 11.3) is *1-methyl-4-phenyl-1,2,3,6-tetrahydropyridine (MPTP)* (Fig. 11.9). A characteristic of the disease is a damaged nigrostriatal dopamine system with a loss of cell bodies in the substantia nigra and a corresponding depletion of transmitter in the varicosities innervating the corpus striatum. Parkinson's disease is characteristically one of later life; it was therefore unexpected when, in 1976, a group of young Californian drug addicts was admitted to hospital with symptoms characteristic of Parkinson's disease. The patients had all injected the same batch of a 'designer drug' analogue of the narcotic meperidine (pethidine). Analysis of the sample revealed the presence of several pyridine impurities, including MPTP. Positron emission tomography showed a dramatic destruction of dopaminergic neurones in the substantia nigra similar to that in Parkinson's disease.

[^{3}H]-MPTP binds to brain MAO-B which oxidizes it by a two-electron step to a dihydro form and then subsequently to the four-electron oxidation product, MPP^+, which is toxic. Thus MAO-B inhibitors confer protection against MPTP toxicity in primates. Rodents are less

Box 11.3 Parkinson's disease

Parkinson's disease affects some 0.5% of the population over 50 and manifests itself as a tremor of the limbs, jerky 'cogwheel' movement, and problems with speech and cognition. Postmortem examination shows a specific degeneration of the substantia nigra which is rich in dopaminergic cell bodies with a loss of the dopaminergic pathways to the striatum and globus pallidus. The disease is progressive and it is estimated that an 80% loss of dopamine from striatal varicosities can occur before clinical symptoms are seen. In order to increase the efficiency of the residual varicosities L-DOPA can be administered orally (together with a peripheral DOPA decarboxylase inhibitor) to bypass the rate-limitating enzyme in dopamine synthesis, tyrosine hydroxylase. L-DOPA therapy does have problems, including a loss of efficacy with time and psychiatric complications. Dopamine receptor agonists can also be used to control symptoms — alone or in combination with L-DOPA.

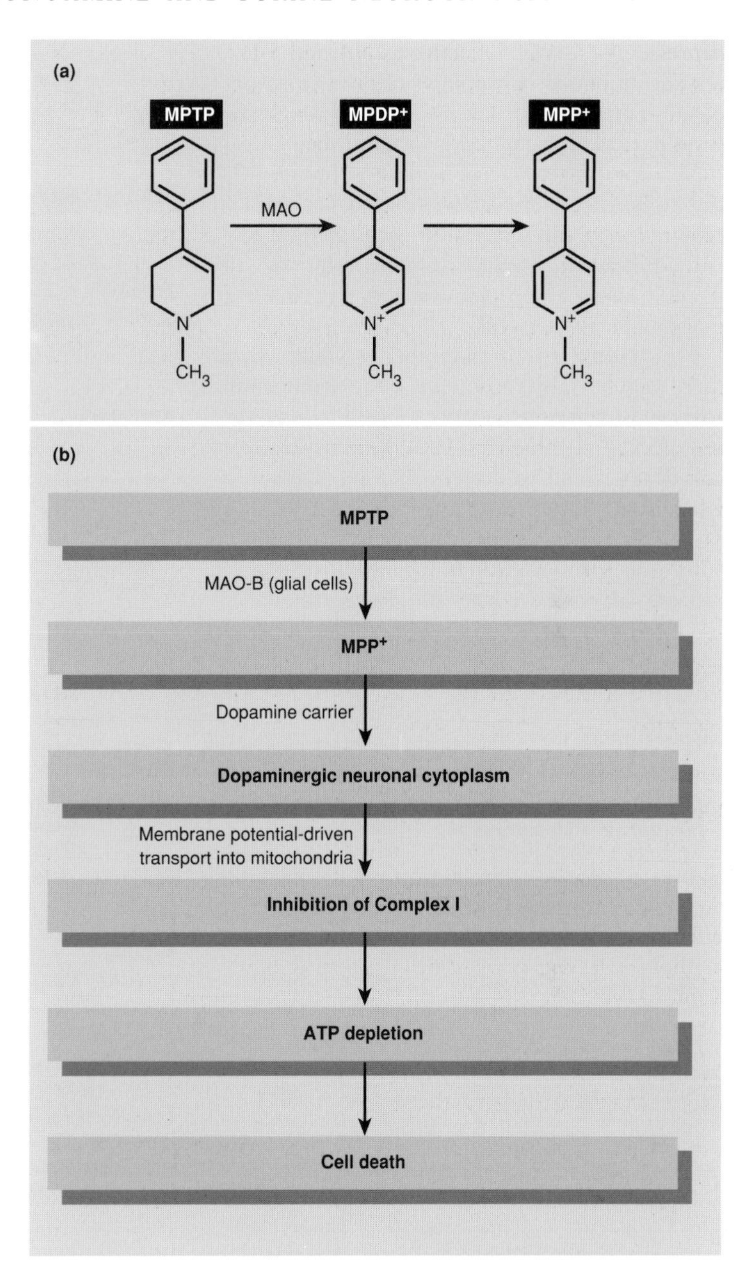

Fig. 11.9 The metabolism of 1-methyl-4-phenyl-1,2,3,6-tetrahydropyridine (MPTP). (a) MPTP is oxidized to the neurotoxic MPP^+ by MAO-B. (b) MPP^+ is a substrate for the plasma membrane dopamine carrier specific to dopaminergic neurones. Once in the cytoplasm MPP^+ acts as a lipophilic cation and is accumulated within the mitochondrial matrix by the negative interior membrane potential where it inhibits mitochondrial Complex I preventing oxidative phosphorylation.

sensitive than primates to MPTP, probably because they possess less of the B isoform of MAO; however, once this stage is bypassed, by direct intracranial administration of MPP^+, the characteristic damage is seen also in the rat.

The location of the MAO-B responsible for producing the MPP^+ is puzzling since central dopaminergic neurones contain very little of the enzyme. As extraneuronal MAO-B is widespread, MPP^+ may be produced outside the terminals. Once produced, MPP^+ is accumulated into dopaminergic terminals although the pathway is contentious. MPP^+ in the cytoplasm of dopaminergic terminals can be further accumulated across the mitochondrial inner membrane in response to the membrane potential. This behaviour would be predicted of a hydrophobic cation but there is also the possibility that a specific carrier exists. MPP^+ accumulated into the mitochondrion inhibits Complex I of the respiratory chain. Thus the cell will be deprived of energy from oxidative phosphorylation and die.

This two-stage concentration is supported by the finding that the concentrations of MPP^+ required to inhibit respiration decrease from several millimolar (sonicated mitochondria) to 0.5 mM (intact respiring mitochondria) to a few micromolar (intact cells).

It is natural to consider the possibility that there may be natural products or environmental toxins which mimic the action of MPP^+, however any mechanism which results in the same selective cell death would produce symptoms similar to Parkinson's disease. One possibility is that autoxidation of dopamine with specific cells in the substantia nigra can lead to the formation of free radicals which cause neuronal degeneration.

Further reading

Monoaminergic pathways

Parnavelas, J.G. & Papadopoulos, G.C. (1989) The monoaminergic innervation of the cerebral cortex is not diffuse and nonspecific. *Trends Neurosci.* **12**, 315–315.

Ridet, J.-L., Rajaofetra, N., Teilhac, J.-R., Geffard, M. & Privat, A. (1993) Evidence for nonsynaptic serotonergic and noradrenergic innervation of the rat dorsal horn and possible involvement of neuron–glia interactions. *Neurosci.* **52**, 143–157.

Vesicular storage of catecholamines

Klein, R.L., Duncan, R.W., Selva, T.J., Kong, J.Y., Clayton, W.E., Liaw, Y.L., Rezk, N.F. & Thureson-Klein, A. (1988). Sympatho-adrenal co-storage, release and synthesis of enkephalins and catecholamines induced by acute CNS ischaemia in the pig. In *Cellular and Molecular Basis of Synaptic Transmission*, NATO ASI Series H, Cell Biology, Vol. 21, H. Zimmermann (ed). Berlin: Springer, pp. 377–393.

Njus, D., Kelley, P.M. & Harnadek, G.J. (1986) Bioenergetics of secretory vesicles. *Biochim. Biophys. Acta* **853**, 237–266.

Exocytosis of biogenic amines and ATP

Barrie, A.P. & Nicholls, D.G. (1993) Adenosine A1 receptor inhibition of glutamate exocytosis and protein kinase C-mediated decoupling. *J. Neurochem.* **60**, 1081–1086.

Fiedler, J.L., Pollard, H.B. & Rojas, E. (1992). Quantitative analysis of depolarization-induced ATP release from mouse brain synaptosomes: external Ca-dependent and independent processes. *J Membrane Biol.* **127**, 21–33.

Schizophrenia and affective disorders

Strange, P.G. (1992) *Brain Biochemistry and Brain Disorders*. Oxford: Oxford University Press.

Reynolds, G.P. (1992) Developments in the drug treatment of schizophrenia. *Trends Pharmacol. Sci.* **13**, 116–121.

Receptors

Arriza, J.L., Dawson, T.M., Simerly, R.B., Martin, L.J., Caron, M.G., Snyder, S.H. & Lefkowitz, R.J. (1992) The G-protein-coupled receptor kinases βARK1 and βARK2 are widely distributed at synapses in rat brain. *J. Neurosci.* **12**, 4045–4055.

Birdsall, N.J.M. (1991) Cloning and structure-function of the H2 histamine receptor. *Trends Pharmacol. Sci.* **12**, 9–10.

Bobker, D.H. & Williams, J.T. (1990) Ion conductances affected by 5-HT receptor subtypes in mammalian neurons. *Trends Pharmacol. Sci.* **13**, 169–173.

Christian, E.P. & Weinreich, D. (1992) Presynaptic histamine H_1 and H_3 receptors modulate sympathetic ganglionic synaptic transmission in the guinea-pig. *J. Physiol. (Lond.)* **457**, 407–430.

Fredholm, B.B. & Dunwiddie, T.V. (1988) How does adenosine inhibit transmitter release? (TIPS Review). *Trends Pharmacol. Sci.* **9**, 130–135.

Harrison, J.K., Pearson, W.R. & Lynch, K.R. (1991) Molecular characterization of α1 and α2-adrenoceptors. *Trends Pharmacol. Sci.* **12**, 62–67.

Hen, R. (1992) Of mice and flies: commonalities among 5-HT receptors. *Trends Pharmacol. Sci.* **13**, 160–165.

Hibert, M.F., Trumppkallmeyer, S., Bruinvels, A. & Hoflack, J. (1991) Three-dimensional models of neurotransmitter G-binding protein-coupled receptors. *Mol. Pharmacol.* **40**, 8–15.

Lefkowitz, R.J., Hausdorff, W.P. & Caron, M.G. (1990) Role of phosphorylation in desensitization of the β-adrenoceptor. *Trends Pharmacol. Sci.* **11**, 190–194.

Linden, J.L., Tucker, A.L. & Lynch, K.R. (1991) Molecular cloning of adenosine A1 and A2 receptors. *Trends Pharmacol. Sci.* **12**: 326–328.

Lledo, P.M., Homburger, V., Bockaert, J. & Vincent, J.D. (1992) Differential G protein-mediated coupling of D2 dopamine receptors to K^+ and Ca^{2+} currents in rat anterior pituitary cells. *Neuron* **8**, 455–463.

Rudolphi, K.A., Schubert, P., Parkinson, F.E. & Fredholm, B.B. (1992) Neuroprotective role of adenosine in cerebral ischaemia. *Trends Pharmacol. Sci.* **13**, 439–445.

Sibley, D.R. & Monsma, F.J. (1992) Molecular biology of dopamine receptors. *Trends Pharmacol. Sci.* **13**, 61–69.

Stiles, G.L. (1992) Adenosine receptors. *J. Biol. Chem.* **267**, 6451–6454.

Strange, P.G. (1993) New insights into dopamine receptors in the central nervous system. *Neurochem. Int.* **22**, 223–236.

Summers, R.J. & McMartin, L.R. (1993) Adrenoceptors and their second messenger systems. *J. Neurochem.* **60**, 10–23.

Tota, M.R., Candelore, M.R., Dixon, R.A.F. & Strader, C.D. (1991) Biophysical and genetic analysis of the ligand-binding site of the β-receptor. *Trends Pharmacol. Sci.* **12**, 4–6.

Ionotropic receptors for 5-HT and ATP

Bean, B.P. (1992) Pharmacology and electrophysiology of ATP-activated ion channels. *Trends Pharmacol. Sci.* **13**, 87–90.

Maricq, A.V., Peterson, A.S., Brake, A.J., Myers, R.M. & Julius, D. (1991) Primary structure and functional expression of the $5HT_3$ receptor, a serotonin-gated ion channel. *Science* **254**, 432–435.

Neurotoxins acting on catecholamine transmission

Singer, T.P. & Ramsay, R.R. (1990) Mechanism of the neurotoxicity of MPTP — an update (Review Letter). *FEBS Lett.* **274**, 1–6.

12

Neuropeptides

12.1 Introduction

Neuropeptides and their receptors have a widespread distribution both within and outside the CNS. They are frequently co-stored in the same terminals as classical transmitters but segregated into large dense-core vesicles (LDCVs), allowing for their frequency-dependent differential release as discussed in Chapter 8. Most of our understanding of peptide action comes from studies of the peripheral nervous system, and despite the enormous proliferation of putative neuropeptides in the CNS it has been, until recently, difficult to demonstrate direct physiological roles for peptide transmitters in the CNS.

Neuropeptides can be classified into *opioid-* and *non-opioid* peptides (Table 12.1). The former comprises a group of small peptides which act at a group of receptors which are also responsive to plant opiate alkaloids, while the latter includes neuropeptides which act either within the CNS or else are secreted into the bloodstream by neurosecretory neurones with terminals in the pituitary gland. There are numerous neuropeptides, but we shall concentrate upon the opioid peptides and a few of the major non-opioid peptides whose study has thrown light on release or postsynaptic mechanisms.

As discussed in Chapter 8, the presynaptic mechanisms associated with neuropeptide synthesis, storage and release are profoundly different from those for the Type I or Type II transmitters. Neuropeptides are synthesized in the neuronal cell body as large precursor peptides which are then sorted into LDCVs and are subsequently processed within the vesicles by specific proteases. The processed neuropeptides are typically present in very low concentrations (femtomoles to picomoles per gram of brain) compared with Type I or Type II transmitters (*see* Table 1.1) and frequently occur as 'cotransmitters' in the same terminals as Type I or II transmitters. They are released from sites other than active zones by frequency-dependent stimulation and may act as local hormones as well as at specific synapses. Finally, neuropeptides are not retrieved from the synapse by re-uptake, but are degraded by extracellular peptidases.

The cell bodies of neurones secreting neuropeptides such as *arginine-vasopressin* (AVP) and *oxytocin* (OXY) are concentrated in the hypothalamus. Cell bodies synthesizing other prominent peptides such as *cholecystokinin* (CCK) and *vasointestinal peptide* (VIP) are found largely in the neocortex. Cell bodies synthesizing the precursor proteins for two different groups of opioid peptides, the enkephalins derived from proenkephalin (ProEnk) and the endorphins derived from proopiomelanocortin (POMC) show dramatically different distributions. ProEnk cell bodies are found in almost all brain regions and typically project short axons, while POMC-synthesizing neurones are concentrated almost entirely in the pituitary and nearby regions of the hypothalamus, sending out long axons to diverse brain areas as well as axons specialized for neurosecretion from the pituitary.

Table 12.1 Primary sequences of some major neuropeptides.

Peptide	Chain length	Sequence
Opiate peptides		
[Met]-enkephalin	5	YGGFM
[Leu]-enkephalin	5	YGGFL
β-endorphin	31	YGGFM TSEKS QTPLV TLFKN AIVKN AHKKG Q
Non-opiate peptides		
Arg-vasopressin	9	CYFQN CPRG (SS bond C1–C6)
Oxytocin	9	CYIQN CPLG (SS bond C1–C6)
Substance-P	11	RPKPQ QFFGL M
Endothelin-1	21	CSCSS LMBKE CVYFC HLBII W (SS bonds C1–C15 and C3–C11)
Galanin-29 (rat)	29	GWTLN SAGYL LGPHA IDNHR SFSDK HGLT
Somatostatin	14	AGCKN FFWKT FTSC (SS bond C3–C14)

12.2 Synthesis and processing: generic aspects

Most secreted peptides are formed by posttranslational modification of larger precursor proteins. The initial 'pre-propeptide' is synthesized in the cell body on the rough endoplasmic reticulum (ER) and directed into the lumen of the ER by a signal sequence which is then cleaved by an endopeptidase. The cDNA for a number of pre-propeptides indicate some consistent features in the amino acid sequence: an N-terminal signal sequence of some 20 residues is followed by a peptide separated from the first neuropeptide at a Lys-Arg, a second inactive peptide separates the second neuropeptide and another Lys-Arg separates the C-terminal region.

Initial glycosylation of the proteins is carried out in the ER. The proteins destined for the secretory vesicles must then be transported from the ER to the Golgi apparatus by a process which is regulated and may be rate limiting. Within the Golgi the proteins may be phosphorylated, and the glycosylation modified by the removal of mannose units and the addition of galactose and sialic acid units. Neuropeptide precursors, the enzymes required to process them and vesicular membrane proteins such as dopamine β-hydroxylase are directed to the regulated secretory pathway for processing into secretory vesicles, transport to the terminals and storage prior to release.

The Golgi is responsible for the sorting of secreted proteins between these two pathways. However, there appears to be no information contained within the sequence of the propeptide to direct it to one or other of the pathways. One possibility is that the proteins destined to supply the contents of the secretory vesicles aggregate within the Golgi (perhaps under the influence of internal pH or Ca^{2+} concentration) and that the regulated pathway recognizes such aggregates. The *secretogranins* (or chromogranins) which are found in the lumen of LDCVs (*see* Section 8.2) may play a role in assisting this aggregation. A further problem, which is currently unresolved, is how the integral proteins of the secretory vesicle membrane (H^+-ATPases, transporters, cytochromes, etc.) are segregated from similar plasmalemmal transporters destined for the terminal plasma membrane via the constitutive pathway.

During transport to the terminal, the propeptide is further processed to give the final neuropeptide by specific peptidases (Fig. 12.1). In addition to proteolytic cleavage, processing may involve amidation of the C-terminus (e.g. CCK), cyclization of an N-terminal glutamate to pyroglutamate (e.g. thyrotropin-releasing hormone, TRH), acetylation of the terminal amine (e.g. α_1-melanocyte-stimulating hormone, α-MSH, and β-endorphin) and sulphation of tyrosyl-OH groups (e.g. CCK). LDCVs must therefore additionnally contain the enzymes for these processes.

Precursor proteins are very versatile, and may contain the sequences for a number of neuropeptides or for multiple copies of a given neuropeptide such as [Met]-enkephalin (Fig. 12.1) as well as sequence for carrier molecules which are co-

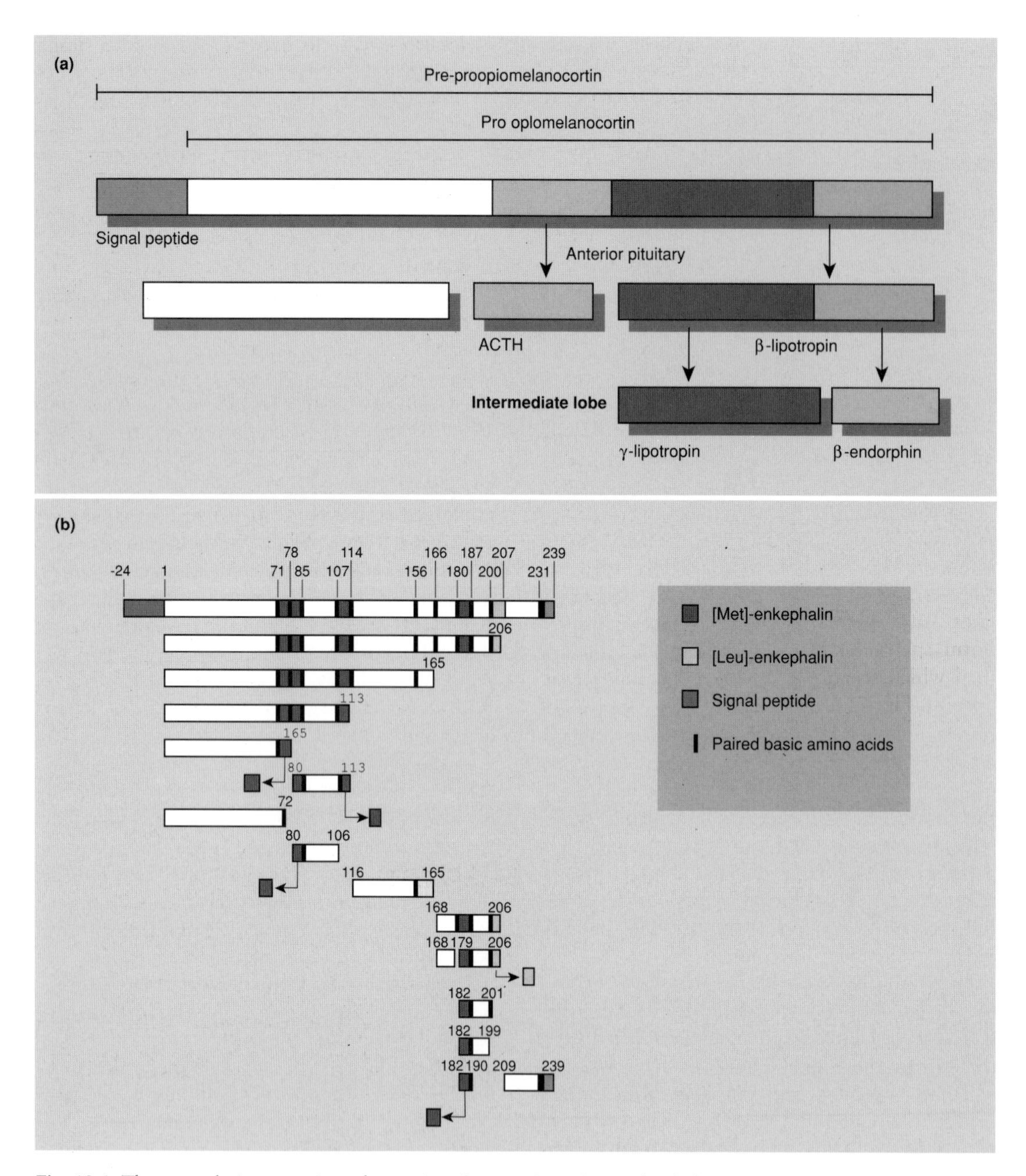

Fig. 12.1 The proteolytic processing of proopiomelanocortin and proenkephalin A. (a) Pre-proopiomelanocortin is the precursor of adrenocorticotrophic hormone (ACTH), lipotropins, melanocyte-stimulating hormones and endorphins. (b) Bovine pre-proenkephalin A generates a complex series of proteolytic fragment *en route* to the pentapeptide [Leu]- and [Met]-enkephalins. Cleavage sites are located at paired basic residues (bars). From Dillen *et al.* (1993).

released with the neuropeptides but which do not in themselves possess biological activity, such as the cysteinerich neurophysins which are co-released with AVP and OXY.

In contrast to small synaptic vesicles, there is no localized recycling and refilling of vesicles within the terminal, instead new vesicles must be transported from the cell body. Thus the amplification step inherent in the synthesis of Type I and II transmitters, where one molecule of enzyme transported down the axon can direct the synthesis of thousands of transmitter molecules, is lost in the case of the neuropeptides. The need to utilize neuropeptides more efficiently is reflected in the extremely high affinity of most peptide receptors, and in the need for specific patterns of stimulation in order to evoke release (*see* Fig. 8.3).

12.3 Non-opioid neuropeptides

Oxytocin and vasopressin. The first peptidergic neurones to be described were those with large *magnocellular* cell bodies in specific areas of the hypothalamus (Fig. 12.2) which synthesize AVP and OXY and project both to conventional synapses within many regions of the CNS and also to terminals in the posterior lobe of the pituitary (also called the *neurohypophysis*) which form specialized synapses with epithelial cells and which secrete the peptides into the bloodstream (Fig. 12.2). A rat has about 4000 OXY-releasing cells in its hypothalamus and a similar number of AVP neurones.

Both of these nonapeptides (*see* Table 12.1) are cyclized via disulphide bonds and contain C-terminal amide groups; the latter is a common feature of many neuropeptides. Amidation is performed by an enzyme within the secretory vesicle which at the same time cleaves off a C-terminal extension from the active peptide on the N-terminal side of a sequence 'Gly-basic-basic'. AVP is found in all mammals except the pig where the N-terminal amino acid is Lys.

Homogenization of the posterior pituitary leads to the formation of neurosecretosomes which behave as synaptosomes secreting AVP and OXY (*see* Section 2.6.1). Some neurosecretosomes are sufficiently large to permit patch clamping and electrophysiological determination of secretion and the Ca^{2+}-channels coupled to peptide exocytosis (Fig. 12.3).

The tachykinins (or neurokinins). There are five peptides in this family: substance P (SP), neurokinin A (or substance K), neurokinin B, neuropeptide K and neuropeptide γ. They have the

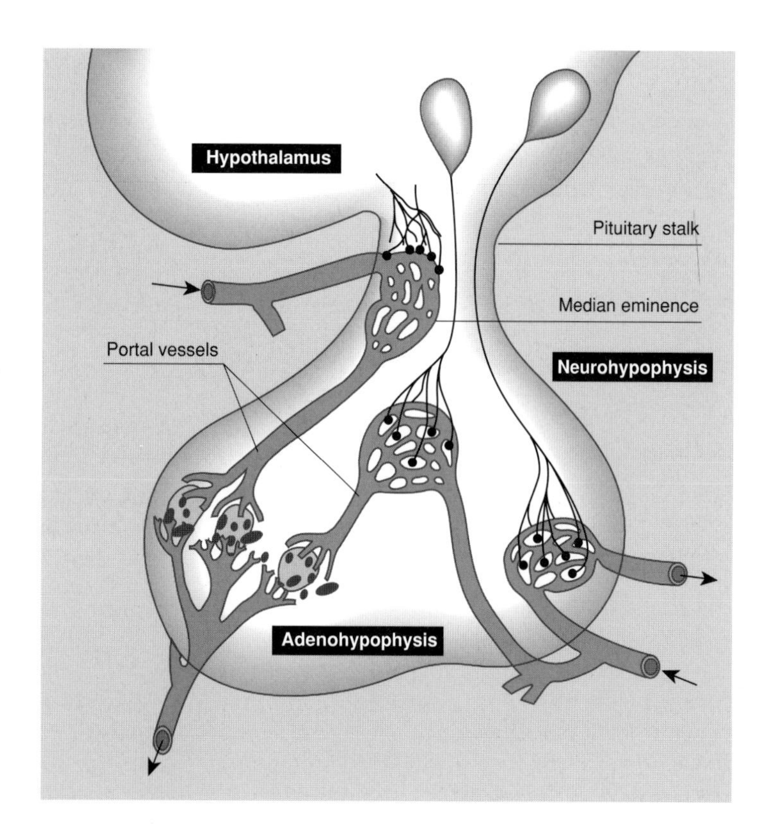

Fig. 12.2 The pituitary and its control by the hypothalamus. The hypothalamus controls the endocrine system in two ways. Magnocellular neurones project axons to the posterior pituitary, or neurohypophysis, where their large peptidergic terminals release *oxytocin* or *vasopressin*. The terminals may be isolated as *neurosecretosomes*. Other hypothalamic neurones secrete *releasing hormones* into the portal system which drains into the anterior pituitary, or adenohypophysis, where they control the secretion of anterior pituitary hormones.

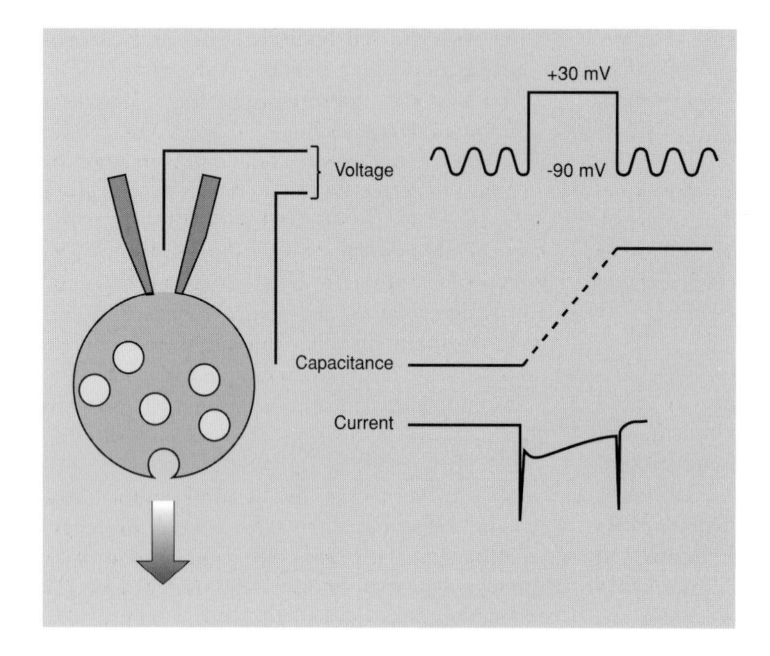

Fig. 12.3 Monitoring Ca^{2+} entry and peptide exocytosis in single neurosecretosomes. Neurosecretosomes (posterior pituitary peptidergic terminals) were immobilized on coverslips and patch clamped in 'whole terminal' mode. Capacitance was measured from the change in phase when a sinusoidal voltage (insufficient to cause exocytosis) was applied about a holding potential of −90 mV. The terminal was then depolarized to +30 mV for 80 ms before being repolarized. The new phase change indicated an increase in membrane capacitance of 50 fF, equivalent to the fusion of 20–80 vesicles. The current during the depolarization was largely due to the opening of Ca^{2+}-channels. Adapted from Lim *et al.* (1990).

common C-terminal sequence Phe-X-Gly-Leu-Met-NH_2 and are involved in sensory transmission and pain perception. SP (*see* Table 12.1) was the first neuropeptide to be discovered, in 1931, although its structure as an 11-residue peptide was not determined until much later. It originates from cell bodies in the caudate-putamen with terminals in the globus pallidus and the substantia nigra (where its concentration exceeds 1 nmol · g^{-1} wet weight).

One subclass of neurokinin receptor, NK3, potentiates the release of dopamine from striatal terminals, while SP enhances the $[Ca^{2+}]_c$ responses to NMDA in dorsal horn neurones of the spinal cord, probably via activation of PKC. SP plays a major role in the dorsal horn and in sympathetic ganglias a primary sensory neurotransmitter. *Capsaicin*, from hot peppers, depletes SP immunoreactivity in the spinal cord, although its effect does not seem to be directed purely towards these terminals since levels of CCK (see below) are also decreased.

The endothelins. Endothelin-1, -2 and -3 each possess 21 amino acid residues with two internal disulphide bridges (*see* Table 12.1). As well as being potent vasoconstrictors, the endothelins interact with neuronal receptors. cDNA clones have been isolated for two endothelin receptors, ET_A and ET_B with 426 and 441 residues respectively. Both receptors are coupled to phosphoinositide turnover *in situ*, but little is currently known about the neuronal action of the endothelins apart from their vasoconstrictive activity. The snake venom toxin *sarafotoxin S6b* possesses similar CNS activity.

Vasointestinal peptide. VIP functions both in the central and peripheral nervous system. The highest concentration of VIP is found in the cerebral cortex. At least 80% of the bipolar cholinergic neurones in rat cortex are also positive for VIP-like immunoreactivity and probably serve as interneurones (i.e. neurones with short axons and with cell body and terminals within a given brain area).

Somatostatin (SOM). SOM can exist as a cyclic 14 and 28 amino acid peptide with no C-terminal amidation (*see* Table 12.1). As with the C-terminal extensions of the enkephalins, it is necessary to decide whether SOM-28 possesses physiological actions of its own, or is merely a precursor of the active SOM-14. A differential effectiveness of the two forms in different cells suggests that SOM-28 may be more than merely a precursor. SOM was originally detected in the hypothalamus as a regulator of pituitary secretion, but SOM-positive neurones are widely distributed, most appearing to act as interneurones, some of which may contain GABA as a cotransmitter (see below). The neuropeptide is involved in cognitive functions and the modulation of locomotor activity.

Neuropeptide Y (NPY). NPY is a 36 residue, C-terminal amidated peptide. The peptide is relatively abundant and NPY-positive cell bodies are widely distributed throughout the cerebral cortex.

Calcitonin gene-related peptide (CGRP). The mRNA transcript of the rat calcitonin gene undergoes differential processing in different tissues: in the thyroid it is processed into the mRNA for a precursor for calcitonin, whereas in nervous tissue it can also be processed into the mRNA for a precursor of CGRP. The presence of the peptide in the CNS was initially detected by synthesizing the predicted C-terminus of the peptide, raising antibodies and determining that CGRP-like immunoreactivity was present throughout brain.

Bradykinin and related peptides. Bradykinin is a nine residue peptide; together with kallidin (lys-bradykinin) and met-lys-bradykinin they comprise the *kinins* which are released from injured tissues and act on either neuronal or non-neuronal receptors. The former can be pre- or postsynaptic.

Cholecystokinin. The total brain content of CCK is far higher than that of any other neuropeptide and is found at highest concentration within the cerebral cortex. CCK can exist in multiple forms with 39, 33, 12, eight, four or three residues depending on the extent of proteolytic cleavage within the vesicles. CCK-8 can additionally exist in sulphated and non-sulphated forms; the former is resistant to cleavage by amidopeptidases and is therefore more potent. CCK has been reported to modulate the release of dopamine and GABA. The CCK_B form of the receptor is predominant in the brain, while CCK_A receptors occur peripherally.

Galanin. Galanin is a 29 amino acid neuropeptide which does not belong to any known family of biologically active peptides. Galanin-like immunoreactivity and receptor binding is widely distributed in the brain stem, frequently in association with the cell bodies of cholinergic, serotonergic or noradrenergic neurones. The receptors are not limited to the CNS, however, and galanin can inhibit insulin secretion from the pancreas. Since galanin is rather resistant to proteolysis thus has a long half-life following secretion. Galanin has predominantly inhibitory hyperpolarizing actions postsynaptically consistent with a coupling of galanin receptors via G_i/G_o to K^+-channel activation, although as so often with this mode of coupling, the alternative inhibition of Ca^{2+}-channels may also occur.

12.4 Opioid peptides

Opioid peptides are the natural endogenous agonists for the opiate receptors which were first identified by their ability to bind plant opiate alkaloids, such as morphine, stereospecifically and with high affinity. The first opioid peptides were identified in pig brain in 1975, and were found to be simple pentapeptides: Tyr-Gly-Gly-Phe-Met ([Met]-enkephalin) and at lower concentrations Tyr-Gly-Gly-Phe-Leu ([Leu]-enkephalin). Following this initial discovery, at least 12 endogenous peptides have been described which have opiate-like effects in the CNS and which originate by differential cleavage of just three precursor polypeptides which possess considerable homology and may have evolved from a common gene.

All the active opioids are C-terminal extensions of either [Leu]- or [Met]-enkephalin (*see* Table 12.1). Their distribution within the CNS depends both on the expression of the precursor proteins and the distribution of the specific proteases responsible for cleavage of the precursors. The extremely low concentrations of the opioid peptides requires the use of sensitive radioimmunoassays using specific monoclonal or polyclonal antibodies raised following cross-linking to larger peptides. This in turn leaves the assays open to the possibility of cross-reaction with other peptides, particularly in view of the close sequence homology of many neuropeptides. This problem is particularly apparent with monoclonal antibodies and must be closely controlled for. For that reason, peptides detected by purely immunological means are referred to as, for example, 'enkephalin-like immunoreactivity' (Enk-LI). *In situ* hybridization of mRNA with DNA probes can be used to detect the presence of mRNA for the specific precursor proteins, but that this reveals only the sites of synthesis and not the storage sites in the terminals.

12.5 Neuropeptide synthesis

Following the determination of the sequence of isolated and purified neuropeptides, oligonucleotide probes can be constructed corresponding to these sequences and used to determine the sequence of cDNAs corresponding to the mRNAs coding for the larger precursor peptides. From the deduced amino acid sequence it is possible to see whether the precursor contains any other candidate peptides which would be likely to be produced by cleavage at pairs of basic residues. In this way β-endorphin and other peptides were originally identified.

Proenkephalin and derived opioids. ProEnk was first detected in the adrenal cortex from the sequence of an isolated mRNA which appeared to contain six copies of the [Met]-enkephalin sequence and one of [Leu]-enkephalin. The CNS form of the protein appears to be identical (*see* Fig. 12.1). From the distribution of pairs of basic amino acid residues (processing signals for the trypsin-like proteases) it was predicted that cleavage of the precursor would yield four copies of [Met]-enkephalin, one copy of [Leu]-enkephalin and in addition two novel opioid peptides with seven and eight residues respectively due to C-terminal extensions of [Met]-enkephalin. Particularly in the CNS the processing proceeds further to give the free enkephalins as the principle products.

The enkephalins are the most widely distributed of the opioids within the CNS and are found both in neurones forming local circuits and those with long axonal projections to other brain areas. The distribution of [Met]-enkephalin and [Leu]-enkephalin is identical, and the ratio of their concentrations reflects the presence of more copies of the former in the precursor protein. Their highest concentration is in the *globus pallidus*.

Pro-opiomelanocortin and derived opioids. POMC is a 263 residue protein which was the first neuropeptide precursor protein to be identified from mRNA using recombinant DNA techniques (*see* Fig. 12.1). POMC undergoes a complex cleavage scheme which can potentially yield not only endorphins, but also the pituitary hormone β-lipoprotein, adrenocorticotrophic hormone (ACTH) and melanocyte-stimulating hormones (MSH), depending on the pattern of processing. Cleavage occurs at pairs of basic amino acid residues (Lys or Arg). β-endorphin is the predominant opioid product of POMC cleavage.

POMC is primarily found in the intermediate lobe of the pituitary gland, but β-endorphin expressing cell bodies are also located in the arcuate nucleus of the hypothalamus, from where β-endorphin secreting nerve fibres project to many areas of brain.

Prodynorphin and derived opioids. Prodynorphin is found both in the CNS and in secretory tissues and was first detected as an mRNA sequence containing the sequence of three opioid peptides with C-terminal extensions from [Leu]-enkephalin, namely dynorphin A (1–17), dynorphin A (1–8) and dynorphin B. The dynorphins are found in highest concentrations in the hypothalamus and the posterior pituitary, but there is a wide distribution within the CNS, although notably both the cerebral cortex and cerebellum have very low concentrations.

Large mossy fibre terminals in the hippocampus provide an excitatory input to CA3 neurones. The terminals release both glutamate and dynorphin, the peptide activating κ_1 opiate receptors (*see* Section 12.7) causing a presynaptic inhibition of glutamate release from neighbouring mossy fibre terminals.

12.6 Coexistence and co-release of neuropeptides with other transmitters

Coexistence of neuropeptides with Type II transmitters in the same terminal may be the rule rather than the exception. Thus small and large dense-cored vesicles can frequently be seen within the same terminal (Fig. 12.4). The ease with which antibodies to the peptides may be prepared has facilitated histochemical localization of peptides, or strictly peptide-like immunoactivity. Other more discriminating techniques involve chemical or surgical lesions (following the simultaneous disappearance of the peptide and another transmitter as the terminals degenerate) and direct determination of

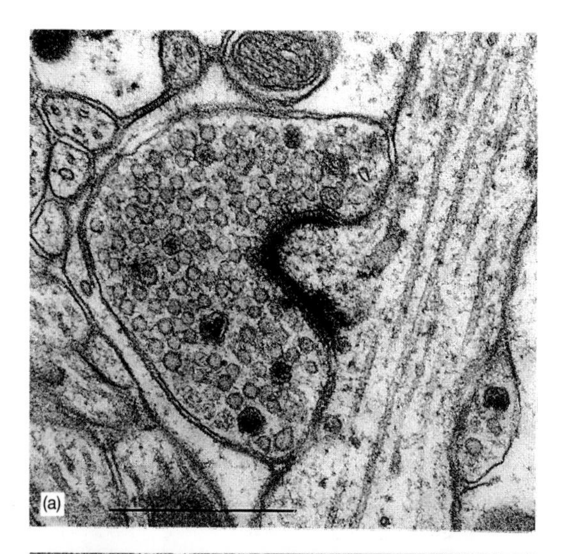

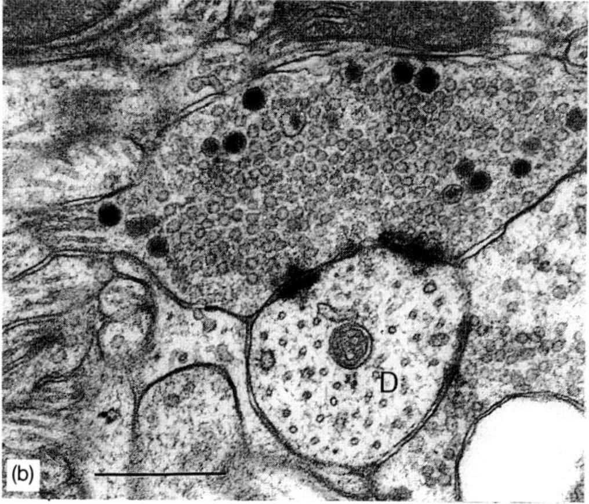

Fig. 12.4 Coexistence of small and large dense-cored vesicles (LDCVs) in the same terminal. (a) Primary afferent nerve terminals in the spinal cord are immunoreactive for the peptide substance P which is localized to LDCVs and glutamate. From De Biasi & Rustoni (1988). (b) A terminal from rat CNS synapsing onto a dendrite (D) and containing both SSVs and LDCVs (SG). From Buma (1989), reproduced with permission from Swets Publishing Service.

co-release from defined terminals. Although more than 100 examples of coexistence have been detected histochemically, the mechanistic relevance has been determined in rather few cases, and then predominantly in the peripheral nervous system, where it is easier to confirm co-release and to document the postsynaptic action.

It is not yet possible to predict which combination of transmitters will be found in a given neurone; since the DNA in all neurones carries the information for the synthesis of any neuropeptide or enzyme for the synthesis of Type I or Type II transmitter, this would require a knowledge of the factors which govern neuronal gene expression. However, some frequently observed combinations are as follows (it should be emphasized that many other combinations may exist):

1 *ACh and VIP.* VIP is present in, and co-released from, particular cholinergic postganglionic neurones such as those of the cat submandibular gland. ACh is present both in small clear vesicles and in LDCVs. On density gradient separation only the heavier fraction is found to contain VIP (strictly VIP-like immunoactivity). However, most cholinergic neurones do not contain VIP.

2 *NA with an opioid peptide or neuropeptide Y.* Catecholamines are stored together with [Leu]-enkephalin and [Met]-enkephalin in chromaffin vesicles. Similarly sympathetic LDCVs contain NA together with NPY, and it is possible that the same vesicles also contain [Met]-enkephalin.

3 *5-HT with substance P.* This may be observed in spinal cord and substantia nigra.

4 *Dynorphin with glutamate.* This occurs in the hippocampal mossy fibre terminals discussed above.

Further types of co-transmission may be seen:

1 Multiple neuropeptides derived from a common precursor peptide, e.g. POMC giving rise to α-MSH and β-endorphin; while ACTH, β-lipotropin and β-endorphin may all be found within the same neurones, these being derived from a common precursor peptide.

2 Multiple neuropeptides which are the products of the same gene, but a result of differential splicing of the primary transcript to yield two different mature mRNAs.

3 Multiple neuropeptides which are the products of distinct genes coexpressed in the same cells, e.g. the coexistence of SOM and NPY.

4 Finally, coexistence of classical neurotransmitters may occur in invertebrates (e.g. 5-HT and ACh in *Aplysia* ganglia).

Although peptides are invariably in LDCVs, Type II transmitters may be in both small and large vesicles. A strict colocalization of enkephalins, catecholamines and ATP in the same vesicles occurs in the chromaffin cell, with resultant parallel co-release.

12.6.1 Regulation of the release of coexisting transmitters

Presynaptic regulation of classical transmitter release, frequently by inhibitory autoreceptors, is well established. In the case of coexisting transmitters a presynaptic receptor will usually regulate the release of both transmitters in the same sense. Thus in terminals where ACh and VIP coexist, presynaptic muscarinic receptors inhibit the release of both transmitters in parallel (Fig. 12.5), while in the vas deferens presynaptic α_2-adrenoceptors inhibit in parallel release of NA and NPY. Conversely presynaptic neuro-

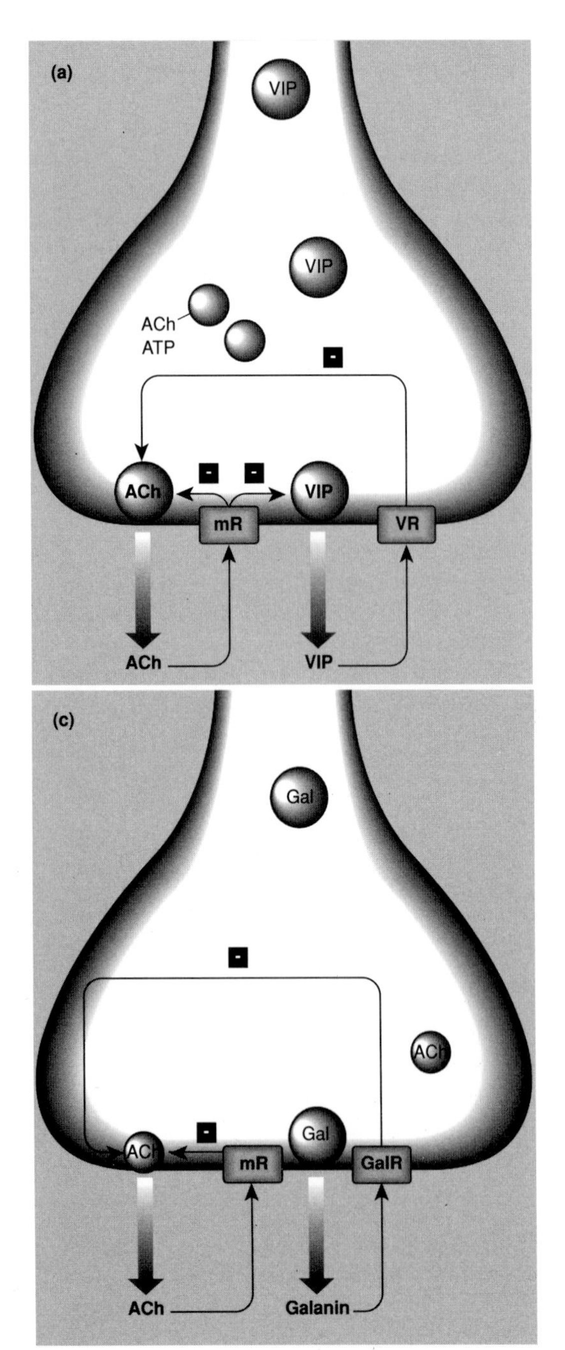

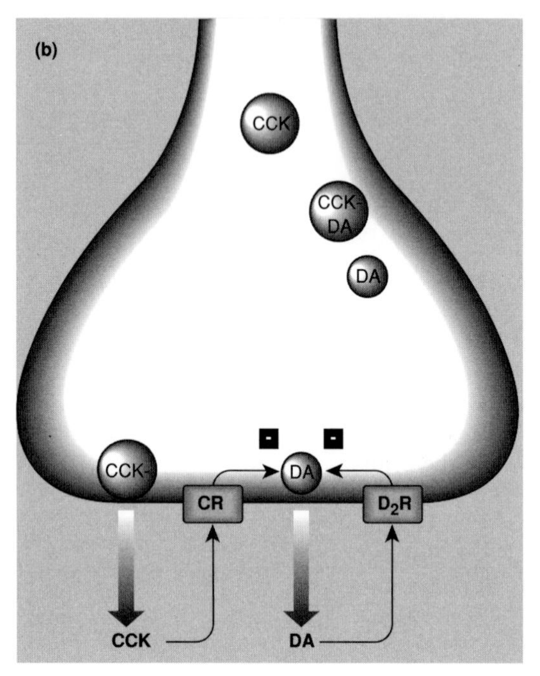

Fig. 12.5 Some examples of presynaptic 'cross-talk' between peptidergic and non-peptidergic transmitters. (a) In the rat cerebral cortex ACh can inhibit the release of both ACh itself and vaso-intestinal peptide (VIP), while VIP can inhibit ACh release. (b) Cholecystokinin (CCK) can inhibit dopamine (DA) release in the rat nucleus accumbans, as can DA itself via D_2 receptors. (c) ACh release can be inhibited in the rat ventral hippocampus by both ACh itself and galanin (Gal). mR, VR, CR, D_2R and GalR are receptors for ACh, VIP, CCK, DA and Gal respectively. Adapted from Bartfai *et al.* (1988).

peptide autoreceptors are capable of regulating the release of the classical transmitter. In combination with the distinctive frequency dependencies for the release of the transmitters (*see* Fig. 8.3), these mechanisms allow for a sophisticated regulation of the proportion of released transmitter (Fig. 12.5).

12.7 Neuropeptide receptors

To date, G-protein-coupled receptors for some 17 different neuropeptides have been sequenced by molecular cloning. Sequence homologies with other metabotropic receptors indicate that they share the common topography of 7-transmembrane α-helices and an extracellular N-terminal which can vary in length from 20 to 100 residues and contain Asn residues for *N*-glycosylation, although there are no obvious common motifs which distinguish peptide receptors from other G-protein-coupled receptors. The transmembrane domains, particularly TM4, TM6 and TM7 show the highest degree of conservation, while the cytoplasmic C-terminus is again variable, varying in length from <30 to >100 residues and frequently with Cys residues capable of palmitoylation.

A number of approaches have been used to clone neuropeptide receptors:

1 Classical protein purification using specific ligand binding as an assay, followed by partial sequencing, construction of oligonucleotides derived from these sequences and screening of a cDNA library, e.g. the endothelin ET_B receptor.

2 Screening for functional expression of RNA derived from *in vitro* transcription of a cDNA library e.g. following injection into *Xenopus* oocytes (*see* Section 2.9) as described for the NMDA receptor (*see* Fig. 9.14).

3 Screening of cDNA libraries with probes against conserved transmembrane regions of other G-proteins, although in this case it is not possible to predict the ligand for any receptor so cloned. In a number of cases 'orphan receptors' have been found with no known ligands.

Since the natural neuropeptides are rapidly degraded by extracellular proteases (see below), much use is made of synthetic agonists which are resistant to proteolysis, frequently by the substitution of a D-amino acid for the natural stereoisomer (for example, DADL — [D-Ala^2-D-Leu^5]Enk as an enkephalin analogue).

The pharmacological effects of the opioids are very similar to those of plant-derived and synthetic alkaloid opiates, being broadly associated with pain perception, tolerance, physical dependence and stress mechanisms. Opioid peptides interact with the endocrine system and increase release of peptide hormones such as growth hormone, prolactin, ACTH while decreasing follicle-stimulating hormone, luteinizing hormone and thyrotropin levels.

Opioid receptors are G-protein linked. Their original classification was pharmacological, based on binding specificities. μ Receptors have uniquely high affinity for morphine, while δ receptors have less affinity for morphine and more affinity for enkephalins, κ receptors have little affinity for either morphine or enkephalins, but bind dynorphins.

μ-opioid receptors are present on many nerve terminals where they inhibit the release of a number of neurotransmitters, including adrenaline, DA, ACh and SP. As with other presynaptic inhibitory receptors, their mechanism of action has to be extrapolated from electrophysiological studies at the cell soma. Assuming that the same mechanisms function at the presynaptic terminal a generalization is that agonists cause a direct inhibition of Ca^{2+}-channels and/or an activation of K^+-channels, leading to a shortening of action potential duration.

δ-opioid receptors are largely presynaptic and are found at high concentration in deep nuclei of the cerebral cortex, hypothalamus and substantia nigra. They are potent inhibitors of transmitter release, for example, in the large mossy fibre terminals of the hippocampus (*see* Fig. 1.10) where they are co-stored with glutamate and inhibit exocytosis by an autoreceptor effect on Ca^{2+}-and/or K^+-channels.

So-called sigma (σ) receptors do not fall into any clear category. Benzomorphan derivatives which were developed in an attempt to produce less addictive opiate analgesics were found to be of limited use due to psychomimetic side-effects. The site to which these compounds bound was termed the σ receptor. A number of binding proteins have been partially purified, but it is not clear that they represent true receptors, since there is not unambiguous information on any modulation of known second messenger systems.

The postsynaptic effects of neuropeptides are largely defined by the G-protein with which their receptors interact. Thus a family of *SOM* receptors has been cloned, termed SSTR1–3, which show distinct but overlapping patterns of expression in the brain. Cultured cell lines stably expressing the recombinant receptors indicates that SSTR2 and SSTR3 can be coupled via pertussis toxin-sensitive G-proteins to the classic responses of inhibited adenylyl cyclase (via $G_{i\alpha 3}$, SSTR3), Ca^{2+}-channel inhibition (via $G_{o\alpha}$, SSTR2) and K^{+}-channel activation (via $G_{1\alpha 3}$, SSTR2). The role of SSTR1 is currently unclear.

12.7.1 Peptide transmitter inactivation

In the absence of re-uptake mechanisms, neuropeptides are inactivated by a variety of cell surface associated peptidases, which do not appear to be peptide specific. *Endopeptidase-24.11*, also known as enkephalinase, is an integral membrane protein localized to neurones, but not glia, and is localized both pre- and postsynaptically. The endopeptidase possesses a short N-terminal cytoplasmic domain, a single transmembrane domain, and a large (700 amino acid) extracellular domain which contains the active site allowing it to hydrolyse released neuropeptides including SP, [Leu]- and [Met]-enkephalin and OXY (but not vasopressin) on the amino side of a hydrophobic residue (Fig. 12.6). The enzyme is found in peptide-rich brain regions such the striatum and the substantia nigra where it colocalizes with SP. A second peptidase, *peptidyl dipeptidase A*, also known as angiotensin-converting enzyme (170 000 kDa) is found associated with dendrites. *Aminopeptidase N* can hydrolyse N-terminal residues of neuropeptides.

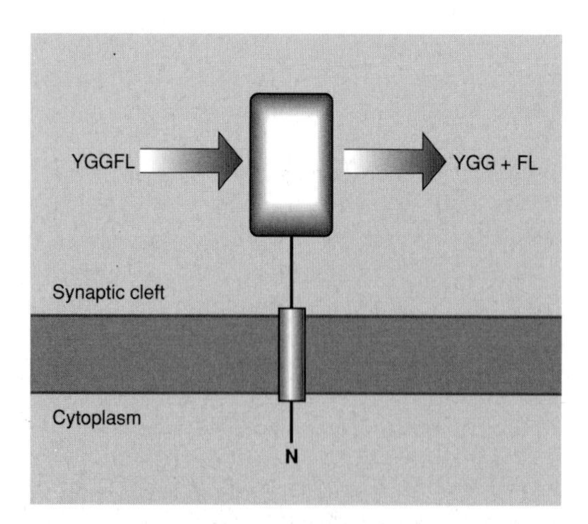

Fig. 12.6 Endopeptidase-24.11. The endopeptidase 24.11 hydrolyses released neuropeptides such as [Leu]-enkephalin (YGGFL) as well as tachykinins and dynorphin. The enzyme has a single transmembrane domain and a large extracellular domain containing the catalytic site. Adapted from Barnes *et al.* (1992).

Further reading

Synthesis and processing

Dillen, L., Miserez, B., Claeys, M., Aunis, D. & De Potter, W. (1993) Posttranslational processing of proenkephalins and chromagranins/secretogranins. *Neurochem. Int.* **22**, 315–352.

Coexistence and release

Artaud, F., Baruch, P., Stutzmann, J.M., Saffroy, M. & Godeheu, G. (1989) Cholecystokinin: co-release with dopamine from nigro-striatal neurons in the cat. *Eur. J. Neurosci.* **1**, 162–171.

Bartfai, T., Fisone, G. & Langel, Ü. (1992) Galanin and galanin antagonists: Molecular and biochemical perspectives. *Trends Pharmacol. Sci.* **13**, 312–317.

Bartfai, T., Iverfeldt, K., Fisone, G. & Serfözö, P. (1988) Regulation of the release of coexisting neurotransmitters. *Annu. Rev. Pharmac. Toxicol.* **28**, 285–310.

Battaglia, G. & Rustioni, A. (1988) Coexistence of glutamate and substance P in dorsal root ganglion neurons of the rat and monkey. *J. Comp. Neurol.* **277**, 302–312.

Buma, P. (1989) Synaptic and non-synaptic release of neuromediators in the central nervous system. *Acta Morphol. Neerl. Scand.* **26**, 81–113.

De Biasi, S. & Rustioni, A. (1988) Glutamate and substance P coexist in primary afferent terminals in the superficial laminae of spinal cord. *Proc. Natl. Acad. Sci. USA* **85**, 7820–7824.

Hoekfelt, T. (1991) Neuropeptides in perspective: The last ten years. *Neuron* **7**, 867–879.

Iverfeldt, K., Serfozo, P., Diaz-Arnesto, L. & Bartfai, T. (1989) Differential release of coexisting neurotransmitters: frequency dependence of the efflux of substance P, thyrotropin releasing hormone and [^{3}H] serotonin from tissue slices of rat ventral spinal cord. *Acta Physiol. Scand.* **137**, 63–71.

Lim, N.F., Nowycky, M.C. & Bookman, R.J. (1990) Direct measurement of exocytosis and calcium currents in single vertebrate nerve terminals. *Nature* **344**, 449–451.

Nordmann, J.J., Dayanithi, G. & Lemos, J.R. (1987) Isolated neurosecretory nerve endings as a tool for studying the mechanism of stimulus-secretion coupling. *Biosci. Rep.* **7**, 411–425.

Zhu, P.C., Thureson-Klein, Å. & Klein, R.L. (1986)

Exocytosis from large dense cored vesicles outside the active synaptic zones of terminals within the trigeminal subnucleus caudalis: a possible mechanism for neuropeptide release. *Neurosci.* **19**, 43–54.

Neuropeptide receptors

Burbach, J.P.H. & Meijer, O.C. (1992) The structure of neuropeptide receptors. *Eur. J. Pharmacol. Mol. Pharmacol.* **227**, 1–18.

Fletcher, G.H. & Chiappinelli, V.A. (1993) The actions of the kappa, opioid agonist U-50 488 on presynaptic nerve terminals of the chick ciliary ganglion. *Neurosci.* **53**, 239–250.

Izquierdo, I. (1990) Acetylcholine release is modulated by different opioid receptor types in different brain regions and species. *Trends Pharmacol. Sci.* **11**, 179–180.

Jin, W., Lee, N.M., Loh, H.H. & Thayer, S.A. (1992) Dual excitatory and inhibitory effects of opioids on intracellular calcium in neuroblastoma × glioma hybrid NG108-15 cells. *Mol. Pharmacol.* **42**, 1083–1089.

Loh, H.H. & Smith, A.P. (1990) Molecular characterization of opioid receptors. *Annu. Rev. Pharmac. Toxicol.* **30**, 123–47.

Taussig, R., Sanchez, S., Rifo, M., Gilman, A.G. & Belardetti, F. (1992) Inhibition of the omega-conotoxin-sensitive calcium current by distinct G proteins. *Neuron* **8**, 799–809.

Toru, T., Konno, F., Takayanagi, I. & Hirobe, M. (1989) Kappa-receptor mechanisms in synaptosomal Ca uptake. *Gen. Pharmacol.* **20**, 249–252.

Wollemann, M. (1990) Recent developments in the research of opioid receptor subtype molecular characterization. *J. Neurochem.* **54**, 1095–1101.

Peptide transmitter inactivation

Barnes, K., Turner, A.J. & Kenny, A.J. (1992) Membrane localization of endopeptidase-24.11 and peptidyl dipeptidase A (angiotensin converting enzyme) in the pig brain. *J. Neurochem.* **58**, 2088–2096.

13

Molecular aspects of neuronal development and plasticity

13.1 Introduction

Neuronal development involves several processes including cell birth and differentiation, cell migration and survival, directed neurite (axon and dendrite) outgrowth and growth cone targetting and finally synaptogenesis. Three classes of molecules have been associated with neuronal development and plasticity: (i) *growth factors* which induce neuronal differentiation and neurite extension; (ii) neural cell *adhesion molecules* which act in concert with growth factors to guide developing axons to their targets; and (iii) *growth-associated proteins* or GAPs which are found in specific areas of developing neurones such as the growth cone at the tip of outgrowing neurites, but whose functions largely remain to be clarified.

Neuronal plasticity is the process whereby specific patterns of input into mature synapses can cause long-lasting changes in the magnitude of the postsynaptic response to subsequent stimulation. This is generally believed to be the essence of the learning and memory processes.

13.2 Neuronal growth factors and neuronal survival and differentiation

Nerve growth factor (NGF) permits the survival and supports the development and differentiation of neurones both in the PNS (particularly sympathetic neurones) and in the CNS where it is predominantly expressed by hippocampal pyramidal and granular neurones and supports cholinergic neurones in the basal forebrain. NGF acts in a retrograde manner, being secreted by the postsynaptic cell onto receptors on the presynaptic terminal.

PC12 cells (*see* Section 2.3), derived from the adrenal tumour phaeochromocytoma, originally secreted NA, although most strains have lost dopamine β-hydroxylase and so store and secrete DA. PC12 cells are frequently exploited for the study of NGF since they do not require the peptide for survival, but differentiate from a characteristic round cell to a neuronal phenotype when exposed to nanomolar NGF (Fig. 13.1).

NGF binds to two classes of receptors on PC12 cells: a low-affinity (K_d *c.* 10^{-9} M) receptor termed $p75^{NGFR}$ (Fig. 13.2) and a 140 000 kDa high-affinity (K_d *c.* 10^{-11} M) receptor, $gp140^{trk}$. The latter, which is present in very low amounts, is the product of a proto-oncogene *trk*. $gp140^{trk}$ is a single transmembrane receptor containing an intrinsic protein–tyrosine kinase activity towards its C-terminus (Fig. 13.2). *Brain-derived neurotrophin factor*, BDNF, and *neurotrophins 3, 4* and *5* display similar biological effects to NGF but each act on characteristic populations of cells by binding to receptors distinct from, but homologous to $gp140^{trk}$.

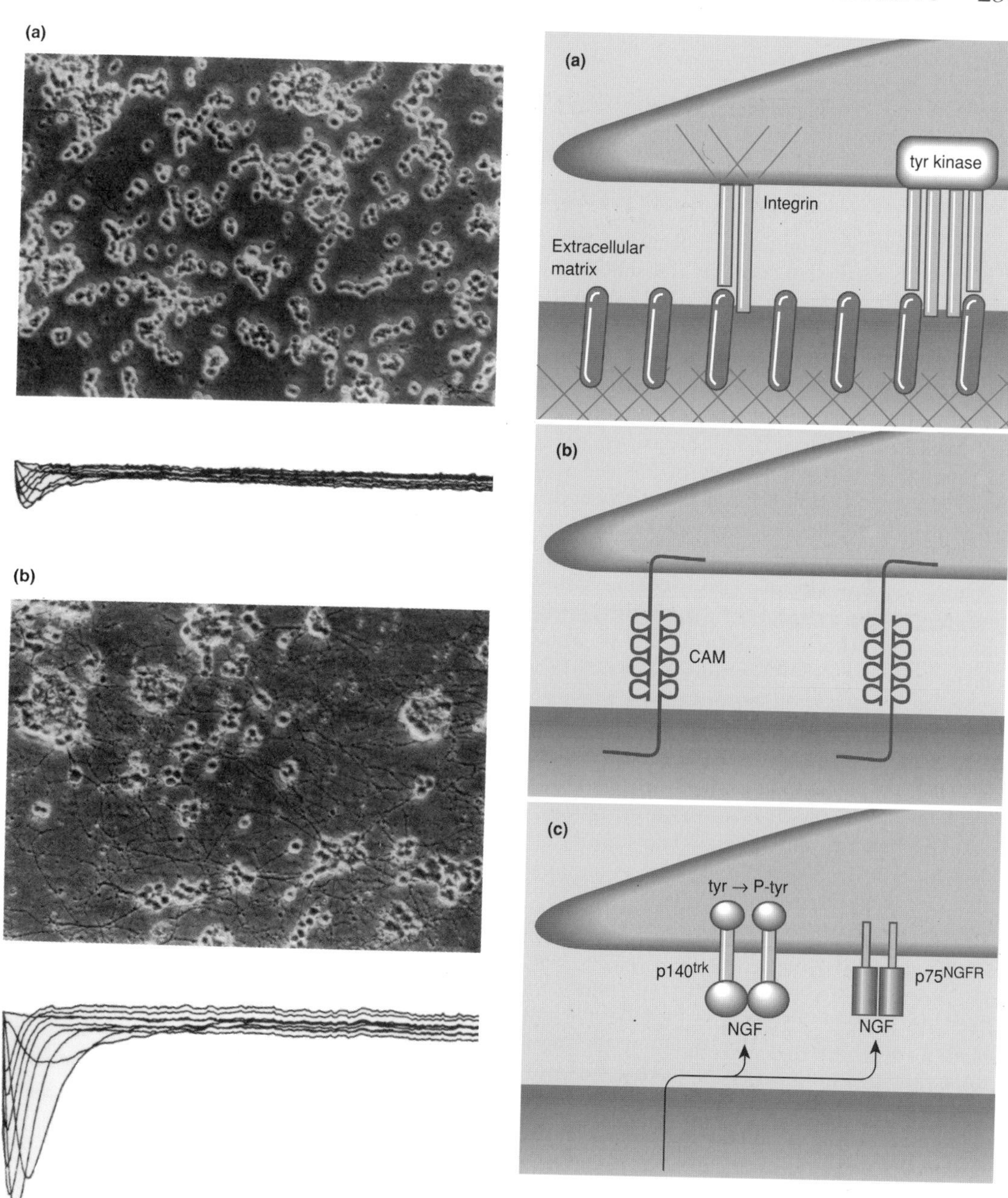

Fig. 13.1 NGF induces PC12 cell differentiation. (a) PC12 cells grown in culture show a rounded morphology and possess few voltage-activated Na^+-channels (shown as whole-cell patch-clamp currents). (b) after 10 days treatment with NGF the cells have extended long neurites and their cell bodies show a six-fold increase in Na^+-channel density. From Pollock *et al.* (1990), by permission of the Society for Neuroscience.

Fig. 13.2 Integrins, cell adhesion molecules and growth factor receptors. (a) The integrin family of cell surface receptors recognize components of the extracellular matrix and may either act as 'nucleation sites' for the cytoskeleton or, after clustering, may trigger a tyrosine kinase cascade. (b) Cell adhesion molecules (CAM) form homomeric attachments between two cells. (c) Many growth factor receptors, such as those for nerve growth factor (NGF), are tyrosine receptor kinases and can undergo autophosphorylation. NGF also has a lower affinity receptor whose role is currently unclear.

13.2.1 Cytosolic and receptor tyrosine kinases

Tyrosine kinases in the nervous system are either cytosolic (including those anchored to the cytoplasmic face of the plasma membrane via N-terminal myristoylation) or integral plasma membrane glycosylated tyrosine receptor kinases with a single transmembrane segment (Fig. 13.2). The best characterized cytosolic tyrosine kinase is the product of the c-*src* gene: mRNA from the widely expressed proto-oncogene c-*src* is differentially spliced in neurones to give a product, $pp60^{c\text{-}src+}$, which is localized throughout the neurone, including synaptic vesicles and growth cones. A target of $pp60^{c\text{-}src+}$ in growth cones is tubulin, which plays a role in extension of filopodia (*see* Section 13.3) while phosphorylation may affect neurite migration. In synaptic vesicles, the integral protein p29 and the homologous synaptophysin (*see* Section 7.4.4) undergo tyrosine phosphorylation although the functional significance is not currently clear.

The receptor tyrosine kinases which are expressed in the nervous system include those for insulin, epidermal growth factor (EGF), platelet-derived growth factor (PDGF), fibroblast growth factor (FGF), and for a family of neurotrophins (including NGF). Each possesses a cytoplasmic catalytic domain and a single transmembrane segment. However the extracellular N-terminal domains are heterogeneous.

Binding of ligand to receptor tyrosine kinases results in dimerization of the receptor and trans-phosphorylation between the monomers on multiple tyrosine residues in the cytoplasmic domain (Fig. 13.2c). The resulting phosphoprotein domains allow specific interactions with enzymes which contain Src-homology in two (SH_2) domains. These include the Src tyrosine kinase and PLC-γ_1, the latter being responsible *in vitro* for the production of phosphatidylinositol 3,4,5-trisphosphate. This in turn can activate the atypical PKC-ζ (*see* Section 5.5.2). Phosphatidylinositol 3-kinase is activated indirectly via a p85 *adapter* subunit. $p21^{ras}$ GTPase-activating protein (GAP, regulating the de-activation of the GTP-activated small G-proteins and not to be confused with GAP-43), is a substrate for the NGF receptor, or via a SH_2-containing intermediate GRB2.

There is currently a limited knowledge of the immediate target proteins for the receptor tyrosine kinases, although Src and the Ras GTP-binding protein are involved sequentially; thus microinjection of antibodies against $p21^{ras}$ can block the transforming effects of NGF while expression of mutated or oncogenic viral forms of the *src* and *ras* genes can mimic NGF-induced neurite extension in the absence of NGF. Two members of the Raf serine/threonine protein kinase family, Raf-1 and B-Raf are hyperphosphorylated by NGF addition to PC12 cells. The kinases are downstream of the Ras protein, although it is not yet established that they are obligatory intermediates in the NGF signalling pathway.

In addition to Raf, NGF stimulates indirectly (via Ras) a number of additional cytoplasmic protein kinases including mitogen-activated protein (MAP) kinases, MAP kinase activators and the *rsk*-encoded kinase (RSK). The first nuclear event which can be detected following NGF addition to PC12 cells is the transient activation of the *immediate early genes*, *fos*, *myc* and *jun*. The protein products fos and jun form a heterodimeric transcription factor which recognizes a seven base-pair enhancer element found upstream of many genes. Neuronal *fos* and *jun* activation is not restricted to NGF, since elevation of $[Ca^{2+}]_c$ by NMDA receptor activation has a similar effect. Interestingly, elevation of $[Ca^{2+}]_c$ by other means is not effective.

The components discussed above appear to be ordered: NGF receptor → Src → Ras → Raf (in fibroblasts but perhaps not in PC12 cells) → MAP kinase activator → MAP kinase → RSK → gene transcription. However, it must not be assumed that this is a simple linear pathway, since NGF receptor, Src, Ras and Raf mediate the expression of progressively fewer genes, implying the existence of branching.

Clearly tyrosine kinases would be unable to function in signal transduction without the presence of tyrosine phosphatases. Tyrosine phosphatases show no homology with serine/threonine phosphatases and exist as an extensive family in two forms: putative receptor tyrosine phosphatases possess a single transmembrane segment, extracellular N-termini homologous to tyrosine receptor kinases and cell adhesion molecules (CAMs, *see* Section 13.3) and cytoplasmic domains which provide a tandem repeat of a catalytic tyrosine phosphatase motif. A second class of tyrosine phosphatases is soluble (or associated via a C-terminal hydrophobic anchor with the endoplasmic reticulum). Specific phosphatases may act as the counter to each tyrosine kinase in the growth factor signal cascade.

13.2.2 Non-tyrosine kinase growth factor receptors

The low-affinity NGF receptor, p75^{NGFR} (*see* Fig. 13.2), is more abundant than gp140^{trk} in most cells and is expressed in all NGF-responsive cells. The transmembrane glycoprotein does not possess a tyrosine kinase domain and many functions have been proposed for the receptor, including coupling via G-proteins to second messenger cascades, involvement in retrograde transport of NGF and association with gp140^{trk} in order to achieve high-affinity binding. However gp140^{trk} can function as a homodimer in the absence of the low-affinity receptor. The nature of any second messenger which might be generated by p75^{NGFR}, if any, is not firmly established, although both arachidonate production and PI turnover is enhanced. However, clearly additional factors must be involved, since other agonists which do increase these messengers do not induce PC12 differentiation.

13.3 Growth cones and neurite extension

Each developing axonal or dendritic neurite has at its leading edge a growth cone (Fig. 13.3). This amoeboid structure is highly dynamic, sending out *filopodia* which in tissue culture respond to the nature of the extracellular matrix upon which they are grown. The receptors on the surface of the growing neurite (and other developing cells) which recognize and interact

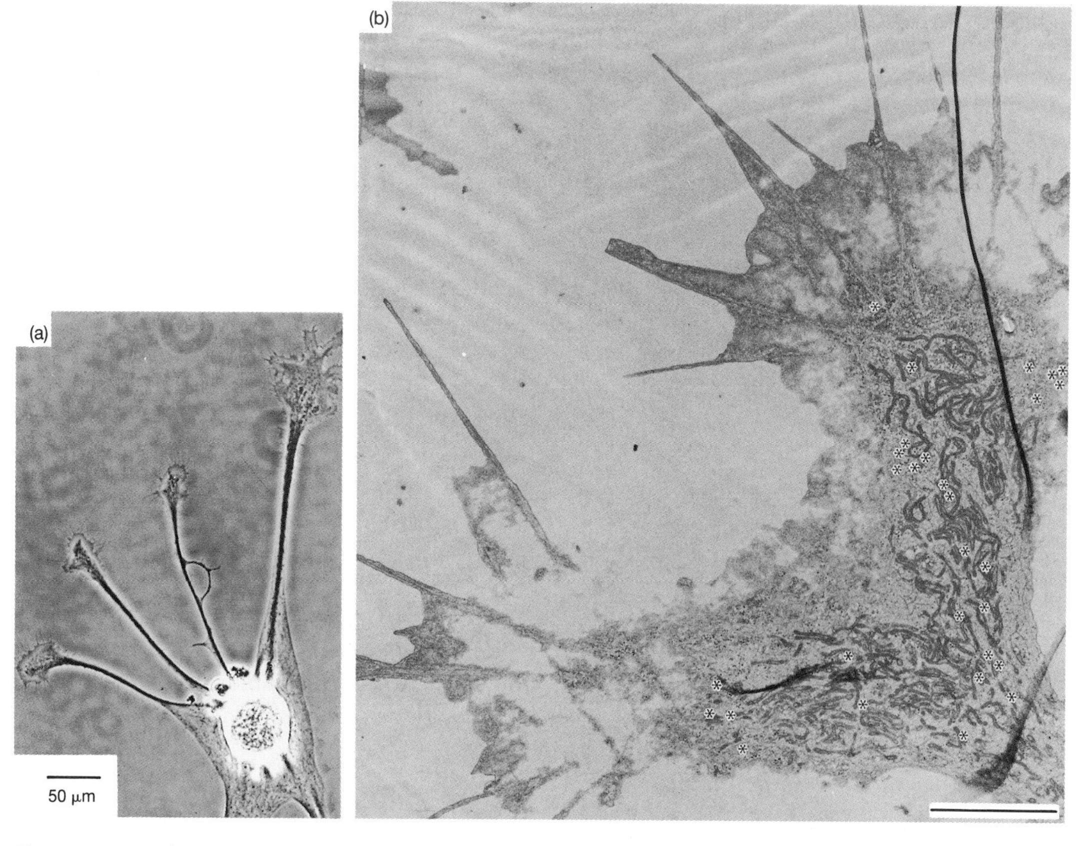

Fig. 13.3 Growth cones. (a) A cultured snail neurone extending large growth cones. (b) Growth cones are capable of independent protein synthesis (unlike mature terminals) and contain polyribosomes (asterisks). Bar = 5 μm. From Davis *et al.* (1992), by permission of the Society for Neuroscience.

with components of the extracellular matrix are known as *integrins*. These are a family of non-covalently linked glycoprotein heterodimers with M_rs of some 280 kDa. The mammalian family of integrins is made up of at least 12 α- and eight β-subunits that can associate to give a large number of combinations. Each subunit possesses a large extracellular domain, a single membrane-spanning region and a short cytoplasmic domain of 20–60 residues (*see* Fig. 13.2a).

A wide variety of these two subunit receptors have been identified with specificities for different extracellular matrix proteins. Thus α5/β1 integrin will specifically bind fibronectin, while others such as α3/β1 will bind fibronectin, laminin and some collagens. Several integrins recognize the sequence Arg-Gly-Asp present in fibronectin. It is evident that integrins are able to act as receptors transducing signals into the cell interior since they are capable of affecting ion transport and gene expression when bound to their appropriate ligand. The mechanism of this transduction is still being investigated. The short cytoplasmic domains of the integrins have no homology with tyrosine receptor kinases or G-protein-coupled receptors but do interact with cytoskeletal components, and it has been proposed that integrin binding to the extracellular matrix may serve as a 'nucleation site' for the organization of the cytoskeleton. Alternatively, integrins may be true receptors, since clustering of integrins as occurs during matrix adhesion can trigger a tyrosine kinase cascade (*see* Fig. 13.2a).

As well as this interaction with the extracellular matrix, neurones express integral plasma membrane proteins which can interact with identical molecules on adjacent cells. The most characterized are the *immunoglobulin (Ig)* and *cadherin (CAD)* families which show Ca^{2+}-independent and Ca^{2+}-dependent homophilic interactions respectively. The Igs possess characteristic C2 domains, which are loops of some 50 amino acids linked by disulphide bonds. CAMs are members of the Ig superfamily (*see* Fig. 13.2b); the first to be described in brain, *neural CAM (N-CAM)*, contains five C2 domains, while other members of the family contain from one to six. There are many different spliced variants of N-CAM, and varied glycosylation may control the strength of homophilic interaction. The Ca^{2+} -dependent *neural CAD (N-CAD)*, not to be confused with N-CAM, is found in the brain and N-CAD expressing cells will only bind to other cells expressing the same protein.

While N-CAM and N-CAD are widely distributed throughout the nervous system, other CAMs are expressed transiently and on specific cells and may play a major role in axonal guidance. N-CAMs and cadherins stimulate neurite growth in PC12 cells and elevate $[Ca^{2+}]_c$. This is blocked by inhibitors of voltage-dependent Ca^{2+}-channels and by pertussis toxin, suggesting the involvement of trimeric G-proteins, however the mechanism of G-protein activation remains to be elucidated.

Neurite extension and the mobility of growth cones requires a close regulation of $[Ca^{2+}]_c$ in the range 100–300 nM. If $[Ca^{2+}]_c$ deviates from this, due to electrical activity of receptor activation, then growth and motility ceases and the neurones may begin a process of programmed cell death (apoptosis). Many cultured neurones, such as cerebellar granule cells, require an elevated Ca^{2+}, either created by partial depolarization with KCl or by activation of Ca^{2+}-transporting NMDA receptors, for optimal growth and neurite extension. Elevated $[Ca^{2+}]_c$ can cause the expression of immediate-early genes (*see* Section 13.2.1) and partially depolarizing conditions can remove a requirement for nerve growth factor.

13.4 Molecular models of memory and learning

The physical basis of learning (the permanent storage of information within the brain) and memory (its retrieval) presents the greatest challenge facing neuroscience. Vivid *cognitive* memories, or *motor* skills learnt in childhood can persist for the lifetime of the individual, and certainly for much longer than the lifetime of individual proteins in the CNS. While cognitive and motor memory and learning involve different areas of brain, the same type of mechanism based upon stable alterations of synaptic efficiency appears to be involved, rather than the synthesis of specific 'memory molecules' as occurs in immunological 'memory'. In the necessarily restricted scope of this book, three models will be discussed which may throw some light on mechanisms of memory and learning: (i) *long-term-potentiation* (LTP) in the CA1 region of the mammalian hippocampus, where a specific pattern of stimulation leads to a persistent enhancement of synaptic efficiency; (ii) *long-term depression* (LTD) at the parallel fibre/Purkinje cell synapse in the cerebellum; and (iii) the biochemical changes which occur in defined neurones of the sea hare *Aplysia* when it is taught to modify a simple reflex. It must be emphasized that each of these models has their limitations: the mammalian models are so complex that the biochemical mechanisms are only partially understood, while although the *Aplysia* model is characterized in much more detail the molecular mechanisms may diverge from those in the mammalian CNS.

13.5 Long-term potentiation

In 1949 the psychologist Donald Hebb proposed that repeated or persistent firing of a synapse may cause some stable change which increases the efficiency with which it thereafter excites the postsynaptic cell, and that in order to induce this change simultaneous presynaptic and postsynaptic activity is required. Hebb's proposals were an attempt to put into neuronal terms psychological theories of learning which can be traced back as far as Descartes.

One mammalian model which is actively exploited as a model for a Hebbian process is hippocampal LTP. The hippocampus may be responsible for the initial processing and storing of memory before permanent deposition in the cerebral cortex. LTP is a long-lasting increase in synaptic strength (i.e. postsynaptic response relative to presynaptic stimulus) which may be induced *in vitro* in a number of hippocampal regions by a brief high-frequency ('tetanic') stimulation of afferent fibres. The potentiation outlasts other short-term forms of synaptic enhancement following tetanic stimulation which may be related to the temporary accumulation of Ca^{2+} within the terminal, and may last for several weeks *in vivo*.

Although LTP is not restricted to the hippocampus, the brain region has a number of experimental advantages. Firstly it is possible to prepare brain slices which maintain the lamellar structure of the hippocampus with the main synaptic connections intact (*see* Fig. 1.10). Secondly there are a number of synaptic connections within the hippocampus which are sufficiently simple in anatomical and electrophysiological terms to facilitate their investigation, the most actively investigated of which has been the synaptic contacts which the Schaffer collateral axons from neurones in the CA3 layer make with the dendrites of pyramidal cells in the CA1 layer (Fig. 13.4). It should be emphasized, however, that LTP can also be observed at other hippocampal synapses, notably the mossy fibre/CA3 synapse and that the mechanism differs from that to be discussed below by being independent of NMDA receptor activation. Long-term potentiation at this synapse is believed to be predominantly presynaptic in both induction and expression.

Each side of a rat hippocampus contains some 330 000 CA1 cells; a single CA1 cell has 25 000 spines spread along a dendritic tree and a single Schaffer fibre will make one or two synaptic contacts with an individual CA1 cell. About 200 synapses must be activated simultaneously by a single electrical stimulus to a bundle of Schafer collateral axons in order to fire a typical CA1 pyramidal neurone. The excitatory postsynaptic potential (epsp) can be detected by either an intracellular electrode implanted in a representative postsynaptic cell or by an extracellular electrode placed in close proximity to a field of synapses. Although intracellular recordings of LTP are performed *in vitro* on slices, extracellular recording techniques can also be utilized for *in vivo* recordings, allowing LTP to be monitored for extended periods not limited by the viability of the slice preparation.

The electrophysiological criterion that LTP has been induced is a long-lasting increase in the epsp in response to the test pulse (Fig. 13.4). LTP can be induced in two ways. The first is by a short, high-frequency ('tetanic') stimulus (e.g. 100 Hz for 1 s) can be applied to the afferent pathway. A second means of inducing LTP provides more information as to the mechanism. Two independent bundles of fibres, A and B, which innervate the same neurones are selected. A weak, W, tetanic electrical stimulus to bundle A fails to induce LTP, however the same weak stimulus will produce LTP between bundle A and the CA1 neurone if it is paired with a parallel 'strong', S, stimulus to bundle B (Fig. 13.4d). Such *associative* LTP has some of the characteristics of models for learning: it is induced rapidly, it persists and it only occurs if there is a close temporal relationship between

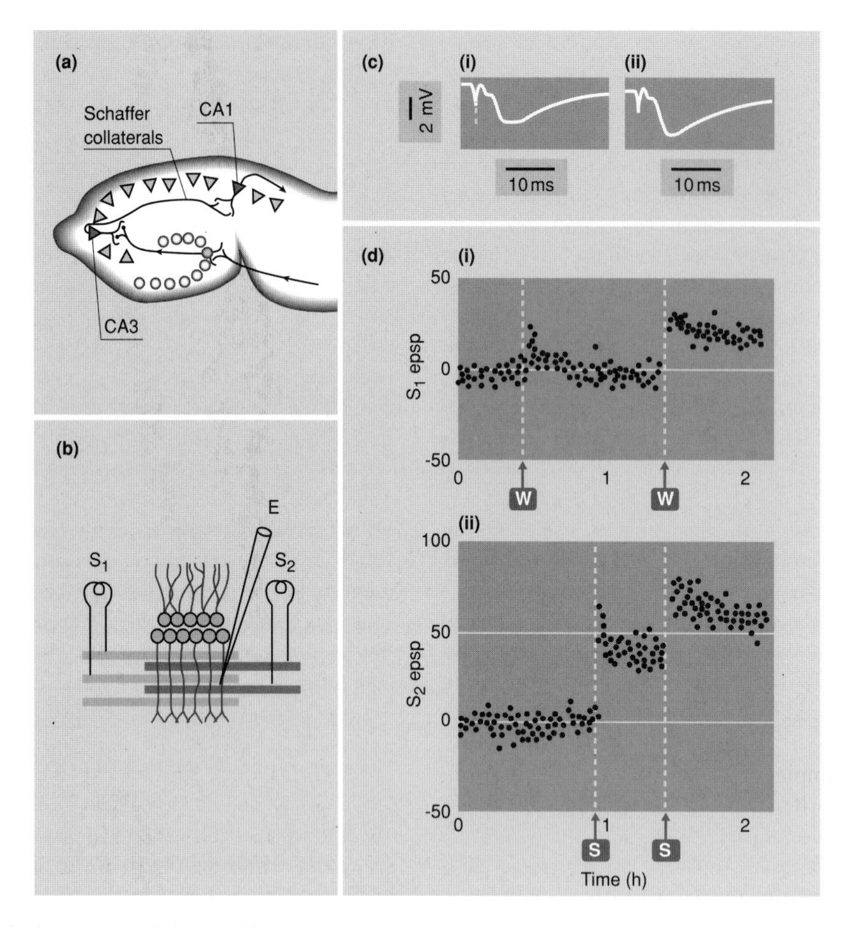

Fig. 13.4 The hippocampal slice and long-term potentiation. (a) Transverse section through rat hippocampus. Shaded areas represent the location of the cell bodies of dentate gyrus granule cells and CA3 and CA1 pyramidal neurones. The main excitatory pathways are the input to the dentate gyrus granule cells via the *perforant pathway*; the mossy fibres from the granule cells to the CA3 pyramidal cells and the *Schaffer collaterals* from the CA3 neurones to the CA1 cells. (b) Two sets of afferent fibres innervating a common population of CA1 neurones in a hippocampal slice are independently stimulated via electrodes S_1 (providing a weak stimulus) and S_2 (providing a strong stimulus). (c) The extracellular electrode (E) responds to the depolarization of the dendrites with an *excitatory postsynaptic potential* (epsp) (i) before and (ii) after long-term potentiation (LTP). (d) The slope of the epsp in response to a single test pulse delivered via S_1 or S_2 is a measure of the postsynaptic effect. Single test pulses delivered via either stimulating electrode produces a given epsp slope (dots). (i) A weak tetanus (W) via S_1 produces a short-lived enhancement, but no LTP. (ii) In contrast, a strong tetanus (S) via S_2 causes a long-lasting potentiation of the response to subsequent single test stimuli, which is further enhanced by a second tetanus. If the weak tetanus through S_1 is synchronized with the strong stimulus via S_2, however, long-term potentiation is induced between the weakly stimulated fibres and the dendrites of the CA1 neurones. Adapted from Bliss & Collingridge (1993), reproduced with permission from *Nature*. © 1993 Macmillan Magazines Ltd.

the inputs from bundles A and B, indeed the **W** stimulus of bundle A can even preceed the **S** stimulus to bundle B by as much as 40 ms.

The interpretation of these results is that the **S** input from bundle B generates a critical level of postsynaptic depolarization in the target cell at exactly the right time when the **W** input from bundle A is occurring. In support of this, LTP cannot be induced by tetanic stimulation if depolarization of a postsynaptic cell is prevented by electrophysiological means, on the other hand LTP *can* be induced when presynaptic stimulation is paired with postsynaptic depolarization. The simple tetanic stimulation discussed above achieves the same ends by providing both synaptic activity and depolarization.

13.5.1 Receptor involvement in the induction of LTP

The excitatory inputs to the pyramidal cells of the hippocampus release glutamate, and AMPA receptors are responsible for synaptic transmission both before and after the induction of LTP. However the hippocampus is very rich in NMDA receptors and NMDA receptor antagonists block the *induction* of LTP.

AMPA and NMDA receptors coexist on individual dendritic spines. Glutamate would be released onto both receptors in response to low frequency, **W**, electrical stimulation; however the depolarization due to the AMPA receptor activation would be rapidly diluted due to electronic spreading of the depolarization, preventing the NMDA receptors from relieving their Mg^{2+} block and conducting Ca^{2+}. Only during tetanic stimulation of the **S** input, or electrophysiological depolarization by a microelectrode, is the delocalized depolarization sufficient so that a parallel **W** input can induce Ca^{2+} entry through the NMDA receptors.

In this simple form, the hypothesis does not account for the specificity of LTP to individual synapses, since the NMDA receptors on all the spines participating in the **W** impulse would also be activated. Specificity may depend on a precise time interval between the firing of a specific synapse at which LTP is to be induced and the non-specific firing of many synapses to produce the necessary delocalized depolarization. This is consistent with the efficacy of high-frequency stimulation in inducing LTP; the NMDA receptors activated by one pulse being able to conduct Ca^{2+} when depolarized by the next stimulus. Thus the synapses stimulated by a single **S** pulse may fail to induce LTP because their NMDA receptors fail to activate before the single AMPA receptor mediated depolarization is over (*see* Fig. 13.4).

Once LTP has been induced the NMDA receptors appear to play no further role in transmission. Thus NMDA antagonists such as AP5 block the induction of LTP but have no effect on the expression of LTP which has already been induced, or on normal synaptic transmission.

There are at least three possible mechanisms which might underlie the synaptic strengthening observed in LTP: (i) the postsynaptic membrane may increase its sensitivity to released glutamate; (ii) the dendritic spines may change their morphology to alter their electrical coupling with the dendrite itself; and (iii) the locus may be presynaptic in that more transmitter may be released for a given stimulus. All three possibilities have their proponents, and the evidence to date is sufficiently ambiguous to prevent any option from being eliminated.

13.5.2 Postsynaptic induction and maintenance of LTP

As discussed in Section 13.5.1, there is compelling evidence that the NMDA receptor is involved in the induction of LTP. Since the characteristic of NMDA receptor activation is that it induces a large and sustained elevation in postsynaptic $[Ca^{2+}]_c$, most working hypotheses of LTP assume that an increase in postsynaptic $[Ca^{2+}]_c$ plays an essential role in LTP. Since Ca^{2+} entry through voltage-activated Ca^{2+}-channels does not induce LTP the NMDA receptors must either induce an increase in $[Ca^{2+}]_c$ which differs in extent, location or time dependency, or the receptors may also generate an unidentified additional message. If the critical NMDA receptors are located on the heads of dendritic spines, this will serve to restrict the transient increase in $[Ca^{2+}]_c$ to the immediate region of the receptor and limit its diffusion away into the dendrite.

A number of possible links between this elevated $[Ca^{2+}]_c$ and the induction of LTP have been proposed (Fig. 13.5). The elevated $[Ca^{2+}]_c$ may activate postsynaptic protein kinases: a remarkable feature of the postsynaptic density is the extraordinary concentration of Ca^{2+}-calmodulin protein kinase II (CaMKII) which can account for 20% of the protein of the density. CaMKII can be activated *in vitro* by an NMDA receptor mediated increase in Ca^{2+}; this results in autophosphorylation of the kinase, converting it to a constitutively active form in which it no longer requires Ca^{2+} (*see* Section 5.9.2). Evidence for the involvement of postsynaptic CaMKII in LTP was until recently speculative. However a function of CaMKII has been revealed most dramatically by the genetic engineering of a mutant mouse deficient in the expression of the α-subunit of CaMKII. The mice have no obvious neuroanatomical defects and show normal neurotransmission, but are deficient in their ability to induce LTP. While it is not clear to

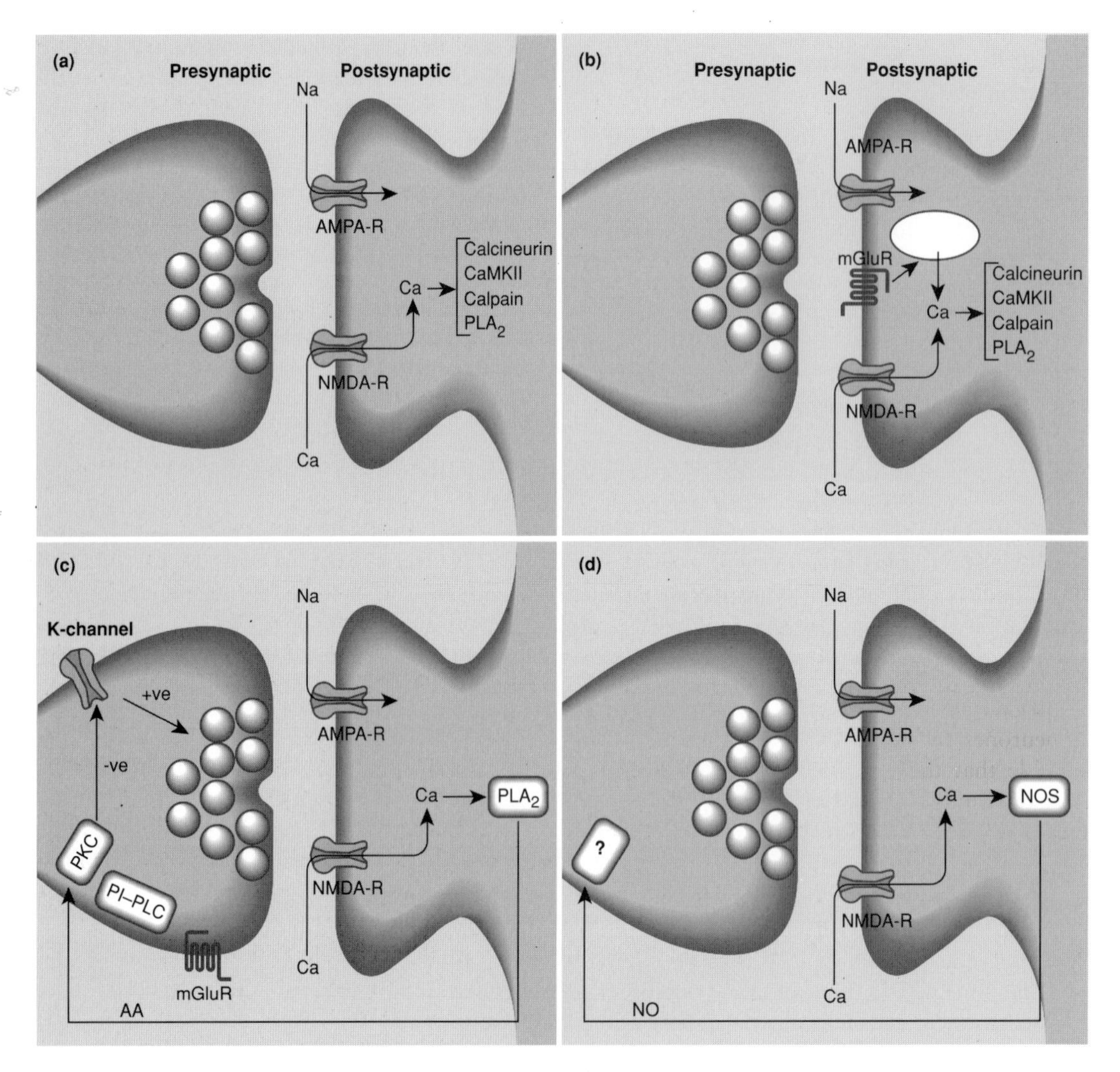

Fig. 13.5 Possible molecular mechanisms in the induction of hippocampal long-term potentiation (LTP). During tetanic stimulation, released glutamate depolarizes the postsynaptic membrane by activating AMPA receptors, allowing NMDA receptors to activate and conduct Ca^{2+} into the cytoplasm. (a) Possible targets of the elevated $[Ca^{2+}]_c$ include PLA_2, calpain (a Ca^{2+}-dependent protease), the phosphatase calcineurin and CaMKII. This last has been reported to activate AMPA receptors. (b) A postsynaptic mGluR may enhance the Ca^{2+} elevation by releasing internal stores. (c) Ca^{2+} may activate PLA_2 leading to the generation of arachidonate (AA) as a retrograde messenger. At the presynaptic membrane AA together with DAG from a *presynaptic* mGluR may activate PKC which inhibits a K^+-channel prolonging action potentials and enhancing glutamate release. (d) Nitric oxide generated by the Ca^{2+}/calmodulin-dependent NO-synthase (NOS) may also act as a retrograde messenger.

what extent the β-subunit can substitute for the deficient subunit to maintain activity, these findings do indicate that the extremely high concentration of CaMKII in postsynaptic densities plays an important role in this learning process. It has recently been shown that CaMKII can phosphorylate and activate AMPA/kainate (KA) receptors which would provide a plausible mechanism for a sustained potentiation.

PKC may also be implicated in the postsynaptic maintenance of LTP. PKC can be cleaved to a constitutively active form which remains catalytically active in the absence of Ca^{2+} and diacylglycerol. Non-specific protein kinase inhibitors, such as sphingosine, prevent the induction of LTP. Conversely, inhibitors such as H-7 which block catalytic activity (and so inhibit both the original and activator-independent forms of the kinases) also block the maintenance of LTP. However, it should be emphasized that the specificities of most protein kinase inhibitors leaves much to be desired, and that careful controls are essential to avoid artefacts, for example, effects on the energetics of the synapse.

Brain contains a group of Ca^{2+}-activated proteases called *calpains*, and it is possible that the NMDA receptor mediated increase in $[Ca^{2+}]_c$ in the postsynaptic spine may be sufficient to activate calpain and initiate a permanent change in the postsynaptic cytoskeletal architecture, either mobilizing extra glutamate receptors or increasing the electrical coupling between the spine and the dendrite. One problem with this theory is that the $[Ca^{2+}]_c$ required to activate the calpains, including the so-called high Ca^{2+}-sensitivity forms, lies in the region 5–30 μM, which is far above the levels reported in neurones by fura-2 imaging. It is however possible that the calpains may be localized in close proximity to the NMDA receptor channels, such that they are exposed to high localized Ca^{2+} concentrations in much the same way as excoytosis of fast neurotransmitters appears to respond to very high localized $[Ca^{2+}]_c$ in the region of presynaptic voltage-activated Ca^{2+}-channels.

13.5.3 Presynaptic aspects of LTP

It has been assumed so far that the induction and maintenance of hippocampal LTP can be described as a purely postsynaptic event. However, from first principles, synaptic strengthening could be due to an increased release of transmitter glutamate instead of a potentiation of postsynaptic events.

Distinct biochemical and electrophysiological approaches can be taken to investigate for possible enhanced presynaptic release. As an example of the former, Lynch and Bliss have reported an increased release of labelled glutamate from hippocampal slices following induction of LTP, and this has been confirmed by monitoring endogenous glutamate release. The electrophysiological approach involves *quantal analysis*, which attempts to distinguish elementary quantal events at single CA3/CA1 synapses and to determine whether LTP is associated with an increase in the number of quanta released for a given stimulus (indicating a presynaptic locus) or whether the same number of quanta is released but their postsynaptic effect is enhanced. In two studies where this analysis was performed presynaptic enhancement was favoured, although it must be borne in mind that a very sophisticated analysis is required and that the validity of quantal analysis in the CNS is actively debated.

13.5.4 Retrograde messengers

An enhanced presynaptic release of glutamate can be reconciled with a postsynaptic NMDA receptor mediated primary event by invoking a *retrograde messenger* capable of diffusing from the postsynaptic membrane and causing a stable enhancement of transmission (*see* Fig. 13.5). Various putative messengers have been proposed, including *arachidonic acid* which could be released postsynaptically in response to a Ca^{2+}-activated phospholipase A_2 and *nitric oxide* produced by the Ca^{+}/calmodulin activated NO-synthase (*see* Section 5.10).

The elevated Ca^{2+} resulting from NMDA receptor activation appears to be particularly effective in activating NO-synthase. This has excited considerable interest in the possibility that NO might function as a retrograde messenger in LTP. In favour of a role for NO is the finding that NO-synthetase inhibitors such as L-N^{G}-nitroarginine (L-NOARG) and the NO scavenger haemoglobin inhibit the induction of hippocampal LTP while the NO generator hydroxylamine induces an LTP-like state. However a major problem with this hypothesis is that NO-synthase may not be present in hippocampal pyramidal cells. Very recently a third retrograde messenger has been proposed, carbon monoxide (CO). CO can be produced physiologically during the degradation of porphyrin rings by haem oxygenase. Inhibition of this enzyme by zincprotoporphyrine IX has been reported to block, or even reverse, LTP. However, there are a number of unanswered problems: there is currently no mechanism for coupling this enzyme to Ca^{2+} or NMDA receptor activation; there is no mechanism for degrading CO; CO is an extremely potent mitochondrial inhibitor; and finally it remains to be proven that the inhibitor is without action on other enzymes in the slice preparation.

It is not intuitive that the very persistence of LTP should depend on the continuous production of a retrograde messenger such as arachidonate or NO. An alternative is that an enhanced presynaptic release is only required during the *establishment* of LTP. If synaptic plasticity is an all-or-nothing event at a given synapse it is important once a synapse becomes committed to potentiation, for example, by NMDA receptor activation, that the release of glutamate remains maximal for the period needed to establish plasticity. However, the release of many transmitters is subject to a negative feedback by autoreceptors activating K^+-channels and/or inhibiting Ca^{2+}-channels which would seem to directly oppose this. Both inhibitory and facilitatory autoreceptor mechanisms have been described at glutamatergic terminals, probably acting respectively via the 'L-AP4' (probably mGluR4) receptor and a phospholipase C-coupled (mGluR1) receptor (*see* Section 9.7.1).

As has been discussed (*see* Section 9.7.1) this positive feedback pathway has been investigated in cerebral cortical synaptosomes, where a synergistic activation of presynaptic PKC by diacylglycerol (generated by the metabotropic glutamate receptor) and arachidonic acid is required to potentiate release (*see* Fig. 9.18), the probable mechanism being an inhibition of K^+-channels controlling the duration of the presynaptic action potentials. Since arachidonic acid has been hypothesized to be a retrograde messenger produced in response to Ca^{2+} entering via the NMDA selective glutamate receptor it could serve to 'couple' this positive feedback precisely during the establishment of potentiation (*see* Fig. 13.5).

13.6 Long-term depression

The components of the inositol phospholipid signal transduction mechanism are extraordinarily enriched in the Purkinje cells of the cerebellum (*see* Fig. 5.21). Autoradiography of $[^3H]$-(1,4,5)IP_3 binding sites several extremely high concentrations of receptors, while the βII and γ isoforms of PKC are similarly enriched. The dendritic spines of Purkinje cells contain a structure, the spine apparatus, which resembles smooth endoplasmic reticulum and is the site of the (1,4,5)IP_3 receptor.

Purkinje cells receive a GABAergic inhibitory input from basket and stellate cells (*see* Fig. 1.9) and two excitatory inputs: from the climbing fibres projecting from the inferior olives of the brain stem and from the parallel fibres originating from granule cells. The parallel fibre/Purkinje cell synapse is the most abundant in brain. Firstly cerebellar granule cells are the most numerous neurones, numbering 100 000 000 in the rat; secondly each parallel fibre innervates several hundred thousand Purkinje cells, and conversely each Purkinje cell possesses an immense dendritic tree and receives a converging input from some 100 000 parallel fibres.

Long-term depression involves a long-lasting inhibition of synaptic transmission. The cerebellar form of LTD can be induced in cerebellar Purkinje cells by coordinated activation of parallel fibres and climbing fibres (which, unusually, are believed to employ aspartate rather than glutamate as transmitter) or *in vitro* by simultaneous depolarization and glutamate application (Fig. 13.6) and results in a persistent depression in the activity of the parallel fibre/Purkinje synapse. LTD requires depolarization of the Purkinje neurone by the climbing fibre (which results in a prolonged depolarization and elevation in dendritic $[Ca^{2+}]_c$) at the same time that both AMPA/KA and metabotropic receptors are activated at the parallel fibre/Purkinje synapse. The process seems to be predominantly postsynaptic, since the sensitivity of cultured Purkinje cell to applied AMPA can be depressed when the cells are depolarized at the same time that glutamate (or a cocktail of AMPA and the metabotropic agonist tACPD) is iontophoresed onto the cell (Fig. 13.6). The nature of the modification to the AMPA receptor is not established, although the involvement of the metabotropic receptor could be consistent with a PKC-mediated phosphorylation of the AMPA receptor.

13.7 *Aplysia* as a model for learning

The Californian sea hare, *Aplysia*, is a large marine gastropod mollusc which has simple and well mapped neuronal pathways, capable of directing a limited number of behavioural responses. Individual neuronal cell bodies within defined ganglia can be identified and are sufficiently

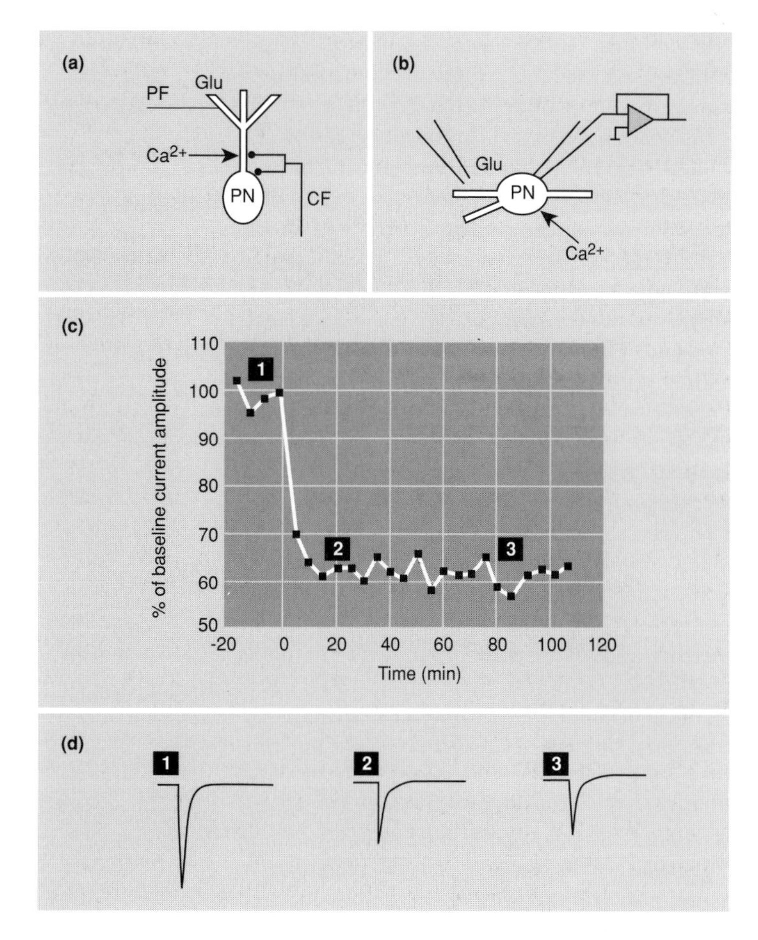

Fig. 13.6 Cerebellar long-term depression (LTD). (a) LTD of the parallel fibre (PF)/Purkinje cell synapse is caused by simultaneous firing of climbing fibres and parallel fibres innervating the Purkinje cell (PN). (b) LTD can be induced *in vitro* in Purkinje cell cultures in the absence of synaptic transmission by mimicking the climbing fibre (CF) depolarization of the Purkinje cell with a nystatin perforated patch electrode at the same time that glutamate (Glu) is iontophoresed onto the dendrites to mimic PF firing. (c) The patch electrode is then used to monitor postsynaptic currents in response to glutamate application before and after induction of LTD. Individual glutamate evoked currents are shown in (d). The activation of both AMPA and metabotropic glutamate receptors is required for the induction of LTD. Data from Linden *et al.* (1991).

large (up to 1 mm diameter) not only for patch clamping but also for the introduction of biochemical agents. The mollusc can be induced to undergo two defensive reflex responses. The *tail–siphon withdrawal reflex* occurs in response to touching the tail and elicits a coordinate withdrawal of tail and siphon. The *siphon–gill withdrawal reflex* is largely controlled by neurones in the abdominal ganglion and occurs when the mantle or siphon is touched, resulting in a withdrawal of siphon and gill. The minimal circuitry required to effect the siphon–gill withdrawal reflex comprises a sensory neurone from the siphon or mantle to detect the touch forming a monosynaptic (direct) contact onto a motor neurone which controls the withdrawal of the siphon or gill (Plate 5, facing page 133).

A painful stimulus to the mantle or siphon causes *sensitization* of this reflex, with the result that the withdrawal in response to subsequent stimuli becomes progressively more vigorous. At the level of the synapse, sensitization is caused by release of 5-HT from terminals of *facilitator neurones* onto the sensory nerve terminals. 5-HT receptor activation is coupled via a G_s to increased cAMP. It is possible to visualize this increase in cAMP by confocal fluorescence microscopy by injecting PKA labelled on the C-subunits with fluorescein and on the R-subunits with rhodamine (Plate 5b). The two probes are sufficiently close in the holoenzyme to allow fluorescent energy transfer, but when the kinase dissociates in response to an increase in cAMP this is lost. The enzyme thus becomes a fluorescent indicator of cAMP levels.

The activation of PKA causes phosphorylation which directly or indirectly inhibits a presynaptic K^+-channel termed K_s (S for serotonin). Prolongation of presynaptic action potentials then results in an increased release (*facilitation*) of transmitter and an enhanced motor response in the mantle and siphon. Direct injection of catalytic subunits of PKA into the soma of sensory neurones to allow their diffusion to the terminal mimics the effect of applied 5-HT.

Facilitation can be studied *in vitro* by co-culturing *Aplysia* sensory and motor neurones (Plate 5c). Direct measurement of Ca^{2+} current in whole-cell patched sensory neurones shows that 5-HT greatly enhances the measured Ca^{2+} current, and that this is largely due to dyhydropyridine (DHP)-sensitive channels analogous to mammalian L-channels. However neither normal transmission nor facilitation is sensitive to DHP inhibitors such as nitrendipine, and direct fura-2 imaging of sensory neurones terminal regions shows that DHP-insensitive Ca^{2+}-channels are present in the terminals, that the increase in free $[Ca^{2+}]$ resulting from a train of action potentials is enhanced by simultaneously applied 5-HT and that it is this nitrendipine-insensitive component which is responsible for the facilitation. Thus somatic whole-cell patching largely fails to detect these presynaptic Ca^{2+}-channels. The DHP-sensitive channels may help in maintaining the level of release during prolonged activity.

The sensitization discussed above only lasts a few minutes; however repeated noxious stimuli can result in a much more persistent sensitization lasting 24 h or more. This latter can be modelled in dissociated cell culture by repeated pulses of 5-HT to the sensory neurone/motor neurone synapse. This results in proteolytic cleavage of the regulatory R-subunit of PKA, resulting in a constitutively active kinase. The PKA fluorescent labelling technique discussed above can be used to show that PKA can translocate into the nucleus of the sensory cell, where it can phosphorylate nuclear transcription factors including CREB. The activation of immediate early genes may switch on longer acting structural genes, and one morphological response to multiple stimuli which is seen is an increase in the number of varicosities. Interestingly, one effect is the internalization of the *Aplysia* version of CAM (*see* Section 13.2.2) which would decrease cell–cell adhesion and aid morphological plasticity.

Further reading

Growth factor receptors

Keegan, K. & Halegoua, S. (1993) Signal transduction pathways in neuronal differentiation. *Curr. Op. Neurobiol.* **3**, 14–19.

Kornberg, L. & Juliano, R.L. (1992) Signal transduction from the extra-cellular matrix: the integrin-tyrosine kinase connection. *Trends Pharmacol. Sci.* **13**, 93–95.

Landis, S.C. (1990) Target regulation of neurotransmitter phenotype (Review). *Trends Neurosci.* **13**, 344–344.

Morgan, J.I. & Curran, T. (1989) Stimulus-transcription coupling in neurones: role of cellular immediate-early genes. *Trends Neurosci.* **12**, 459–462.

Nakanishi, H., Brewer, K.A. & Exton, J.H. (1993) Activation of the zeta isozyme of protein kinase C by phosphatidylinositol 3,4,5-trisphosphate. *J. Biol. Chem.* **268**, 13–16.

Pazin, M.J. & Williams, L.T. (1992) Triggering signalling cascades by receptor tyrosine kinases. *Trends Biochem. Sci.* **17**, 374–378.

Wagner, K.R., Mei, L. & Huganir, R.L. (1991) Protein tyrosine kinases and phosphatases in the nervous system. *Curr. Op. Neurobiol.* **1**, 65–73.

Growth cones and neurite extension

Bamburg, J.R. (1988) The axonal cytoskeleton: stationary or moving matrix. *Trends. Neurosci.* **11**, 248–249.

Davis, L., Dou, P., Dewit, M. & Kater, S.B. (1992) Protein synthesis within neuronal growth cones. *J. Neurosci.* **12**, 4867–4877.

Goldberg, D.J. & Burmeister, D.W. (1989) Looking into growth cones. *Trends Neurosci.* **12**, 503–506.

Gordon-Weeks, P.R. (1989) GAP-43–what does it do in the growth cone? *Trends Neurosci.* **12**, 363–365.

Pollock, J.D., Krempin, M. & Rudy, B. (1990) Differential effects of NGF, FGF, EGF, cAMP and dexamethasone on neurite outgrowth and sodium channel expression in PC12 cells. *J. Neurosci.* **10**, 2626–2637.

Vaughn, J.E. (1989) Fine structure of synaptogenesis in the vertebrate central nervous system (Review). *Synapse* **3**, 255–255.

Wheatley, S.C., Suburo, A.M., Horn, D.A., Vucicevic, V., Terenghi, G., Polak, J.M. & Latchman, D.S. (1992) Redistribution of secretory granule components precedes that of synaptic vesicle proteins during differentiation of a neuronal cell line in serum-free medium. *Neurosci.* **51**, 575–582.

Long-term potentiation

Baudry, M. & Davis, J.L. (1991) *Long-Term Potentiation: A Debate of Current Issues.* Cambridge, Mass.: Bradford.

Bliss, T.V.P. & Collingridge, G.L. (1993) A synaptic model of memory: long-term potentiation in the hippocampus. *Nature* **361**, 31–39.

Bliss, T.V.P. & Dolphin, A.C. (1982) What is the

mechanism of long-term potentiation in the hippocampus. *Trends Neurosci.* **5**, 289–290.

Fazeli, M.S. (1992) Synaptic plasticity: on the trail of the retrograde messenger. *Trends Neurosci.* **15**, 115–117.

Fukunaga, K., Soderling, T.R. & Miyamoto, E. (1992) Activation of Ca^{2+}/calmodulin-dependent protein kinase II and protein kinase C by glutamate in cultured rat hippocampal neurones. *J. Biol. Chem.* **267**, 22 527–22 533.

Herrero, I., Miras-Portugal, M.T. & Sanchez-Prieto, J. (1992) Positive feedback of glutamate exocytosis by metabotropic presynaptic receptor stimulation. *Nature* **360**, 163–166.

Korn, H. & Faber, D.S. (1991) Quantal analysis and synaptic efficiency in the CNS. *Trends Pharmacol. Sci.* **12**, 439–445.

Malenka, R.C., Kauer, J.A., Perkel, D.J. & Nicholl, R.A. (1989) The impact of postsynaptic calcium on synaptic transmission — its role in long-term potentiation (Review). *Trends Neurosci.* **12**, 444–450.

Malinow, R. & Tsien, R.W. (1990) Presynaptic enhancement shown by whole-cell recordings of long-term-potentiation in hippocampal slices. *Nature* **346**, 177–180.

Morris, R. & Collingridge, G. (1993) Neuroscience: expanding the potential. *Nature* **364**, 104–105.

Siegelbaum, S.A. & Kandel, E.R. (1991) Learning-related synaptic plasticity: LTP and LTD. *Curr. Op. Neurobiol.* **1**, 113–120.

Swope, S.L., Moss, S.J., Blackstone, C.D. & Huganir, R.L. (1992) Phosphorylation of ligand-gated ion channels: a possible mode of synaptic plasticity. *FASEB J.* **6**, 2514–2523.

Thomson, A.M. (1992) Quantal analysis of quantal analysis. *Trends. Neurosci.* **15**, 167–168.

Long-term depression

Daniel, H., Hemart, N., Jaillard, D. & Crepel, F. (1992) Coactivation of metabotropic glutamate receptors and of voltage-gated calcium channels induces long-term depression in cerebellar Purkinje cells *in vitro*. *Exp. Brain Res.* **90**, 327–331.

Hirano, T. (1990) Depression and potentiation of the synaptic transmission between a granule cell and a Purkinje cell in rat cerebellar culture. *Neurosci. Lett.* **119**, 141–144.

Ito, M. (1986) Long-term depression as a memory process in the cerebellum. *Neurosci. Res* **3**, 531–539.

Ito, M. (1989) Long-term depression. *Ann. Rev. Neurosci.* **12**, 85–102.

Linden, D.J., Dickinson, M.H., Smeyne, M. & Connor, J.A. (1991) A long-term depression of AMPA currents in cultured cerebellar Purkinje neurons. *Neuron* **7**, 81–89.

Aplysia as a model for learning

Altman, J.S. & Kien, J. (1990) Highlighting *Aplysia*'s networks. *Trends Neurosci.* **13**, 81–82.

Bacskai, B.J., Hochner, B., Mahaut-Smith, M., Adams, S.R., Kaang, B.-K., Kandel, E.R. & Tsien, R.Y. (1993) Spatially resolved dynamics of cAMP and protein kinase A subunits in *Aplysia* sensory neurones. *Science* **260**, 222–226.

Eliot, L.S., Kandel, E.R., Siegelbaum, S.A. & Blumenfeld, H. (1993) Imaging terminals of *Aplysia* sensory neurones demonstrates role of enhanced Ca^{2+} influx in presynaptic facilitation. *Nature* **361**, 634–637.

Index

Page numbers in *italic* refer to figures, in **bold** to tables.